AF838553

Emerging Innovations in Wireless Networks and Broadband Technologies

Naveen Chilamkurti
La Trobe University, Australia

A volume in the Advances in Wireless Technologies and Telecommunication (AWTT) Book Series

Information Science REFERENCE
An Imprint of IGI Global

Published in the United States of America by
Information Science Reference (an imprint of IGI Global)
701 E. Chocolate Avenue
Hershey PA 17033
Tel: 717-533-8845
Fax: 717-533-8661
E-mail: cust@igi-global.com
Web site: http://www.igi-global.com

Copyright © 2016 by IGI Global. All rights reserved. No part of this publication may be reproduced, stored or distributed in any form or by any means, electronic or mechanical, including photocopying, without written permission from the publisher.
Product or company names used in this set are for identification purposes only. Inclusion of the names of the products or companies does not indicate a claim of ownership by IGI Global of the trademark or registered trademark.

Library of Congress Cataloging-in-Publication Data

Names: Chilamkurti, Naveen, 1974- editor.
Title: Emerging innovations in wireless networks and broadband technologies / Naveen Chilamkurti, editor.
Description: Hershey, PA : Information Science Reference, 2016. | Includes bibliographical references and index.
Identifiers: LCCN 2015046871| ISBN 9781466699410 (hardcover) | ISBN 9781466699427 (ebook)
Subjects: LCSH: Wireless communication systems--Technological innovations.
Classification: LCC TK5103.2 .E494 2016 | DDC 621.382--dc23 LC record available at http://lccn.loc.gov/2015046871

This book is published in the IGI Global book series Advances in Wireless Technologies and Telecommunication (AWTT) (ISSN: 2327-3305; eISSN: 2327-3313)

British Cataloguing in Publication Data
A Cataloguing in Publication record for this book is available from the British Library.

All work contributed to this book is new, previously-unpublished material. The views expressed in this book are those of the authors, but not necessarily of the publisher.

Advances in Wireless Technologies and Telecommunication (AWTT) Book Series

ISSN: 2327-3305
EISSN: 2327-3313

Mission

The wireless computing industry is constantly evolving, redesigning the ways in which individuals share information. Wireless technology and telecommunication remain one of the most important technologies in business organizations. The utilization of these technologies has enhanced business efficiency by enabling dynamic resources in all aspects of society.

The **Advances in Wireless Technologies and Telecommunication Book Series** aims to provide researchers and academic communities with quality research on the concepts and developments in the wireless technology fields. Developers, engineers, students, research strategists, and IT managers will find this series useful to gain insight into next generation wireless technologies and telecommunication.

Coverage

- Wireless Sensor Networks
- Wireless Broadband
- Network Management
- Mobile Web Services
- Grid Communications
- Mobile Communications
- Wireless Technologies
- Broadcasting
- Virtual Network Operations
- Mobile Technology

IGI Global is currently accepting manuscripts for publication within this series. To submit a proposal for a volume in this series, please contact our Acquisition Editors at Acquisitions@igi-global.com or visit: http://www.igi-global.com/publish/.

The Advances in Wireless Technologies and Telecommunication (AWTT) Book Series (ISSN 2327-3305) is published by IGI Global, 701 E. Chocolate Avenue, Hershey, PA 17033-1240, USA, www.igi-global.com. This series is composed of titles available for purchase individually; each title is edited to be contextually exclusive from any other title within the series. For pricing and ordering information please visit http://www.igi-global.com/book-series/advances-wireless-technologies-telecommunication-awtt/73684. Postmaster: Send all address changes to above address. Copyright © 2016 IGI Global. All rights, including translation in other languages reserved by the publisher. No part of this series may be reproduced or used in any form or by any means – graphics, electronic, or mechanical, including photocopying, recording, taping, or information and retrieval systems – without written permission from the publisher, except for non commercial, educational use, including classroom teaching purposes. The views expressed in this series are those of the authors, but not necessarily of IGI Global.

Titles in this Series

For a list of additional titles in this series, please visit: www.igi-global.com

Critical Socio-Technical Issues Surrounding Mobile Computing
Norshidah Mohamed (Prince Sultan University, Kingdom of Saudi Arabia and Universiti Teknologi Malaysia, Malaysia) Teddy Mantoro (USBI-Sampoerna University, Indonesia) Media Ayu (USBI-Sampoerna University, Indonesia) and Murni Mahmud (International Islamic University Malaysia, Malaysia)
Information Science Reference • copyright 2016 • 357pp • H/C (ISBN: 9781466694385) • US $210.00 (our price)

Handbook of Research on Next Generation Mobile Communication Systems
Athanasios D. Panagopoulos (National Technical University of Athens, Greece)
Information Science Reference • copyright 2016 • 604pp • H/C (ISBN: 9781466687325) • US $370.00 (our price)

Game Theory Framework Applied to Wireless Communication Networks
Chungang Yang (Xidian University, China) and Jiandong Li (Xidian University, China)
Information Science Reference • copyright 2016 • 502pp • H/C (ISBN: 9781466686427) • US $235.00 (our price)

Enabling Real-Time Mobile Cloud Computing through Emerging Technologies
Tolga Soyata (University of Rochester, USA)
Information Science Reference • copyright 2015 • 399pp • H/C (ISBN: 9781466686625) • US $195.00 (our price)

Emerging Perspectives on the Design, Use, and Evaluation of Mobile and Handheld Devices
Joanna Lumsden (School of Engineering and Applied Science, Aston University, UK)
Information Science Reference • copyright 2015 • 334pp • H/C (ISBN: 9781466685833) • US $205.00 (our price)

Technological Breakthroughs in Modern Wireless Sensor Applications
Hamid Sharif (University of Nebraska – Lincoln, USA) and Yousef S. Kavian (Shahid Chamran University of Ahvaz, Iran)
Information Science Reference • copyright 2015 • 417pp • H/C (ISBN: 9781466682511) • US $200.00 (our price)

IGI GLOBAL
DISSEMINATOR OF KNOWLEDGE
www.igi-global.com

701 E. Chocolate Ave., Hershey, PA 17033
Order online at www.igi-global.com or call 717-533-8845 x100
To place a standing order for titles released in this series,
contact: cust@igi-global.com
Mon-Fri 8:00 am - 5:00 pm (est) or fax 24 hours a day 717-533-8661

Table of Contents

Detailed Table of Contents

In the unlicensed band, the notion of primary user and secondary user (to implement cognitive radio) is not explicit. By dynamic priority assignment we propose to implement cognitive radio in the unlicensed band. In time critical events, the data which is most important, has to be given the time slots. Wireless Sensor nodes in our case are considered to be mobile, and hence make it difficult to prioritize one over another. A node may be out of the reach of the cluster head or base station by the time it is allotted a time slot and hence mobility is a constraint. With the data changing dynamically and factors such as energy and mobility, which are major constraints, assigning priority to the nodes becomes difficult. In this chapter, we have discussed how Wireless Sensor Networks are able to allocate priorities to nodes in the unlicensed band with multiple parameters being posed. We have done simulations on NS-2 and have shown the implementation results.

Wireless sensor networks are deployed in unattended and hostile environment for many applications such as battlefield surveillance. The WSN applications may require knowing the locations of the node in the network to assist in neighbour discovery, selective information sharing and so on. The trivial approach to node localization is to equip each node with GPS. However, the cost and size of GPS enabled nodes make it impractical for resource-constrained, low cost WSNs. GPS-free node localization has been addressed using two different approaches - Beacon Based (BB) and Without Beacon Based (WBB). In BB approach, few nodes aware of their locations serve as beacons to help other nodes in the network localize themselves. In WBB approach, nodes need to localize themselves with the help of their neighbours only. Although, knowledge of nodes' location within network is desirable, exposure of node location information to adversary may lead to undesirable consequences, such as ease of planning for node capture attack, and hence the need of secure localization. The BB approach has been studied extensively under adversarial model and many algorithms based on BB approach have been proposed in literature in order to localize nodes in a secure manner. In contrast, WBB approach for node localization under adversarial model has not received substantial attention from researchers. In this chapter, we discuss static and dynamic key settings for node localization using WBB for node localization under adversarial model. We consider the Localized Combinatorial Keying (LEAP) and Localized Encryption and Authentication Protocol (LEAP) as the building block and propose a protocol for pair-wise key establishment and key revocation to facilitate secure node localization without using beacon nodes in mobile sensor networks, aiming at providing resilience against node impersonation attack and thus minimizing the impact of node capture threats. We provide a comparison of the improved protocol with other related protocols. We show that the improved protocol provides effective node localization in a secure manner with minimal node capture threats.

Avoiding from congestion and provision of reliable communication characterising the low energy consumption and high data rate is one of the momentous challenges at Media Access Control (MAC) layer. This become more difficult to achieve when there is energy constraint mixed with mobility of nodes. Same issue is addressed in this underlying paper. Here we have proposed a Time-Sharing Energy Efficient Congestion Control (TSEEC) technique for Mobile Wireless Sensor Networks. Time Division Multiple Access Protocol (TDMA) and Statistical Time Division Multiple Access Protocol (STDMA) are major constituents of this technique. These helps in conserving the energy by controlling the sleeping, waking up and listening states of sensor nodes. Load Based Allocation and Time Allocation Leister techniques further helps in conserving the network energy minimizing the network congestion. First mentioned technique is designed on the basis of STDMA Protocol and uses the sensor node information to dynamically assign the time slots while later said technique is does the job of mobility management of sensor node. This Time Allocation Leister techniques further comprises of Extricated Time Allocation (ETA), Shift Back Time Allocation (SBTA), and eScaped Time Allocation (STA) sub techniques for managing the joing and leaving of nodes to cluster and redundant\absence of data for communication respectively. To control the movement of mobile sensor nodes, we have also introduced mobility pattern as part of TSEEC that helps in making the protocol adaptive to traffic environment and to mobility as well. A comparitive analysis of proposed mechanism with SMAC is performed in NS2 along with mathematical anslysis by considering energy consumption, and packet deliver ratio as performance evaluation parameters. The results for the former outperforms to that of later. Moreover, comparative analysis of the proposed TSEEC with other MAC protocols is also presented.

Wireless sensor networks are greatly habituated in widespread applications but still yet step behind human intelligence and vision. The main reason is constraints of processing, energy consumptions and communication of image data over the sensor nodes. Wireless sensor network is a cooperative network of nodes called motes. Image compression and transmission over a wide ranged sensor network is an emerging challenge with respect to battery, life time constraints. It reduces communication latency and makes sensor network efficient with respect to energy consumption. In this paper we will have an analysis and comparative look on different image compression techniques in order to reduce computational load, memory requirements and enhance coding speed and image quality. Along with compression, different transmission methods will be discussed and analyzed with respect to energy consumption for better performance in wireless sensor networks.

Network layer functionalists are of core importance in the communication process and so the routing with energy aware trait is indispensable for improved network performance and increased network lifetime. Designing of protocol at this under discussion layer must consider the aforementioned factors especially for energy aware routing process. In wireless sensor networks there may be hundreds or thousands of sensor nodes communicating with each other and with the base station, which consumes more energy in exchanging data and information with the additive issues of unbalanced load and intolerable faults. Two main types of network architectures for sensed data dissemination from source to destination exist in the literature; Flat network architecture, clustered network architecture. In flat architecture based networks, uniformity can be seen since all the network nodes work in a same mode and generally do not have any distinguished role.

Wireless Sensor Nodes (WSNs) are small in size and have limited energy resources. Recent technological advances have facilitated widespread use of wireless sensor networks in many real world applications. In real life situations WSN has to cover an area or monitor a number of nodes on a plane. Sensor node's coverage range is proportional to their cost, as high cost sensor nodes have higher coverage ranges. The main goal of this paper is to minimize the node placement cost with the help of uniform and non-uniform 2D grid planes. Authors propose a new algorithm for data transformation between strongly connected sensor nodes, based on graph theory.

In the last few years Intelligent Transportation Systems (ITS) based on wireless vehicular networks have been attracting interest, since they can contribute towards improving road transport safety and efficiency and ameliorate environmental conditions and life quality. In order to widely spread these technologies, standardization at each layer of the networking protocol stacks has to be done. Therefore, a suite of protocols along with the architecture for the wireless environments with vehicles has been developed and standardized. Both in the US as well as in Europe the selected wireless communication protocol has been the 802.11p protocol developed by the IEEE. In this chapter, we discuss the potential impact of ITS towards achieving the above targets of improving road safety and traffic control. We review the overall architecture and the protocol functionality and present in detail a number of applications that have been developed demonstrating selected use-cases on an 802.11p compliant system prototype. Specifically, we discuss the implementation of selected applications on the NEC's Linkbird-MX platform, which supports IEEE 802.11p based communications, showing how its functionality can be exploited to build efficient road safety and traffic management applications, and evaluate the performance of the system using an experimental testbed.

Mobile Cloud Computing (MCC) paradigm includes all the emerging technological advances, mechanisms and schemes for the efficient resource offloading and the energy-efficient provision of services to mobile users. In addition, the mobile nodes will act as flexible networking points in emerging mobile networking architectures, where several challenges have to be addressed, like the high energy consumption and the data packets transmission failure, under a Mobile Peer-to-Peer (MP2P) approach. Towards addressing such challenges, several factors that contribute to the increased consumption of the energy, have to be investigated, as well as issues related with the loss of data during the provision of services. In this framework, a Traffic-based S-MAC scheme is proposed in this chapter, towards increasing the data exchange and minimize the energy consumption, between mobile nodes operating under an Ad-Hoc approach. The performance of the proposed scheduling scheme was thoroughly evaluated, through a number of simulation experiments.

The overlay networks composed of residential gateways (i.e. home-box) leverage their storage and upload capacity to achieve scalable and cost-efficient content distribution. In this chapter, we present the architecture of the home-box overlay for video on demand services, with the network-aware request redirection and content caching strategy that optimizes the resource usage at both network and client side, for reducing the overall distribution cost. The proposed system is compared with existing solutions through comprehensive simulations. The results demonstrate the advantage of network-aware and popularity-based caching strategy, with reduced the overall cost of the VoD services.

Opportunistic network applications are usually assumed to work only with unordered immutable messages, like photos, videos or music files, while applications that depend on ordered or mutable messages, like chat or shared contents editing applications, are ignored. In this chapter, we examine how causal and total ordering can be achieved in an opportunistic network. By leveraging on existing dissemination algorithms, we investigate if causal order can be efficiently achieved in terms of hit rate and latency compared to not using any order. Afterwards, we propose a Commutative Replicated Data Type algorithm based on Logoot that uses the nature of opportunistic networks to its advantage. Finally, we present the results of the experiments for the new algorithm by using an opportunistic network emulator, mobility traces and chat traces.

Preface

In recent years, Long Term Evolution (LTE), now the new 5th Generation telecommunication standard has been well and truly evolved. Due to this new standard, new multiple interconnected communication standard, ranging from wireless metropolitan area networks down to wireless personal networks have been deployed (Chin, Fan, & Haines, 2014). Ad-hoc network are portrayed as the arrangement of remote wireless sensors that make use of wireless multi-hop radio link to relay between the nodes. Wireless Sensor Network (WSN) comprises an extensive number of Ad-hoc networks that are designed to sense the phenomena. As these WSN's are powered by small batteries, low power usage in sensor is required to engage long lifetime by promising low commitment cycle operation and low signal processing. On top of this, sensor localization is a vital and basic issue for network management and operation. Such knowledge of the node location and territory can be used to execute energy capable messages controlling protocols in sensor networks (Alwageed, 2015). At the network layer, we need to find distinctive networks for discovering energy capable protocols to increase the lifetime of the network. Thus requiring multi-path routing protocols that are capable of using multiple short routes, increasing reliability and thus requiring less energy.

WSN is a promising technology with a great potential to transform our world. But dynamic network topologies and harsh environmental conditions may cause sensor node failures and performance degradation. This requires WSN to support adaptive network operations, including signal-processing algorithms, communication and routing protocols to cope with dynamic wireless-channel conditions and varying connectivity (Rawat, Singh, Chaouchi, Bonnin, 2013). In WSN, some sensors can able to equip with cameras and microphones. Such WSNs are capable of storing, processing, and retrieving multimedia data such as video, audio and images. They must cope with various challenges such as high bandwidth demand, high energy consumption, and Quality of Service (QoS), compression and cross-layer design. Given limited energy resources, bandwidth and radio access, keeping QoS at high level is very difficult, but a certain level of QoS must be achieved for reliable content delivery, which is highly challenging.

Bandwidth demanding P2P (Peer-to-Peer) overlay communications strongly require scalable and differentiated provisioning of network resources to enable and deliver demand based contents (Logota, Marques, & Rodriguez, 2013). A major challenge in such systems is to assure a proper synchronization, at the end-user, of the content received and to provide best-effort services such as Internet traffic. But, with Unknown peers and unpredictable network behaviors and other factors that threaten stability in P2P systems and the content delivery may not be guaranteed.

Video distribution over the Internet has been growing every year. Although today there exist many approaches for P2P live streaming over IP networks, P2P Video on Demand (VoD) service provisioning is essentially a different problem as it involves streaming of pre encoded content and, as such, it adds the whole content availability dimension to the problem (Muñoz-Gea et al., 2012).

Besides the client-server mode, the use of content delivery network and the Peer-to-Peer approach is also popular (Yang, Zhou, Chen, Fu, & Chiu, 2015). A view-upload decoupled approach is one of the popular choices in distributing video over P2P networks. The major advantage of this approach is that cross group load balancing can more effectively utilize peer's serve capacity and Video-on-demand can be easily achieved.

In Chapter 1, dynamic priority scheduling of critical data in wireless sensor networks is investigated. In Section I of this chapter, a brief background on cognitive radio concepts such primary and secondary users are discussed. Next, a proposed dynamic priority assignment techniques and mobility issues in unlicensed band are presented. With the data changing dynamically and factors such as energy and mobility, which are major constraints, assigning priority to the nodes becomes difficult. In Section 2, related works are thoroughly investigated. In Section III, we shall discuss about the improvements and the modifications to the existing algorithm. The final sections deal with proposed algorithm analysis and implementation, and the chapter concludes with future research directions

Chapter 2 investigates secure node localization in clustered sensor networks with effective key revocation. Sensor nodes in a WSN are typically deployed in unattended, hostile terrains and therefore, are mostly not aware of their location. The WSN applications may require knowing the locations of the node in the network to assist in neighbour discovery, selective information sharing and so on. The approaches include GPS, Beacon Based (BB) and Without Beacon Based (WBB). GPS enabled nodes are too heavyweight to suit WSNs. In BB approach, few nodes aware of their locations serve as beacons to help other nodes in the network localize themselves. In WBB approach, nodes need to localize themselves with the help of their neighbours only. Although, knowledge of nodes' location within network is desirable, exposure of node location information to adversary may lead to undesir able consequences, such as ease of planning for a node capture attack, and hence the need of secure localization. The BB approach has been studied extensively under adversarial model and many algorithms based on BB approach have been proposed in literature in order to localize nodes in a secure manner. In contrast, the WBB approach for node localization under adversarial model has not received substantial attention from researchers. In this chapter, static and dynamic key settings for node localization using WBB for node localization under the adversarial model are discussed. A protocol for the pairwise key establishment and key revocation to facilitate secure node localization without using beacon nodes in mobile sensor networks is adopted from pre-existing protocols such as LEAP, aiming at providing resilience against node impersonation attack and thus minimizing the impact of node capture threats. The proposed protocol is compared with related studies for performance gain and security.

Chapter 3 discusses energy efficient congestion control in wireless sensor networks. Firstly, the background of the existing schemes and their challenges with regard to their unique design level constraints because of the limited signal range, processing power, storage capabilities, as well as communication pattern are investigated. As avoiding congestion and provision of reliable communication in these resource constrained systems are challenging tasks at the MAC layer, a novel design of MAC protocol is required that focuses on energy efficiency as well as congestion control. The congestion issue is also aggravated when constraint in energy requirements is coupled with the mobility of nodes. Here, we propose a Time-Sharing Energy Efficient Congestion Control (TSEEC) technique for Mobile Wireless Sensor Networks, which also includes Time Division Multiple Access Protocol (TDMA) and Statistical Time Division Multiple Access Protocol (STDMA) as major constituents. These techniques help in conserving the energy by managing the sleep, wake up and listen states of sensor nodes. The proposed scheme is composed of two main strategies that is dealing with the delay in a network; Load Based Allocation and Time Allocation Leister. The first technique is designed on the basis of STDMA Protocol and uses the sensor node information to dynamically assign the time slots while later technique handles the job of mobility management of sensor nodes. Both techniques further help in conserving the network energy minimizing the network congestion. The Time Allocation Leister technique further comprises of Extricated Time Allocation (ETA), Shift Back Time Allocation (SBTA), and eScaped Time Allocation (STA) sub techniques for managing the joining and leaving of nodes to the cluster and redundant/absence of data for communication. Mobility pattern as part of TSEEC has been introduced to control the movement of mobile sensor nodes making the protocol adaptive to traffic environment and mobility. A simulation and analytical analysis of the proposed mechanism with SMAC has been performed using

NS2 and mathematically by considering energy consumption, and packet delivers ratio as performance evaluation parameters. The proposed system outperforms the existing system. The chapter concludes by suggesting future studies such as finding the neighbouring mobile sensor nodes

Chapter 4 investigates the techniques and analysis of energy efficient image compression and transmission in wireless sensor networks. The chapter discusses the challenges of Image compression and transmission over a wide ranged sensor network with respect to battery, life time constraints. This chapter looks at analysis and compares different image compression and transmission techniques with respect to computational load, memory constraints and transmission speed.

Chapter 5 explores the issue of novel energy aware algorithm to design a multilayer architecture for dense wireless sensor networks. The chapter outlines the importance of energy aware network layer routing functionalities ass it is indispensable for improved network performance and increased network lifetime. This chapter reviews the existing research work, and proposes innovative design technique named Extended Multilayer Cluster Designing Algorithm (E-MCDA) for Lifetime Improvement of wireless sensor network. We improved the idea of MCDA with novel algorithms for time slot allocation at network setup phase to make the cluster designing process more energy efficient, energy efficient cluster head selection, and 'Required Node Degree' based cluster member selection for near equal size clusters. The chapter proposes that the work could be extended in the future employing various techniques.

In Chapter 6, cost minimization of sensor placement and routing in wireless sensor networks with regard to placement and routing issues in a random plane are explained. The chapter underlines the importance of sensor placement from the point of view of cost minimization, and node coverage and connectivity. The sensor node's coverage range is proportional to their cost, as high cost sensor nodes have higher coverage ranges. The main goal of this chapter is to minimize the node placement cost with the help of uniform and non-uniform 2D grid planes. A new algorithm for data transformation between strongly connected sensor nodes, based on graph theory has been proposed. It was found to be that t k-mean clustering can solve the problem of deploying sensors on random points on plane. Furthermore, modifying the problem to graph theory's strongly connected components ensures the transmission of data from one node to another without interruption. The future research dire has also been suggested.

Chapter 7 explores the application of 802.11p Based VANET for improving road safety and traffic management. In this chapter, the emphasis was given to the contribution of Intelligent Transportation Systems (ITS) based on wireless vehicular towards improving road transport safety and traffic efficiency as well as environmental conditions and life quality. The standardization at each layer of the networking protocol stacks is also dealt with. The chapter also reviews the overall architecture and the protocol functionality, and presents some use cases on an 802.11p compliant system prototype.

Chapter 8 explores a scheduling scheme for throughput optimization in mobile peer-to-peer networks. It explains that mobile cloud computing (MCC) paradigm includes all the emerging technological advances, mechanisms and schemes for the efficient resource offloading and the energy-efficient provision of services to mobile users. The chapter further explains that the mobile nodes will act as flexible networking points in emerging mobile networking architectures, where several challenges have to be addressed, like the high energy consumption and the data packet transmission failure, under a Mobile Peer-to-Peer (MP2P) approach. Towards addressing such challenges, several factors that contribute to the increased consumption of the energy is investigated. The chapter proposes Traffic-based S-MAC scheme towards increasing the data exchange and minimize the energy consumption, between mobile nodes operating under an Ad-Hoc approach. The performance of the proposed scheduling scheme was thoroughly evaluated, through a number of simulation experiments.

Chapter 9 deals with a P2P home-box overlay for efficient content distribution. It outlines that the overlay networks composed of residential gateways (i.e. home-box) leverage their storage and upload capacity to achieve scalable and cost-efficient content distribution. The chapter details the architecture of the home-box overlay for video on demand services, with the network-aware request redirection and content caching strategy that optimizes the resource usage at both network and client side, for reducing the overall distribution cost. The proposed system is then compared with existing solutions through comprehensive simulations. Finally, the results demonstrating the advantage of network-aware and popularity-based caching strategy, with reduced the overall cost of the VoD services are presented.

Chapter 10 pinpoints how causal and total ordering can be achieved in an opportunistic network. In the recent years, Opportunistic Networks (ONs) have become an important research field as a viable solution for mobile networks. These network resorts towards unordered immutable messages, like photos, videos or music files. By leveraging on existing dissemination algorithms, we investigate if causal order can be efficiently achieved in terms of hit rate and latency. The chapter proposes a Commutative Replicated Data Type algorithm based on Logoot that uses the nature of opportunistic networks to its advantage. Finally, the results of the experiments for the new algorithm by using an opportunistic network emulator, mobility traces and chat traces are presented.

REFERENCES

Alwageed. (2015). Prospective research challenges in wireless sensor networks. *International Journal of Applied Information Systems, 9*(6), 17-27.

Chin, Fan, & Haines. (2014). Emerging technologies and research challenges for 5G wireless networks. *IEEE Wireless Communications, 21*(2), 106-112.

Logota, Marques, & Rodriguez. (2013). *A cross-layer resource over-provisioning architecture for P2P networks*. Paper presented at the 2013 18th International Conference on Digital Signal Processing (DSP), Fira.

Muñoz-Gea, et al.. (2012). *Design and analysis of a peer-assisted VoD provisioning system for managed networks*. Springer Multimedia Tools and Applications Journal.

Rawat, S., Singh, K. D., Chaouchi, H., & Bonnin, J. M. (2013). Wireless sensor networks: A survey on recent developments and potential synergies. *The Journal of Supercomputing, 68*(1), 1–48. doi:10.1007/s11227-013-1021-9

Yang, Z., & Chen, F. (2015). Turbocharged video distribution via P2P. *IEEE Transactions on Circuits and Systems for Video Technology, 25*(2), 287–299. doi:10.1109/TCSVT.2014.2351093

Acknowledgment

The editor would like to acknowledge the help of all the people involved in this project and, more specifically, to the authors and reviewers that took part in the review process. Without their support, this book would not have become a reality.

Our sincere gratitude goes to the chapter's authors who contributed their time and expertise to this book.

I wish to personally thank IGI Global book publication team for their immense patience and help to make this book completed and published.

Chapter 1
Dynamic Priority Scheduling of Critical Data in Wireless Sensor Networks

Arvind Viswanathan
McMaster University, Canada

Garimella Rama Murthy
IIIT Hyderabad, India

Naveen Chilamkurti
La Trobe University, Australia

ABSTRACT

In the unlicensed band, the notion of primary user and secondary user (to implement cognitive radio) is not explicit. By dynamic priority assignment we propose to implement cognitive radio in the unlicensed band. In time critical events, the data which is most important, has to be given the time slots. Wireless Sensor nodes in our case are considered to be mobile, and hence make it difficult to prioritize one over another. A node may be out of the reach of the cluster head or base station by the time it is allotted a time slot and hence mobility is a constraint. With the data changing dynamically and factors such as energy and mobility, which are major constraints, assigning priority to the nodes becomes difficult. In this chapter, we have discussed how Wireless Sensor Networks are able to allocate priorities to nodes in the unlicensed band with multiple parameters being posed. We have done simulations on NS-2 and have shown the implementation results.

DOI: 10.4018/978-1-4666-9941-0.ch001

Copyright ©2016, IGI Global. Copying or distributing in print or electronic forms without written permission of IGI Global is prohibited.

INTRODUCTION

Wireless Technologies play an integral role in our day to day lives. Though the channels are reusable, due to increasing number of users, we have to turn to the unlicensed spectrum as there is a limited licensed spectrum. The unlicensed spectrum (Akyildiz, Lee, Vuran, & Mohanty, 2006) is one where any user is allowed to access as there is no priority mechanism to allow the more important events to be given higher priorities (Yang, & Vaidya, 2002) for communication. This can be disastrous in the long run as the events of higher importance are being starved of the spectrum due to lack of a priority mechanism. We can assume that there are multiple networks competing for the spectrum and hence there is a need to assign a priority for the networks for higher importance during the occurrence of a critical event. Each network itself will have multiple nodes which will also be competing for the channels (Sriporamanont, 2006). These nodes will also have to be allotted a certain priority.

To decide how networks are given different priorities and how nodes in each network are allocated priorities we can resort to a combined TDMA – FDMA access mechanism.

To reach our goal we need to follow the following steps:

- Select a network.
- Select a node.

Here we are taking into consideration the following assumptions:

- Mobile base stations, Cluster Heads and sensors (all are mobile).
- Controlled Mobility versus Uncontrolled Mobility.

This research chapter is organized as follows. In Section I, we shall give a brief background on cognitive radio. In Section II, a discussion on the work previously done is reported. In Section III, we shall discuss about the improvements and the modifications to the existing algorithm. In Section IV and V, we will talk about the new proposed algorithm and the implementations of the algorithm. We talk about the future prospects of the idea developed and finally, we will conclude the chapter in Section VII.

I. BACKGROUND STUDY

The concept of cognitive radio is to access the available channel in the wireless spectrum in such a way that more wireless communications are able to run in parallel, at a given point of time, to maximize the spectrum utilization. The spectrum is broadly divided into the licensed and the unlicensed band. The licensed band is further subdivided into primary and secondary users. Primary users are given highest priority and when they are not accessing a channel then it may be allocated to a secondary user. However in the unlicensed band there is no such distinct division between the users and hence any one may use it. Hence, many events of importance such as fire in a building may not be given priority as the priority in the unlicensed band is given randomly. We have used the concept of Cognitive Wireless Sensor Networks (Kumar, Singhal, & Garimella, 2012) in this chapter.

II. PREVIOUS WORK DONE

Modified Distributed Laxity-based Priority Scheduling scheme (MDLPS) (RamMurthy, Reddy, Varma, 2011) is a packet scheduling scheme that improves the average end to end delay and the packet delivery ratio within the deadline when compared with the Distributed Laxity-based Priority Scheduling scheme (DLPS) and the Distributed Priority Scheduling (DPS). The DLPS is a scheduling scheme where the state of the neighboring nodes and the feedback regarding the packet loss from the nodes nearby are taken into consideration.

Previously, priority was given with the help of parameters such as battery power, battery threshold level and mobility in the MDLPS scheme. The Priority Index (PI) was computed with the help of the Packet Delivery Ratio (PDR), Uniform Laxity Budget (ULB), mobility of the node (v), and desired PDR for the flow defined by the parameter M.

$$\mathrm{PI} = \frac{PDR}{M} * ULB * \frac{1}{V} * X \tag{1}$$

We have also seen that priority index is high when the priority is low and the priority index is low when the priority is high. Hence, it has an inverse relation with the priority.

$$PI \propto \frac{1}{priority\_of\_the\_node\_or\_network} \tag{2}$$

Figure 1. Execution of nodes based on mobility as a priority (MDLPS)

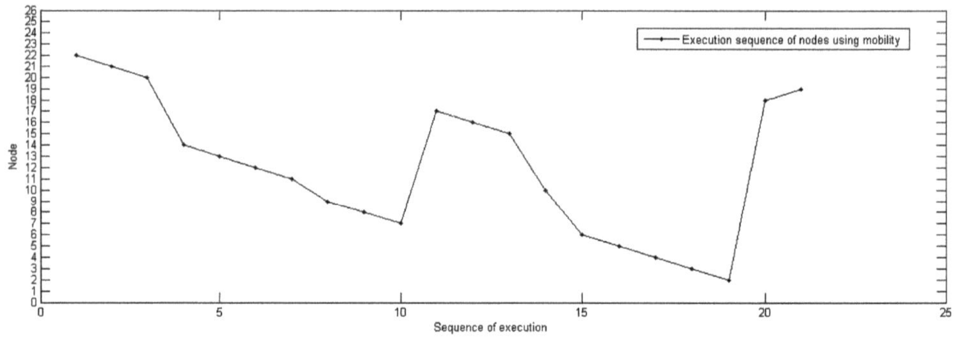

The priority of the nodes were computed on the basis of battery power and an associated threshold. An inverse relation was stated because, if the battery power was below a certain threshold, it would not have enough power to send the data. Hence, the nodes which had energy lower than the threshold were given a higher PI. The relationship between Battery Power and the threshold were as follows:

$$BP \propto \frac{1}{PI} \text{ if } BP \leq BPth \tag{3}$$

$$BP \propto \text{PI if } BP \geq BPth \tag{4}$$

BPth, is the threshold level of the battery power. Each node has a particular amount of battery power in them. The nodes which had full battery power can be made to wait for some time. Hence, they were not given priority first (Figure 1). However, more modifications were done. The number of threshold levels given for battery power were increased. The nodes which were above the hard threshold were divided into multiple levels. This was done so that the nodes having battery voltage greater than the threshold level could avoid undue penalization.

Another important modification that was made, was w.r.t. Packet Delivery Ratio (PDR). When the PDR was below a certain threshold then the packets were discarded by assigning them a higher PI. This is done to allow the nodes with a higher PDR to transmit first so that more data is transmitted.

With respect to mobility the PI had an inverse relationship, i.e. a direct relation with the priority of the node. This was so as the nodes that were moving faster away from the node had to dump the data first so that they do not move away from the range of the decimation node.

$$PI \propto \frac{1}{v} \tag{5}$$

A modification to the calculations in the Universal Laxity Budget (ULB) was also done. The equation was applicable in the case of highly mobile networks. This ensured that the nodes which are about to leave the coverage area of the decimation nodes would dump the data before they left. The new equation was as follows:

$$ULB = \frac{Deadline - CurrentTime}{2^{remaining\_hopes}} \tag{6}$$

III. MODIFICATIONS AND IMPROVEMENTS

The main objective of the new protocol is to improve in terms of utilizing the unlicensed band efficiently, saving battery power (Nitnaware, & Verma, 2010) giving priority to the critical events, prioritizing the sensors on the basis of their velocities and data being transmitted. Previously we were trying to improve the end to end delay and the throughput of the system. We did not focus on the need to improve the order of execution of the nodes. The nodes were allowed to execute on the basis of battery power and the velocities. The data values were not taken into consideration which is the most important parameter.

The Mobility (Eude, Ducourthial, & Shawky, 2005) was not divided into multiple threshold levels. Mobility (v), has been a factor as the nodes are continuously moving and once they come near the base stations it is necessary that they dump the information. This is so as it is not known when they will be allotted the time slot again and when they will be able to transmit the data. We can segment the mobility into three sections.

- High Mobility (V_H).
- Medium Mobility (V_M).
- Low Mobility (V_L).

Since we are including parameters like data and battery power we have made three different levels. If nodes with data of equal importance and similar battery powers are moving then we will give importance to the node with higher mobility. At the time of the occurrence of a critical event we note the speed of the nodes and assign them to their respective mobility levels. When a priority clash occurs then we will check which level the node is in and assign priorities. This is more efficient than checking it every time a priority clash occurs.

IV. NEW PROPOSED ALGORITHM

- Different wireless networks and nodes will be represented in the form of priority tuples.

Priority tuples= (N1, N2) (7)

N1=Priority index of a network
N2= Priority index of the nodes in the network of priority N1

The tuples N1 and N2 can be computed on the basis of the following parameters:

- N1, is decided on the basis of the critical events reported, importance of the data to be communicated, i.e., urgency of the message to be communicated.
- N2, is decided on factors such as, mobility index, i.e., velocity of the node and battery life

Initially, the priority of the network will be selected on the basis of the outreach of the network in the area where the critical event is taking place. In case of a fire in a room, the network which is connected to the room and the rooms near it should be given priority as that network will be able to inform all the nearby nodes about the critical event i.e., the network with higher node density in the area where the critical event has taken place should be given priority [Figure 2]. The speed of communication of the network must also be considered. A network with a higher bandwidth should be prioritized.

- Layering and sectoring can be applied to the sensor field to extend the battery life of the nodes. This way the nodes will broadcast only in the required direction thereby saving a lot of power. The nodes with a higher battery power should be given a higher priority index as they will be able to wait for a larger time span.

Figure 2. Throughput (kbps) v/s the number of connections

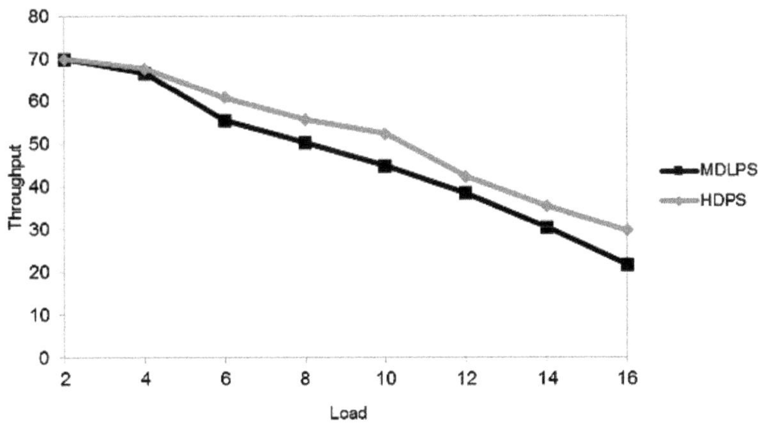

- There are multiple nodes that want access to the spectrum. The spectrum can be divided with the help of TDMA and FDMA. The intersection points of the TDMA/FDMA will be considered as the positions. We are taking as an assumption that in the unlicensed band, the number of sources are larger than the number of positions. When a particular event occurs the nodes which have been given the highest priority are allocated to the positions. This remains static as long as another critical event does not occur. The nodes can be considered as the sources as they are the ones which get the data and they have to send it to the sinks, i.e. the positions on the TDMA/FDMA frame. The sinks here are the base stations.
- In the case of dynamic scheduling, data that is being transmitted can be used as an important parameter to allot priority [Figure 3]. The sensor nodes will be mobile and the base stations will be relatively less mobile. The data will be checked locally by the Cluster Heads (CHs) if it is important. The most important data will be given the lowest priority index, i.e. the highest priority. On a global level all the CHs will check with each other as to which has the most important data to be transmitted and then assign the priorities. Hence, during the occurrence of a critical event such as a fire, the sensor with the most important data in that time slot will be allowed to transmit first to the Base Station (BS). Thus, we can say that the node with the most important data will be given the highest priority.

$$\text{Data Importance} \propto \frac{1}{PI} \tag{8}$$

Figure 3. MDLPS v/s HDPS- Sequence of execution

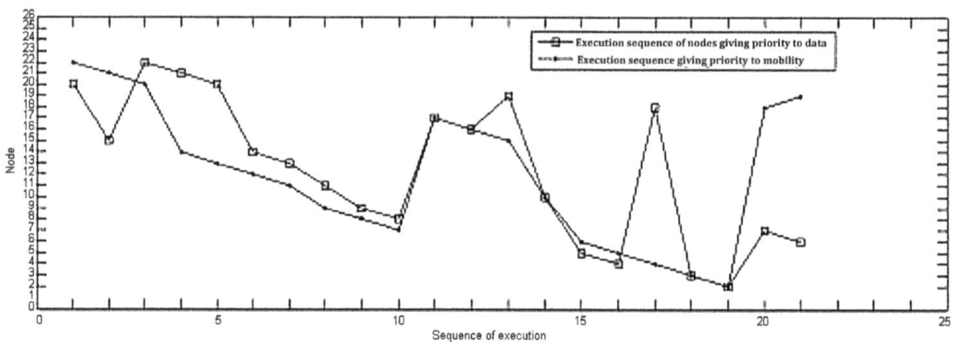

V. IMPLEMENTATION

The nodes are given a random way point motion. The cluster heads and the Base Station are also mobile, though their mobility is controlled and is relatively less compared to the sensor nodes. The nodes have random mobility. They are all connected using the UDP agent and a CBR traffic model. We have taken the Two-Ray ground Reflection Model (Xiuchao, 2004) as the nodes are relatively distant from each other. This model is more realistic and is widely used in comparison to the free space model. The terrain area is large and hence this model has been preferred.

The implementation was done on the Network Simulator 2 -2.34 (NS2-2.34). Some of the parameters taken are given in Table 1.

The energy model was used to implement a threshold level on the battery. The nodes were assigned different speeds so that the priorities could be allotted on the basis of mobility. The nodes were given an interval of 0.5 seconds between transmissions. The simulation graph and the analytical study of the effect of the new protocol has been given.

The above figure is based on the MDLPS algorithm and takes into consideration the mobility and the remaining battery power of the node before allocating priority. Hence we have inferred a simulation study which shows the sequence of execution of the nodes, i.e. the priority allocation of the node- the first executed node is given highest priority and the node executed at the last is given the least priority.

Our proposed algorithm has been termed as HDPS (Heterogeneous Dynamic Priority Scheduling) in the simulation above. In this simulation study we have taken two parameters into consideration, i.e. the throughput and the load on the connection. We are making a comparative analysis between the previously proposed algorithm (MDLPS) and the newly proposed algorithm (HDPS). We notice that HDPS

Table 1. Parameters taken in the NS2 simulation

Parameter	Value
Agent	UDP
Routing Protocol	AODV
Mobility Model	Random Way Point
Data Flow	Constant Bit Rate (CBR)
Node Placement	Random
Terrain Area	2000 x 2000
Session Duration	100s
Queue Size	50
Initial Energy Level	50
Propagation Type	Two-Ray ground Reflection Model
Antenna Type	Omni-Antenna
Number of mobile nodes	22
Packet Size	1000

which incorporates the added concept of checking data in a node for allocating priority and has a better throughput even when the load is increased. Choosing the appropriate nodes helps in obtaining a better throughput. As the data is executed in the correct order the flow of information is continuous and there is no latency. Hence, on increasing the load we are able to achieve a slightly better throughput in comparison to MDLPS.

Though MDLPS was an improvement over DLPS and DPS, the order of execution was not proper. Hence, the critical events were starved by not being allotted the time slots.

VI. FUTURE PROSPECTS

The work that we have done here depends a lot on the data that is being given by the sensor nodes. If this data were to be stored in a particular database with certain constraints and functional dependencies defined in the database architecture, then we have a chance of improving the probability of improving the existing priority model. The model involves assigning a dynamic data, priorities based on the pre-defined constraints and functional dependencies.

Let us consider an example where the building is on fire and we need to find out which rooms to evacuate and in which order so as to minimize the damage done. The assigned functional dependencies would be as follows:

{Room_Temperature, Motion_Sensor} -> {Priority_Assigned}

In this case we have taken the most probable types of sensors available in rooms in a building, and we can see that temperature of the room alone will not be the cause for causing alarm in the room. It is highly probably that the room is empty and hence we are adding a second attribute to that, a motion sensor. This would help in identifying if there are people there in the room. According to this the rooms with maximum fire and people would be prioritized first. This can however be applied in multiple areas such as in areas such as automating air, railway or road traffic by modifying the constraints that we are inputting and the functional dependencies in the database.

VII. CONCLUSION

We have proposed a new algorithm (HDPS) by which we can compute which is the most time critical event. The data in the network is the most important factor and has to be given maximum importance to select any node over the other. The advantage of this over other the previous algorithm is that in time critical events the priority decisions that are taken are more logical to humans and will be helpful in real-time situations such as fire in a building rather than just choosing based on numbers. Choosing a node based on data gives us this advantage. This idea can be implemented in various networks to find out which events are to be executed first in the unlicensed band. The simulation studies have been discussed .We have discussed a few enhancements which could lead to further improve the efficiency of the protocol.

REFERENCES

Akyildiz, I. F., Lee, W. Y., Vuran, M. C., & Mohanty, S. (2006). NeXt generation/dynamic spectrum access/cognitive radio wireless networks: A survey. *Computer Networks: The International Journal of Computer and Telecommunications Networking*, *50*(13), 2127–2159. doi:10.1016/j.comnet.2006.05.001

Eude, N., Ducourthial, B., & Shawky, M. (2005). *Enhancing ns-2 simulator for high mobility ad hoc networks in Car-to-Car communication context.* The 7th IFIP International Conference on Mobile and Wireless Communications Networks, Morocco.

Kumar, S., Singhal, D., & Garimella, R. M. (2012). *Cognitive Wireless Sensor Networks in Intelligent Sensor Networks: The Integration of Sensor Networks, Signal Processing and Machine Learning*. Retrieved from https://books.google.co.in/books?id=ZmLNBQAAQBAJ&printsec=frontcover

Nitnaware, D., & Verma, A. (2010). Energy Constraint Node Cache Based Routing Protocol for AdHoc Network. *International Journal of Wireless & Mobile Networks*, *1*(1), 77–86.

RamMurthy, G., Naveen Reddy, G., & Ravi Shankar Varma, A.RamMurthy. (2011). Modified Distributed Laxity Based Priority Scheduling Scheme. *Wireless Personal Communications*, *58*(3), 627–636. doi:10.1007/s11277-010-0126-1

Sriporamanont, T. (2006). *Wireless Sensor Network Simulator.* (Master's Thesis). Halmstad University, Sweden.

The Network Simulator. (n.d.). *NS-2*. Retrieved from http://www.isi.edu/nsnam/ns

Xiuchao, W. (2004). *Simulate 802.11b Channel within NS2*. Retrieved from http://cir.nus.edu.sg/reactivetcp/report/80211ChannelinNS2_new.pdf

Yang, X., & Vaidya, N. H. (2002). Priority Scheduling in Wireless Ad Hoc Networks. *Proceedings of the 3rd ACM international symposium on Mobile ad hoc networking & computing (MobiHoc'02)*. doi:10.1145/513800.513809

Chapter 2
Secure Node Localization in Clustered Sensor Networks with Effective Key Revocation

Rachit Mittal
Dhirubhai Ambani Institute of Information and Communication Technology, India

Sarita Agrawal
Dhirubhai Ambani Institute of Information and Communication Technology, India

Manik Lal Das
Dhirubhai Ambani Institute of Information and Communication Technology, India

ABSTRACT

Wireless sensor networks are deployed in unattended and hostile environment for many applications such as battlefield surveillance. The WSN applications may require knowing the locations of the node in the network to assist in neighbour discovery, selective information sharing and so on. The trivial approach to node localization is to equip each node with GPS. However, the cost and size of GPS enabled nodes make it impractical for resource-constrained, low cost WSNs. GPS-free node localization has been addressed using two different approaches - Beacon Based (BB) and Without Beacon Based (WBB). In BB approach, few nodes aware of their locations serve as beacons to help other nodes in the network localize themselves. In WBB approach,

DOI: 10.4018/978-1-4666-9941-0.ch002

Copyright ©2016, IGI Global. Copying or distributing in print or electronic forms without written permission of IGI Global is prohibited.

nodes need to localize themselves with the help of their neighbours only. Although, knowledge of nodes' location within network is desirable, exposure of node location information to adversary may lead to undesirable consequences, such as ease of planning for node capture attack, and hence the need of secure localization. The BB approach has been studied extensively under adversarial model and many algorithms based on BB approach have been proposed in literature in order to localize nodes in a secure manner. In contrast, WBB approach for node localization under adversarial model has not received substantial attention from researchers. In this chapter, we discuss static and dynamic key settings for node localization using WBB for node localization under adversarial model. We consider the Localized Combinatorial Keying (LEAP) and Localized Encryption and Authentication Protocol (LEAP) as the building block and propose a protocol for pair-wise key establishment and key revocation to facilitate secure node localization without using beacon nodes in mobile sensor networks, aiming at providing resilience against node impersonation attack and thus minimizing the impact of node capture threats. We provide a comparison of the improved protocol with other related protocols. We show that the improved protocol provides effective node localization in a secure manner with minimal node capture threats.

INTRODUCTION

WSN is a network of sensor nodes that communicate via wireless links to cooperatively monitor the environment or object. Sensor nodes in a WSN are typically deployed in unattended, hostile terrains and therefore, are mostly not aware of their locations. Many WSN applications require the nodes to know their locations after the network deployment. For this purpose, sensor node localization process is used. The advantages of knowing the location information of sensor nodes in WSN include:

- Network coverage checking and location-based information querying.
- Selective forwarding instead of broadcasting information.
- Neighbour discovery.
- Data aggregation after finding neighboring node.

In the localization process, sensor nodes in a WSN determine spatial relationships between themselves (Srinivasan & Wu, 2007; Rasmussen, Capkun, & Cagalj, 2007; Mi, Stankovic, & Stoleru, 2012). The determination of spatial relationship signifies that the nodes are able to assign location coordinates to each other either with reference to a coordinate system localized to a particular network or with reference to a global coordinate system such as GPS (Global Positioning System). Broadly, the approaches used toward localization consist of two types of nodes:

- **Beacon Node / Anchor Node:** Node, who knows its physical location (either via GPS receiver or by manual configuration).
- **Regular Node / Non-Beacon Node:** Node, who does not know its position and no special hardware is required to acquire its location information.

Beacon-Based Approach

Beacon-based (BB) approach uses Beacon node(s) as the basis for computing the location coordinates of other non-Beacon nodes. The idea is to implement the Beacon node with directional antennas and the non-Beacon nodes localizing themselves based on the range of whichever antenna they belong to. Naturally, the security is an important concern in BB approach when an adversary attempts to compromise some Beacon nodes (Liu, Ning, Liu, Wang, & Du, 2008).

However, In the context of node localization in WSN that is inherently distributed in nature, BB approach does not seem to be suitable because (i) in BB approach, a large number of beacon nodes are required to localize all non-beacon nodes; (ii) BB approach is not scalable well in comparison to WBB approach; (iii) in BB approach, localization is completely dependent on Beacon nodes.

Without Beacon-Based Approach

When the network does not have beacon/anchor nodes, the regular nodes may coordinate with their neighbour nodes and build a localized coordinate system wherein a node's location is determined in reference to its neighbour nodes. However, when communication is established with other networks, the local coordinate system can be converted to a global coordinate system with the help of the central authority such as base station. One of the main advantages of WBB approach is scalability. Generally, the WBB approach involves following three steps to establish a localized coordinate system in the network.

1. Discovery of one-hop neighbors and sharing a secret pair wise key with them.
2. Calculation of distance to each neighboring node.
3. Assigning coordinates to neighboring nodes based on the distance measurements.

The main aspect in WBB approach of node localization is the discovery of neighbours and establishing secret pair-wise key with them. Once the communication between a pair of nodes is secured via a pair-wise key, integrity of distance measurement and coordinate assignment is ensured and hence the correctness of the localization process.

Secure Localization

Due to hostile deployment of sensor nodes, the security of node location is very important. The adversary may attempt to attack the network by altering the location information of the node resulting in severe consequences such as wrong decision makings in battle field and false alarm triggers in forest fire or building monitoring applications.

We can consider secure localization of sensor nodes from two perspectives. First aspect is attack on nodes wherein the adversary may compromise nodes to interfere with the localization process by pretending to be an unknown or anchor nodes. To deal with this issue, we need to have secure node authentication process in place. Second aspect is the attacker may attempt to change, delete or replay the location information. Therefore, we need to ensure the integrity of the localization information.

A number of algorithms have been proposed in the literature for secure localization in WSN. Some well-known BB approach based algorithms are SeRLoc (Secure Localization) (Lazos & Poovendran, 2004), ROPE (Robust Position Estimation) (Lazos, Poovendran, & Capkun, 2005), and HirLoc (High Resolution Localization) (Lazos & Poovendran, 2006). These algorithms use Range-free secure localization and uses sectored antennas. Secure Walking GPS based localization suggested in (Mi, Stankovic, & Stoleru, 2012) used master node equipped with GPS, which is present at the time of deployment. In this scheme, the global coordinates received from GPS device are transformed into local coordinates and then local coordinates are broadcasted into the network for the other nodes to localize themselves. Location based key distribution is then carried out. Mobile beacon based secure localization called SLMB is proposed by Zhang, He and Yu (2012) to reduce the energy consumption by sensor nodes. In this scheme, information from unknown sensor node is collected by a mobile beacon which also figures out location relationship with the node. The base station receives this information from mobile beacon to analyze and calculate location of the node. Niu et al. (2012) proposed a localization scheme which utilizes the relation between hop-counts and geographic distance between nodes without using distance measuring devices or complex encrypting algorithms. Recently, Chen et al. (2014) proposed the localization scheme to locate the attackers and remove them from the network.

Using WBB approach, LDAC (Localized-Decentralized Algorithm for Countering Wormholes) is presented in (Giannetsos & Dimitriou, 2012) simply looks for an evidence of non-attack based only on the connectivity information. This is a completely localized method that works for static, dynamic and as well as mobile networks.

Our Contributions

As GPS-based localization is economically costly and GPS signal is not always available, we consider the WBB approach for sensor nodes localization. We discuss related approaches for localizing neighbouring nodes in WSN. We also discuss the trade-off in our choice of LEAP (Localized Encryption and Authentication Protocol) (Zhu, Setia, & Jajodia, 2003) as the key management scheme along with comparisons with schemes and present an improved protocol for pair-wise key establishment. We observed that most of the secure localization schemes deal with securing the WSN from Wormhole attack. In this work, we emphasize on the resilience to node capture attack that is a serious threat with the unattended nature of WSN. We present a key revocation scheme in order to secure the key management process and thus minimizing the impact of node capture attack.

The chapter is organized as follows. Section 2 describes some neighbour discovery algorithms for static and dynamic key management settings. Section 3 presents our protocol for secure key establishment and key revocation in order to facilitate secure node localization in WSN. Section 4 analyzes the proposed protocol. We conclude the paper in section 5.

NEIGHBOUR DISCOVERY AND KEY MANAGEMENT IN WSN

In order to data delivery to the base station, sensor node needs to communicate with the nodes in their neighbourhood. Since the communication channel in a WSN is wireless, there is a need to secure the data exchanged between the nodes. For this purpose, secret key is needed. Simplest way to share a key within WSN is to provide a single network wide key to all the nodes before deployment. However, in such set up, compromise of a single node results in the compromise of the entire network. Another possibility is to assign a unique pair-wise key between each pair of nodes in the network. Now, compromise of a node can only compromise the links associated with that node. However, for a network of n nodes, this approach needs a node to store *n-1* keys (Eschenauer & Gligor, 2002). A number of schemes have been proposed in the literature for discovering neighbours and establishing secure keys for data delivery to the base station. The key distribution schemes used in WSN are broadly classified into *static key management schemes* such as (i) Random key pre-distribution (ii) *Q*-composite random key pre-distribution (Chan, Perrig, & Song, 2003), (iii) Polynomial pool-based key pre-distribution (Liu & Ning, 2003) and (v) General polynomial pool-based key pre-distribution (Liu & Ning, 2003); and *dynamic key management schemes* such as (i) LOCK (Localized Combinatorial Keying) (Eltoweissy, Moharrum, & Mukkamala, 2006) and (ii) LEAP (Localized Encryption and Authentication Protocol) (Zhu, Setia, & Jajodia, 2003). These schemes are gradually build upon the idea of minimizing the impact of node capture on the network that is, increasing the number of nodes that an adversary requires to capture to compromise the whole network. The *dynamic key management schemes* introduce an extra *key revocation* step that allows post deployment renewal of sensor nodes' key either after a predefined time interval or after detection of a node compromise.

Static Key Management Schemes

We briefly explain few approaches for static key management schemes in WSN.

Random Key Pre-Distribution

Random key pre-distribution scheme (Eschenauer & Gligor, 2002) has three basic stages: key pre-distribution, shared key discovery and path key establishment.

- **Key Pre-Distribution:** In this stage, the base station generates a large pool of n keys and their identifiers. Out of these n keys, K ($<n$) keys are randomly picked and preloaded into every node along with their identifiers before nodes get deployed in the field.
- **Shared Key Discovery:** After key pre-distribution, the sensor nodes are deployed in the field, where every node requires discovering its neighbours. If two neighbours share a common key, the key establishment can be done directly; else the key establishment phase should be invoked to establish the shared key. There are ways to detect whether neighbouring nodes share common key(s). The simplest one is to make the nodes broadcast their identifier list to other nodes, and then, other nodes check the list for any common key(s). The problem with this technique is that an adversary can easily get all contents of the key identifier list, which could possibly allow him to impersonate a legitimate node. In order to avoid this threat, an alternate method is provided in the scheme, where every node broadcasts a list $\{x, E_{Ki}(x)\}$ for every i =1, 2,..., k keys in its key ring (here, x acts as a challenge). The

decrypted value of $E_{K_i}(x)$ will enable the recipient to share a common key k, if the common key identifier found between sender and receiver. If a node is compromised, then the communication between nodes with which it shares common keys is affected.

- **Path Key Establishment:** This phase is invoked when two nodes, say u and v, do not share any common keys. In this case, an intermediate node y helps u and v to establish a common key, assuming that both u and v have a common key with y. The node u sends a message to node y; encrypted using their common key, saying that u wants to communicate to node v. Then the node y generates a pair wise key K_{uv} for nodes u and v. In this scenario, the adversary has to compromise all y nodes in between u and v to know the communication between them.

In the unattended deployed WSN, an adversary may capture the nodes and get hold of the keys stored in the node. Using these keys, an adversary may further launch various attacks, such as impersonation attack, selective forwarding and so on. Therefore, the behaviour of the nodes is continuously monitored in the network and if the node is found to be compromised, key revocation process is executed wherein the keys of a compromised node are revoked and also, the keys of all other nodes affected by the compromise of that node are renewed.

Key Revocation: Key revocation process is executed by some Controller nodes which are controlled (and trusted) by the base station. Based on the second assumption, a Controller node has knowledge of the compromised node and it broadcasts a message to other nodes with a list of the key identifiers of the compromised nodes' key ring and new keys. The message is encrypted using K_{ci}, which is the key of the Controller node shared with other nodes at the time of pre-distribution for every node i. Once noticed, the Controller node deletes all keys of the compromised nodes from the entire network.

Recently, Gong et al. (2014) proposed an enhanced Key Management Scheme Based on Random Key Pre-distribution for Wireless Sensor Networks. The initial traditional random key pre-distribution scheme had to tradeoff between network connectivity and security. This enhanced key management scheme proposes the improved connectivity rate between the adjacent sensor nodes and also the security of session key generation is ensured.

Q-Composite Random Key Pre-Distribution

To increase the resilience from node capture, Chan, Perrig, and Song (2003) introduced the q-composite random key pre-distribution scheme that requires two nodes have at least q common keys out of k keys in their key ring to setup a pair wise key. The value of q should be decided optimally such that with a few compromised nodes the adversary cannot compromise the security of the entire network.

Polynomial Based Key Pre-Distribution

Random pre-distribution schemes for key establishment suffer when a part of the network is compromised. Blundo et al. (1993) had introduced the concept of Polynomial based key pre-distribution. The scheme works in two phases:

- **Initialization Phase:** In this phase, the base station randomly generates a symmetric bivariate t-degree polynomial $f(x, y)$, where $f(x, y) = \sum_{i,j=0}^{t} a_{ij} x^i y^j$, and a_{ij} being a scalar. Every node is then provided with its polynomial share, e.g., node u in the network receives $f(u, y)$ share and node v receives $f(v, y)$ as its share.
- **Key Establishment Phase:** For u and v to establish a common key $f(u, v)$, the node u computes $f(u, y)$, that is, replacing y by its value v and node v computes $f(v, u)$.

Since $f(x, y)$ is a symmetric polynomial $f(u, v) = f(v, u)$ becomes the common shared key between u and v.

The above t-degree polynomial based scheme (Blundo et al., 1993) is t-collusion resistant. An adversary needs to compromise $(t+1)$ nodes in order to get the polynomial.

Polynomial Pool-Based Key Pre-Distribution

Liu and Ning (2003) proposed the concept of polynomial pool-based key pre-distribution scheme for establishing pair wise keys between nodes. When the polynomial pool has only one polynomial, then the framework is analogous to the polynomial-based key pre-distribution. In polynomial pool based approach a pool of polynomials is available and the polynomial shares are distributed to the nodes from the pool, either randomly (Random polynomial-pool based approach) or based on the node's position in grid (Grid-based key pre-distribution). The key establishment phase works using the same idea of the key pool-based scheme. Nodes who want to establish a key look for a common polynomial share. If found, a direct key establishment is done; otherwise, Path key establishment phase is invoked for the key establishment. As the attacker needs to capture at least $(t+1)$ nodes to obtain any t-degree polynomial, this scheme has significantly enhanced network resilience to node capture. Later, multiple polynomial share based approach is proposed (Guo, Leung, & Qian, 2010), wherein all the nodes are given the share from all the polynomials, but in different permutation. This scheme ensures that adversary needs to capture at least $m*(t+1)$ nodes, with m polynomials of degree t in the pool.

Recently, Bechkit et al. (2013) proposed a one way hash-chain based method in which a key is concealed in such a way that in case some keys are disclosed, only the derived versions will be disclosed. Thus, the backward keys cannot be used to compromise other links in the network. Any existing pool-based scheme can use this method to improve the network resiliency.

Dynamic Key Management Schemes for WSN

The dynamic key management schemes allow the keys to be updated / renewed during the network operation. The dynamic key update is useful and effective in WSN, in particular, where sensor nodes are mobile in nature.

There are broadly two categories of dynamic key management schemes: Distributed and Centralized. In centralized schemes, a central authority such as base station is responsible for the key management. In distributed schemes, base station involvement is not needed. Polynomial share-based re-keying schemes and EBS based schemes are examples of distributed schemes (He, Niedermeier, & Meer, 2013). In Agrawal et al. (2012) multiple polynomial based approach used for establishing and securely updating the pair wise key between nodes is discussed. Key revocation is another important concept in key management to deal against node capture threat. We briefly explain two widely referred dynamic key management schemes, LOCK (Eltoweissy, Moharrum, & Mukkamala, 2006) and LEAP (Zhu, Setia, & Jajodia, 2003) providing key revocation.

LOCK: Localized Combinatorial Keying

LOCK - Localized Combinatorial Keying (Eltoweissy, Moharrum, & Mukkamala, 2006) is an Exclusion Basis System (EBS) based dynamic key management scheme for clustered sensor networks. EBS was developed by Eltoweissy et al. (2006) with an idea that allows the new keying material after renewal procedure to be securely communicated to nodes. EBS uses *administrative keys* for achieving the desired goal. The basic idea is to have a set of $(k + m)$ keys from which one can form $C(k+ m, k)$ distinct sets and assign each set to a different node. Here, $C(n, r)$ means possible combinations of r objects from n objects. Eviction of a compromised node can be performed by broadcasting "replacement of the k keys" that the evicted node

Table 1. EBS Example - E (10, 3, 2)

	M_0	M_1	M_2	M_3	M_4	M_5	M_6	M_7	M_8	M_9
K_1	1	1	1	1	1	1	0	0	0	0
K_2	1	1	1	0	0	0	1	1	1	0
K_3	1	0	0	1	1	0	1	1	0	1
K_4	0	1	0	1	0	1	1	0	1	1
K_5	0	0	1	0	1	1	0	1	1	1

knows using the m keys that the node does not know. There is a trade-off involved between the number of administrative keys k and the number of re-keying messages m. It is noted that large k increases the storage requirements at the node, while large m increases communication overhead for key management. An EBS, defined as E (N, k, m), is a collection of subsets of K_1, K_2... K_{k+m} each of cardinality k, and an EBS exists when $C\,(k + \mathrm{m}, k) \geq N$.

Example. To construct E (10, 3, 2), the canonical matrix will contain the enumeration of all C (5, 3) ways to form a subset of three keys from five keys (see Table 1). Nodes having 1 as an entry will possess that key.

Suppose M_0 is a compromised node. Since M_0 possesses keys K_1, K_2 and K_3, these keys will have to be renewed using the keys K_4 and K_5 and the following messages are generated and broadcasted for re-keying.

Message 1: $E(K_4(E(K_1(K_1')), E(K_2(K_2')), E(K_3(K_3'))))$
Message 2: $E(K_5(E(K_1(K_1')), E(K_2(K_2')), E(K_3(K_3'))))$

Where, (K_1) is a new key, encrypted with old key K_1. As compromised node is unaware about K_4 and K_5, the adversary cannot decrypt these messages.

It is observed that EBS allows effective re-keying. However, EBS is vulnerable to collusion attack where adversary can compromise selected nodes, say, M_0 and M_5 in the above example, to know all the new keys. If the adversary has no pre knowledge about the keying material stored in a particular node, he cannot obtain K_4 and K_5 from M_5. Therefore, the adversary needs to compromise many nodes before he can get K_4 and K_5. The key management algorithm in LOCK works in the following phases:

- **Key Establishment:** LOCK uses the idea of *polynomial pool based key distribution* to establish key. The process is similar as described in previous section that involves steps where base station preloads nodes with polynomial share and nodes find the common polynomial share to establish key.

- **Key Revocation:** To perform key revocation, the nodes are loaded with administrative keys. LOCK uses two layers of EBS administrative keys, one for cluster heads and other for sensor nodes. The revocation in the cluster heads, EBS_b, is performed by the base station on detection of a node compromise. For sensor nodes, EBS_a is performed by the cluster heads. The eviction of a compromised sensor node is straight forward. However, whenever a cluster head compromise is detected by the base station, it performs two tasks. Firstly, the compromised cluster head is evicted and the keys known to it are renewed in a way such that it cannot know the new keys. Secondly, since the administrative keys of that cluster are known to the cluster head, those keys have to be renewed. The base station can deploy a new cluster head which is authenticated by the sensor nodes using a key shared by the sensors with the base station at the time of pre-distribution or the orphaned sensors (whose cluster head is evicted because of being compromised) are to be redistributed to other uncompromised clusters. The base station performs the task of informing other cluster heads about the identities of orphaned sensors which can then ask other sensors to join.

LEAP: Localized Encryption and Authentication Protocol

LEAP -Localized Encryption and Authentication Protocol (Zhu, Setia, & Jajodia, 2003) is a hierarchical key management scheme that allows nodes to establish a pair wise key with their neighbours. LEAP consists of four steps and works as follows:

- **Key Pre-Distribution:** The base station generates an initial key K_I and loads every node with this key. Node u derives a master key $K_u = f_{KI}(u)$, where $f(.)$ is a pseudo-random function and u is the node's identity.
- **Neighbour Discovery:** When node u is deployed, u starts a timer to keep track of time T_{min}. It then tries to discover its neighbours by broadcasting a *HELLO* message which contains its identity u. Each neighbour v responds with an *ACK* message including its identity and a *MAC* (Message Authentication Code) of their identities.

u →*: *u* (Broadcasts its identity)

$v \rightarrow u$: v, MAC(K_{vu} || v) for every neighbour v

Neighbour v is authenticated, as node u knows K_I and it can derive K_v and then verify node *v's* identity.

- **Pair Wise Key Establishment:** Node u computes its pair wise key with v, $K_{uv} = f_{Kv}(u)$. Node v can also compute K_{uv} in the same way. Node u's authenticity is achieved by K_{uv} used in the subsequent messages.
- **Key Erasure:** When the timer expires, node u erases K_I and all keys K_v of its neighbours, which it derives in the neighbour discovery phase.

Since key K_I is deleted, an adversary, even after capturing node after T_{min}, cannot derive keys of other nodes in the network. After time $T_{min,}$ if the adversary captures a node then he can know that node's key only and its neighbours, if any. Consequently, there will be no impact of capturing of other nodes. The scheme can be further simplified when two neighbouring nodes, say u and v, are added at the same time. For example, if u receives v's response to u's *HELLO* before u responds to v's *HELLO, u* will suppress its own response. However, if u and v finish their neighbour discovery phase separately, in the pair wise key establishment step they will have two different pair wise keys, K_{uv} and K_{vu}. In this case, they may choose K_{uv} as their pair wise key if $u<v$.

The Extended LEAP

If a node capture occurs before T_{min} time, then the dependency of whole network on a single K_I has to be reduced. It is assumed that node addition occurs in successive m time intervals $T_1, T_2 \ldots T_m$, with any number of nodes, in a particular time interval. Now, the base station randomly generates some keys: $K_1, K_2, \ldots, Km$ and these keys serve as initial secrets for the nodes according to the time interval in which they are deployed. A node u deployed in time interval T_i will have secret K_i from which it derives its master key $K_i(u) = f_{Ki}(u)$ for T_i. It is also loaded with master keys $K_j(u) = f_{Kj}(u)$ for all $i < j (= m)$. Here, key $K_j(u)$ for all $i < j (= m)$ is used for establishing pair wise keys with nodes that will be deployed in T_j and $K_i(u)$ to establish pair wise keys with its neighbours. The secret K_i is erased as per the original LEAP scheme.

Example: Suppose a node u is deployed in T_1. Therefore, u will have $K_{1,} f_{K1}(u)$ and $K_j(u)$ for all $2 \leq j (= \text{m})$. Here, $f_{K1}(u)$ is the master key of the node u, and u will delete K_1 after T_{min}. Now, if a node v is deployed in T_2, it will have $K_2, f_{K2}(v)$ and $K_j(v)$ for all $3 \leq j (= m)$. The pair wise key between u and v is $K_{uv} = f_{K2u}(v)$. Also, v can compute $f_{K2u}(v)$ using secret K_2 and can get K_{uv} and u has f_{K2u} loaded in it which it uses to compute the pair wise key.

Observations

The impact of node capture within T_{min} in LEAP is that the single network wide key K_I is revealed to the adversary and thus all the communication links can be captured. The problem in the extended scheme is that the scheme works for a scenario when the node has just been deployed and it has preloaded secret K_i for $1 \leq i \leq m$. However, when a node is mobile and wants to establish a key with its neighbour then since the preloaded secret has been deleted, there is no way for authenticating and establishing the pair wise key.

THE PROPOSED PROTOCOL

We present a secure node localization protocol using WBB approach, aiming at reducing the impact of node capture attack in WSN. The protocol suggests a strong key revocation mechanism in order to achieve this goal.

Network Model

The WSN in the proposed protocol has a three-tier hierarchical structure. The first tier consists of the base station, the second tier consists of some cluster heads, and the third tier consists of the sensor nodes which are aggregated into clusters based on their location. The cluster heads are static and pre-deployed by the base station and the sensor nodes are mobile in nature. The role of these three participants is described as follows:

- **Sensor Node:** At any point of time, a sensor node belongs to a cluster, monitored by the cluster head of that cluster. When a sensor node joins the network, it first performs cluster head discovery to join a particular cluster. After joining a cluster, a sensor node finds its neighbour nodes and establish secure pair-wise key with each neighbour. The unique identity of a sensor node is stored at the base station.
- **Cluster Head:** A Cluster head manages the sensor nodes within its cluster at any given point of time. It provides its cluster nodes with the keying material needed for successful key revocation. A cluster head is solely responsible for performing a key renewal procedure in such a way that the compromised node is not able to get the new keys.
- **Base Station:** Base station is responsible for deploying cluster heads and sensor nodes with initial keying material needed for successful key establishment. The base station is also responsible for detecting any compromised cluster head and if so then, performing a key revocation process at compromised cluster head.

Assumptions

Cluster heads are more powerful in terms of resources such as memory, computation and communication. It is costly to compromise a cluster head as compared to a regular sensor node. Re-keying at cluster head requires the base station to take more strong measures as compared to re-keying at sensor nodes by the cluster heads. The cluster heads can detect a compromised node in their cluster and can then perform key renewal process. An expiry time is associated with each key and after that it is mandatory for a node to renew its key. Failure to do so will lead the cluster head to believe that the node has either left the cluster or is compromised. The underlying assumption involved for successful working of the algorithm is that if a compromised node contacts a cluster head for key renewal, then the cluster head can determine if that node is compromised by analyzing its behavior. The neighbour node discovery and pair-wise key establishment process use LOCK (Eltoweissy, Moharrum, & Mukkamala, 2006) and LEAP (Zhu, Setia, & Jajodia, 2003) as the building blocks.

Adversary Model

We assume the presence of an active adversary who can eavesdrop on the wireless channel, can add/delete/edit packets on the communication channel and can also capture a node to get access of its keys.

Notations Used

Table 2 gives the notations used in the protocol.

Table 2. Notations used in the proposed protocol

Notation	Meaning
T_i	i^{th} Time Interval
K	$=\{K_1, K_2, ..., K_m\}$ - set of initial secrets
K_i	Initial secret key for time interval T_i
$K_i(u)$	Master secret for node u
$f(.)$	Pseudo Random function to generate master secret
CH_i	ID of ith cluster head
A	$=\{A_1, A_2, ..., A_M\}$ Set of administrative keys for performing EBS at cluster heads
K_{CHiB}	Key shared between cluster head CH_i and base station
$h(.)$	Hash function
$E_k(.)$	Encryption under the key k
B_i	$=\{B_{i1}, B_{i2}, ..., B_{iN}\}$ - set of administrative keys assigned to cluster head CH_i
K_{Bu}	Key shared between Base station and node u
MAC (.)	Function to calculate Message Authentication Code
N_x	Nonce selected by entity x

The protocol has three phases: Pair wise key establishment, Distance calculation to each neighbour node and Assigning location coordinates.

Pair Wise Key Establishment

This phase is subdivided into four sub-phases: *Key pre-distribution*, *Cluster formation and Neighbor discovery*, key *establishment,* and *Key revocation.*

Key Pre-Distribution

It is assumed that node addition occurs in successive m time intervals $T_1, T_2, \ldots, T_m$. The base station randomly generates a set K of initial secrets and preloads them in sensor nodes according to the time interval in which nodes are deployed. For example, a node u deployed in time interval T_i has secret K_i which is used by u to derive its master key $K_i(u) = f_{Ki}(u)$. Along with that, node u is also preloaded with master keys $K_j(u) = f_{Kj}(u)$ for all $i < j (= m)$. Here, key $K_j(u)$ for all $i < j (= m)$ is used for establishing pair wise keys with nodes that will be deployed in T_j, and $K_i(u)$ is used for establishing pair wise keys with its neighbours. These pair wise keys are known as *communication keys*.

The base station also generates a set, A, of *Administrator keys* for performing EBS at cluster heads (CHs). A table corresponding to EBS $E(|CH|, k_1, m_1)$ is generated, where $k_1 + m_1 = M$ and $|CH| = C(M, k_1)$, being the total cluster heads that can be uniquely preloaded with their key sets, which are subsets of A. The base station stores the set A to perform key revocation at cluster head. For every cluster head CH_i, the base station generates set B_i and loads CH_i with its corresponding B_i. Each cluster head generates EBS $E(|S|, k_2, m_2)$, where $k_2 + m_2 = N$ and $|S| = C(N, k_2)$, being the total possible sensor nodes in a cluster. Each cluster head is also preloaded with a key K_{CHiB} shared with the base station that is used in the key revocation stage. After that, the base station updates these along all the clusters.

Each cluster head CH_i in the network is pre-loaded with the key K_{CHi} that is used for cluster formation. During cluster formation process, each sensor in a cluster i receives a set of *administrative* keys S_{ij}, where j is the jth subset of B_i. Each sensor in the network is also pre-loaded with the keys for every cluster head CH_i. Besides this, every sensor node u is also preloaded with key K_{Bu} used to authenticate a new cluster head, when revocation takes place. Each key in the network, whether preloaded in a cluster head or in a sensor node has an associated expiry time-stamp given by base station. Base station itself keeps each time stamp (say T) along with its hash $h(T)$, preventing the adversary from changing it. The cluster heads are preloaded with all expiry timestamps of administrative keys of its cluster nodes along with the timestamps for the initial secrets.

Cluster Formation and Neighbor Discovery

After deployment, formation of cluster and discovery of neighbor nodes take place. A Node *u* broadcasts a nonce N_u along with its ID encrypted with key K_{CHi}. The cluster head CH_i in the vicinity of the sensor node *u* replies back with a uni-cast *ACK* message encrypted by the same key. Node *u* stores the identity CH_i of this cluster head and responds to CH_i with an *ACK* message and along with all the keying material, [*K*], stored in node's memory, except key K_{Bu}. The keying material [*K*] helps the cluster head to renew node's key, when needed. After receiving acknowledgement from node *u*, CH_i allocates and sends the subset S_{ij} of administrative keys to node *u*. Since nodes are mobile, for simplicity, we assume that node *u* is in the vicinity of CH_i for a particular time interval *X* after which node u moves and joins another cluster through same process.

After joining a cluster, node *u* discovers its neighbor nodes by broadcasting a *HELLO* message. Each neighbor *v* responds with an *ACK* message including its identity and a MAC of this information generated using key K_v that node *u* could also generate, since node *u* knows K_i and $K_v = f_{Ki}(v)$.

The steps of the protocol are as given below:

$u \rightarrow *: E_{KCHi}\ (u, N_u)$

$CH_i \rightarrow u: E_{KCHi}\ (ACK, N_u)$

$u \rightarrow CH_i: E_{KCHi}\ ([K], ACK, N_u)$

$CH_i \rightarrow u: E_{KCHi}\ (S_{ij}, N_u)$

$u \rightarrow *: u, N^1_u$

$v \rightarrow u: v, MAC\ (K_v, u||v, N^1_u)$ for every neighbour v

Pair Wise Key Establishment

The key establishment is done in the same way as in the extended LEAP scheme (Zhu, Setia, & Jajodia, 2003). The *node discovery* and *key establishment* may be repeated by the node after the time interval *X*, as required.

Key Revocation

The *Key revocation* phase is executed either by base station or cluster head as soon as a node compromise is detected.

- **Key Revocation for Nodes:** The revocation phase renews the node key(s) in such a way that the compromised node cannot learn the new key. The following procedure is followed by a cluster head for the revocation of keys at tier-3.

When CH detects that the node u has been compromised then it searches its database for the keys which are to be renewed. For example, let us assume that the initial secrets K_1, K_2, …, K_n and the administrative keys of node u, i.e., B_{i1}, B_{i2},… ,B_{ik} are to be renewed for the compromised node. The CH first contacts the base station to get new secrets using the key it shares with the base station. The base station broadcasts new secrets to all cluster heads. An update in all the cluster heads is required because a particular sensor node can move from one cluster to another.

CH → Base station: E_{KCHiB} $(K_1, K_2, \ldots, K_n, N_{CH})$

Base station → *: E_{KCHiB} $(E(K_1(K_1')), E(K_2(K_2')), \ldots, E(K_n(K_n')), N_{CH})$

For all cluster heads i; K_i' is the new secret, encrypted with old secret K_i.

Now, CH will renew both set of keys using the remaining m administrative keys $B_{ik+1} \ldots B_{ik+m}$. As the compromised node is unaware about these remaining administrative keys, the compromised node cannot decrypt these messages. Therefore, the keys can be easily renewed. CH does not need to renew the keys shared between sensor nodes and the base station and between base station and the cluster head; because they are specific to a sensor node and other sensor nodes' communication should not get affected. The CH broadcasts m different messages, each encrypted with a different administrative key:

CH → *:

Message 1: $E(B_{ik+1}$ $(E(K_1(K_1')), \ldots, E(K_n(K_n')), E(B_{i1}$ $(B_{i1}'), \ldots, E(B_{ik}$ $(B_{ik}'))\ldots))$

Message 2: $E(B_{ik+2}$ $(E(K_1(K_1')), \ldots, E(K_n(K_n')), E(B_{i1}$ $(B_{i1}')), \ldots, E(B_{ik}$ $(B_{ik}'))$ $\ldots))$

…

Message m: $E(B_{ik+m}$ $(E(K_1(K_1')), \ldots, E(K_n(K_n')), E(B_{i1}$ $(B_{i1}')), \ldots, E(B_{ik}$ $(B_{ik}'))\ldots))$

Node Key Renewal on Request

The CH also performs key renewal when it is contacted by a node whose key is expired. For the key renewal at sensor node, a node u, whose keys are expired, contacts CH by sending a *RENEW* message along with all its keys and hashes of their corresponding expiry times. On receiving, such request from a node u, CH first verifies the expiry timings node u has reported. On successful verification, CH responds back to u with new keys.

$u \rightarrow CH: E_{KCHi}(RENEW, K_1, H(T_1), K_2, H(T_2), \ldots)$

$CH \rightarrow u: E(K_{CHi}(E(K_1(K_1')), \ldots, E(K_n(K_n')), E(B_{i1}(B_{i1}')), \ldots, E(B_{ik}(B_{ik}'))))$

Key Revocation of Cluster Head

The key revocation at tier-2 is done by the base station. It is done after the detection of a compromised CH and involves more work in comparison to tier-3, as sensor nodes under that CH need to be taken care of. When the base station detects a compromised CH_i, it searches its database for the keys which are to be renewed, that is, all the administrative keys of its cluster nodes $B_{i1}, B_{i2}, \ldots, B_{in}$ and the share of administrative keys of CH_i assigned initially by the base station, i.e., $A_1, A_2, \ldots, A_k$. The base station renews both set of keys using the remaining m administrative keys $A_{k+1}, \ldots, A_{k+m}$. Since the compromised CH_i is unaware about these remaining keys, it cannot decrypt these messages. Base station broadcasts m messages, each encrypted with a different administrative key. After receiving the new keys from base station, cluster heads distribute the new administrative keys to their respective cluster nodes. CH broadcasts new administrative keys encrypted by the old ones. Therefore, those who have the previous ones can only decrypt them. Furthermore, as the compromised CH is evicted from the network, the base station has to take care of the cluster nodes under that compromised cluster (orphaned sensor). In this case, either the base station deploys a new CH or redistributes the sensor nodes under the evicted cluster head to other uncompromised clusters. The base station updates the initial secrets in all the CHs to protect the communication in other clusters. Therefore, all uncompromised CHs renew the initial secrets in their clusters to prevent any security breach. The orphaned sensors can start the node discovery phase again in order to redistribute themselves after a time interval X. Alternatively, if a new cluster head is deployed; it is also preloaded with the K_{SiB} which is the key of sensor nodes share with the base station. This would help cluster nodes in authenticating the new CH.

Base Station $\rightarrow$ *:

Message 1: $E(A_{k+1}(E(A_1(A_1')), \ldots, E(A_n(A_k')), E(B_{i1}(B_{i1}')), \ldots, E(B_{ik}(B_{ik}')) \ldots))$

Message 2: $E(A_{k+2}(E(A_1(A_1')), \ldots, E(A_n(A_k')), E(B_{i1}(B_{i1}')), \ldots, E(B_{ik}(B_{in}')) \ldots))$

...

Message m: $E(A_{k+m}(E(A_1(A_1')), \ldots, E(A_n(A_k')), E(B_{i1}(B_{i1}')), \ldots, E(B_{ik}(B_{in}')) \ldots))$

$CH \rightarrow *: E(B_{i1}(B_{i1}')), E(B_{i2}(B_{i2}')), \ldots, E(B_{ik+m}(B_{ik+m}'))$

Base station → *: EK_{CHiB} $(E(K_1(K_1')),E(K_2(K_2')),\ldots,E(K_n(K_n')),N_B)$ (for all CH_i)

Key Renewal for Cluster Head on Request

Base station also performs the key renewal process when it is contacted by a cluster head due to its key being expired. CH contacts base station with a *RENEW* message, all its initial keys and hashes of their expiry timings. Base station verifies the expiry time and on successful verification, base station replies back with its share of administrative keys and new administrative keys for its cluster nodes encrypted by the key of CH shared with the base station. The CH can then distribute these administrative keys to its nodes using the above mentioned CH revocation steps.

CH → Base station: E_{KCHiB} $(RENEW, K_1, H(T_1), K2, H(T_2)\ldots)$
Base station → CH: $E(K_{CHiB}(E(K_1(K_1')),\ldots,E(K_n(K_n'))))$

Calculation of Distance of Neighbouring Node

In the WBB approach of secure node localization, once the nodes securely establish pair wise keys with their neighbours, each node requires to calculate distance to its neighbouring node in a secure way so that it can assign location coordinates to its neighbours and can complete the localization process. In order to calculate the distance between neighbours, one possible approach could be the use of the decrease in power level of RF signal *(Reduced Signal Strength (RSS)* method) and map it to distance. However, an adversary can increase or decrease power level of signal in the pathway, thereby, leading to false distance estimations. Another approach is to send a Radio Frequency (RF) signal to a neighbour node and start a timer (*RF Time of Flight* method). Once a response signal is received from that neighbour, the timer is stopped and the time between sending a message and receiving its response is converted into distance using the speed of light, as the RF signals travel with the speed of light. In this case, the adversary can delay the signal, causing distance enlargements. We discuss the distance bounding protocol and the approaches used to deal with distance enlargement attacks.

Distance Bounding Protocol

Distance bounding protocol (Capkun & Hubaux, 2005) is used to calculate the upper bound of the distance of one device to another. Using this protocol, a sensor node v can verify that a claimant node u at a distance d_{uv} from node v is not at a distance d where $d < d_{uv}$. In this protocol, the claimant u generates a random value N_u, generates a commitment $(c, d) = \text{commit}(N_u)$ and sends the nonce N_u to the

Figure 1. Distance bounding protocol

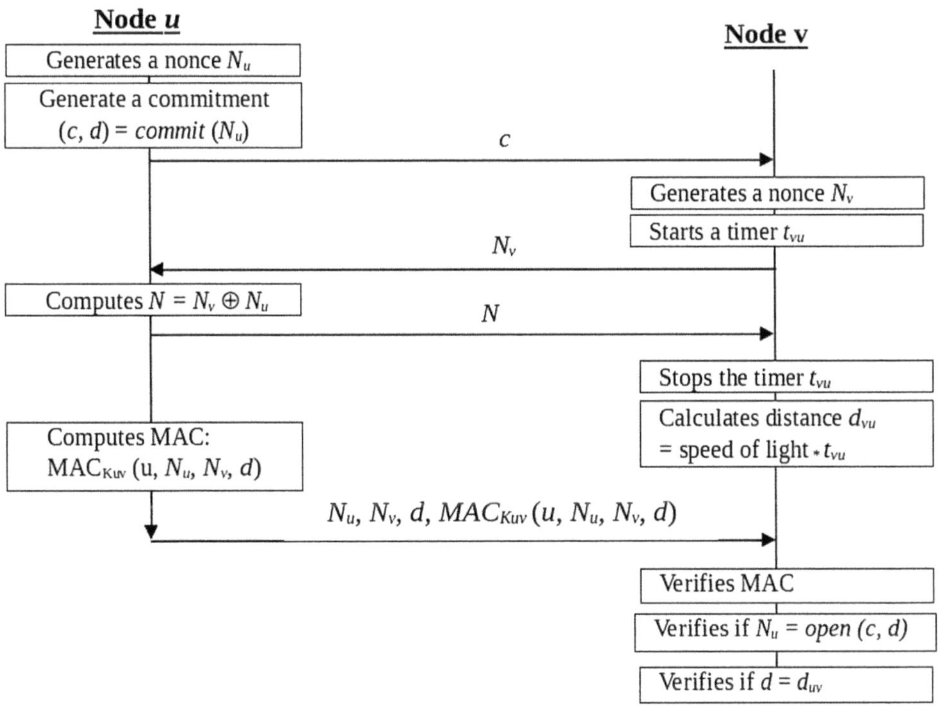

node *v*. Node *v* replies back with a challenge nonce N_v and starts a timer as soon as the challenge has been sent. Upon receiving the challenge from *v*, the claimant *u* responds immediately to *v* with $N_v \oplus N_u$. When *v* receives the response from *u*, *v* stops the timer and converts the challenge-response time t_{vu} to a distance d_{vu}. Now, *u* authenticates itself to *v* and reveals the distance *d*. The authenticity of *d* is ensured with a MAC, using pair wise key K_{uv} shared between them. Finally, *v* verifies if the value N_u received in the time-measuring phase corresponds to the received (*commit*, *open*) pair (*c*, *d*). The protocol steps are as given in Figure 1.

Distance Enlargement Attacks

Trilateration is used for determining the relative positions of objects using the geometry of triangles. It involves gathering a number of reference tuples of the form (*x*, *y*, *z*, *d*), where *d* represents an estimated distance between the source at the location (*x*, *y*, *z*) and the sensor node. The sensor node gathers four such references in case of three dimensional (3D) to accurately and uniquely determine its location.

Verifiable multilateration is an extension to *multilateration*, which is a technique for determining the position coordinates of a node from a set of reference points (four points in 3D and three points in two dimensional (2D)) whose positions are known. It uses the distance between the node and reference points for this computation. For successful computation, the node has to be positioned inside the triangle (2D) or pyramid (3D) formed by the reference points. Because of the distance bounding property, the adversary can only perform distance enlargements. If it increases the distance to one of the reference points, then to keep the position consistent the adversary needs to prove that the distances to any other reference point is shorter than it actually is, which he cannot do because of the distance bounding. Therefore, checking if the node falls within the triangle can lead to detection of enlargements attacks. There are two things that need to be done to perform verifiable multilateration. Firstly, we need four reference points for fifth node to successfully localize it. Secondly, check if the sensor node falls within the pyramid formed by the four reference points. The following test is run to detect if the claimant estimated coordinates (x, y) falls within the verification triangle formed by $((x_i, y_i), (x_j, y_j), (x_k, y_k))$.

$$f_{ij}(u) = (x - y_i)(x_j - x_i) - (x - x_i)(y_j - y_i)$$

$$f_{ki}(u) = (y - y_k)(x_i - x_k) - (x - x_k)(y_i - y_k)$$

$$f_{jk}(u) = (y - y_j)(x_k - x_j) - (x - x_j)(y_k - y_j)$$

If $f_{ij}(u) \,.\, f_{jk}(u) > 0$ and $f_{jk}(u) \,.\, f_{ki}(u) > 0$ then u is in the triangle formed by $((x_i, y_i), (x_j, y_j), (x_k, y_k))$. This test can also be extended to 3D, where the estimated coordinates (x, y, z) can be found by solving four distance quadratic equations using distances computed above.

Assigning Coordinates

Assignment of coordinates is done by the ABC (Assumption Based Coordinate) protocol (Savarese, Rabaey, & Beutel, 2001). The algorithm begins with the assumption that node n_0 is located at (0, 0, 0). The first node to establish communication with n_0, n_1, is assumed to be located at $(r_{01}, 0, 0)$, where r_{01} is the distance between n_0 and n_1. The location of the next node n_2 can then be computed as:

$$x_2 = \frac{r_{01}^2 + r_{02}^2 + r_{12}^2}{2r_{01}}, \; y_2 = \sqrt{r_{02}^2 - x_2^2}, \; z_2 = 0$$

The coordinates for node n_3 is computed as:

$$x_3 = \frac{r_{01}^2 + r_{03}^2 + r_{13}^2}{2r_{01}},\ y_3 = \frac{r_{03}^2 + r_{23}^2 + x_2^2 + y_2^2 - 2x_2x_3}{2y_2},\ z_2 = \sqrt{r_{03}^2 - x_3^2 - y_3^2},$$

where, r_{ij} is the distance between node i and node j.

Using these coordinates for the initial nodes, a node can complete the localization process.

SECURITY ANALYSIS

- **Adversarial Model and Assumptions:** We consider both active and passive adversary, who can launch internal attacks (e.g., node capture) and can perform external attacks (e.g., eavesdropping, replay). The sensor nodes and cluster heads are not considered tamper proof and adversary can get all the keys stored in a node after capturing that node. We assume that the adversary is not aware of what keying material is stored in a node. For example, as in EBS after capturing M_0 node the adversary does not know whether he requires capturing M_5 in order to get keying materials of the network.

Analysis of the Cluster Formation Protocol

The cluster formation protocol works as follows:

1. u → *: $E_{KCHi}(u,N_u)$
2. CH → u: $E_{KCHi}(ACK,N_u)$
3. u → CH: $E_{KCHi}(X,ACK,N_u)$
4. CH → u: $E_{KCHi}(S_{ij}, N_u)$

Node Impersonation attacks:

1. Let us assume that a passive adversary tries to impersonate u in the first step:

$I(u)$ →: $E_{KCHi}(u,N_u)$

As the passive adversary does not know K_{CHi}, he cannot impersonate as u without having K_{CHi}.

2. Assume that an active adversary, who captures a node and acquires K_{CHi}, and then attempts to impersonate as u.

$I(u) \rightarrow *: E_{KCHi}(u, N_u)$

$CH \rightarrow I(u): E_{KCHi}(ACK, N_u)$

$I(u) \rightarrow CH: E_{KCHi}(X, ACK, N_u)$

$CH \rightarrow I(u): E_{KCHi}(S_{ij}, N_u)$

The adversary can also make a successful run of the protocol by compromising a cluster head, thereby, allowing illegitimate nodes to register or knowing the keying material of a node u.

$u \rightarrow *: E_{KCHi}(u, N_u)$

$I(CH) \rightarrow u: E_{KCHi}(ACK, N_u)$

$u \rightarrow I(CH): E_{KCHi}(X, ACK, N_u)$

$I(CH) \rightarrow u: E_{KCHi}(S_{ij}, N_u)$

This attack is mitigated in our protocol. After the expiry time, K_{CHi} has to be renewed and according to our assumption, if a compromised node or cluster head asked for renewal, it will be caught by the corresponding renewing CH/Base station.

Replay attacks: Suppose that an attacker intercepts a previous run of the protocol:

$u \rightarrow *: E_{KCHi}(u, N_u)$

$CH \rightarrow u: E_{KCHi}(ACK, N_u)$

$u \rightarrow CH: E_{KCHi}(X, ACK, N_u)$

$CH \rightarrow u: E_{KCHi}(S_{ij}, N_u)$

Then he replays the same in another session to fool CH into believing that u wants to join the cluster even when u has left the cluster. But, due to nonce validity, the attacker cannot replay the same nonce in other session without being caught. However, the adversary can run a parallel session and can execute following:

1. u → *I*: E_{KCHi} (u,N_u)
   1′. *I*(u) → *: $E_{KCHi}(u,N_u)$
   2′. CH → *I*(*u*): E_{KCHi} (ACK,N_u)
2. CH → *u*: E_{KCHi} (ACK,N_u)
3. *u* → *I*(*CH*): E_{KCHi} (X,ACK,N_u)
   3′. *I*(*u*)→ CH: E_{KCHi} (X,ACK,N_u)
   4′. CH→ *I*(*u*): E_{KCHi} (S_{ij}, N_u)

Though the attacker is able to complete the parallel session with CH (nonce not expired), the adversary is not able to gain anything from this attempt. Furthermore, as the key is renewed after expiry time, it ensures *key freshness*.

Analysis of Neighbour Discovery Phase

- **Node Impersonation Attacks:** Suppose that a passive attacker tries to impersonate *v* and register itself as a neighbour of *u* by the following run of the protocol:

$u \rightarrow *: u, N_u$

$I(v) \rightarrow u$: v, MAC(K_v, $u||v$, N_u) for neighbour v.

As the passive attacker does not possess the secret *k* to construct K_v, he cannot impersonate *v*. An active attacker, on the other hand, can get K_v by capturing the node. In that case, he will be able to convince *u* as a neighbour. However, this will be valid till the time K_v does not get expired. After expiry if the node is still compromised then applying for a renewal procedure can lead to its detection and hence the attack will be caught.

Analysis of the Key Revocation Protocol

Eavesdropping and Replay Attack: A passive adversary may eavesdrop on the wireless communication channel to gain access to some information or he may replay the packets to disrupt the protocol for key revocation. However, the uni-cast messages communicated between a cluster head and the base station are encrypted with the secret key uniquely shared between that cluster head and the base station. Similarly the messages broadcasted by the base station during key revocation/renewal process are protected using the administrative keys not accessible to passive adversary.

CH → Base station: E_{KCHiB} $(K_1, K_2, \ldots, K_n, N_{CH})$

Base station → *: E_{KCHiB} $(E(K_1(K_1')), E(K_2(K_2')), \ldots, E(K_n(K_n')), N_{CH})$

Base Station → *:
Message 1: $E(A_{k+1}(E(A_1(A_1')),..., E(A_n(A_k')), E(B_{i1}(B_{i1}')),..., E(B_{ik}(B_{ik}'))...))$
Message 2: $E(A_{k+2}(E(A_1(A_1')),..., E(A_n(A_k')), E(B_{i1}(B_{i1}')),..., E(B_{ik}(B_{in}'))...))$
...
Message m: $E(A_{k+m}(E(A_1(A_1')),..., E(A_n(A_k')), E(B_{i1}(B_{i1}')),..., E(B_{ik}(B_{in}'))...))$
Base station → *: $EK_{CHiB}(E(K_1(K_1')), E(K_2(K_2')),..., E(K_n(K_n')), N_B)$ for all CH_i.

Similarly, the messages broadcasted by a cluster head to its nodes are encrypted with the administrative keys.

CH → *:
Message 1: $E(B_{ik+1}(E(K_1(K_1')),..., E(K_n(K_n')), E(B_{i1}(B_{i1}'),..., E(B_{ik}(B_{ik}'))...))$
Message 2: $E(B_{ik+2}(E(K_1(K_1')),..., E(K_n(K_n')), E(B_{i1}(B_{i1}')),..., E(B_{ik}(B_{ik}'))...))$
...
Message m: $E(B_{ik+m}(E(K_1(K_1')),..., E(K_n(K_n')), E(B_{i1}(B_{i1}')),..., E(B_{ik}(B_{ik}'))...))$
CH →*: $E(B_{i1}(B_{i1}')), E(B_{i2}(B_{i2}')),..., E(B_{ik+m}(B_{ik+m}'))$

The replay attack is avoided by using nonce N_{CH}, when a cluster head requests key revocation to base station on identifying a node compromise. Also, all the keys in the network are time bound.

- **Node Capture Attack:** Node Major threat due to presence of an active adversary is node compromise or node capture attack. The *node capture attack* is carried out by an adversary in order to obtain the keys stored in the node and use that information to establish links with other nodes in the network for doing malicious activities. After capturing a node, an adversary may launch *node impersonation attack*. However, an active adversary can perform a successful run of this protocol but only till the time K_{CHiB} is not expired. While CH renews keys, the compromised node cannot decrypt messages, as the adversary is unaware about B_{ik+1}, $B_{ik+2}...B_{ik+m.}$ Also at any step, unless CH itself is compromised, an adversary cannot form the messages because of the unavailability of B_{ik+1}, B_{ik+2}, ...,B_{ik+m} . If CH is compromised, the adversary can impersonate till the time these administrative keys are not expired.

Table 3. Comparison of the proposed protocol with related works

Features→ Schemes	Resilient to Node Capture Attack	Supports Key Revocation	Supports Node Mobility
SeRLoc (2004)	No	Not applicable	No
ROPE (2005)	No	Not applicable	Yes
HirLoc (2006)	No	Not applicable	No
Chen et al (2014)	No	Not applicable	No
LDAC (2012)	No	No	Yes
LEAP (2003)	Yes	Yes	No
LOCK (2006)	Yes	Yes	Yes
Proposed protocol	Yes	Yes	Yes

Similarly, key revocation for cluster heads by base station follows similar logic. Therefore, node impersonation attack is resisted in the proposed key revocation protocol. Since all the keys in the network has pre-defined expiry time and are bound to be replaced by new set of keys, before an adversary launch further attacks after compromising a node, any malicious activity done by adversary will be caught and the node keys will be revoked and compromised node will not be provided with new set of keys. Thus, the protocol provides resilience to node capture attack.

- **Forward and Backward Secrecy:** The secrets K_1, K_2,...,K_n used to establish a pair-wise key between neighbor nodes are replaced by new set of keys, whenever key renewal takes place (due to malicious node key revocation or key expiry). Due to this, a node revoked in some time period T_l would not be able to use the keys it holds from previous sessions. Furthermore, any key set used after time period T_l will not be made available to the revoked node. Thus, the key revocation protocol ensures forward and backward secrecy in this group key broadcast scenario.

Comparison of the Proposed Protocol with Related Works

We compare the proposed protocol with some existing schemes relevant to secure node localization and key management protocols and show that the proposed protocol provides secure node localization without using beacon nodes, supports node mobility and key revocation and provides resilience to node capture attack. The comparison is reflected in Table 3.

CONCLUSION

GPS-free secure node localization is an important research topic in WSN, where nodes are mobile in nature. We discussed both static and dynamic key management schemes and proposed an improved protocol for pair-wise key establishment and key revocation based on LEAP and LOCK to facilitate secure node localization. We have addressed secure node localization in WSN without beacon nodes by providing strong key revocation mechanism in the network in order to avoid node impersonation and replay attack. This ensures improved resilience against node capture threat. Effective and timely key revocation is crucial to thwart node capture attacks. In the proposed protocol, we provided the key revocation stages at sensor nodes, cluster heads and base station, which are to be executed as and when a compromise node is detected. We have analyzed the improved protocol and shown that the protocol provides secure node localization with minimal node capture threats. The protocol also ensures forward and backward secrecy for keys broadcasted in the network during key revocation process.

REFERENCES

Agrawal, S., Roman, R., Das, M. L., Mathuria, A., & Lopez, J. (2012). A Novel Key update Protocol in Mobile Sensor Networks. In *Proceedings of International Conference on Information Systems Security* (LNCS), (vol. 7671, pp. 194-207). Berlin: Springer. doi:10.1007/978-3-642-35130-3_14

Bechkit, W., Challal, Y., & Bouabdallah, A. (2013). A New Class of Hash-Chain based Key Pre-distribution Schemes for WSN. *Journal of Computer Communications.*

Blundo, C., De Santis, A., Herzberg, A., Kutten, S., Vaccaro, U., & Yung, M. (1993). Perfectly-secure Key Distribution for Dynamic Conferences. In *Advances in Cryptology* (LNCS), (vol. 740, pp. 471-486). Berlin: Springer.

Capkun, S., & Hubaux, J. (2005). Secure Positioning of Wireless Devices with Application to Sensor Networks. In *Proceedings of the Annual Joint Conference of the IEEE Computer and Communications Societies* (vol. 3, pp. 1917-1928). IEEE. doi:10.1109/INFCOM.2005.1498470

Chan, H., Perrig, A., & Song, P. (2003). Random Key Pre-distribution Schemes for Sensor Networks. In *Proceedings of IEEE Symposium on Security and Privacy*, (pp. 197-213). IEEE.

Chen, H., Chen, W., Wang, Z., Wang, Z., & Li, Y. (2014). Mobile Beacon Based Wormhole Attackers Detection and Positioning in Wireless Sensor Networks. International Journal of Distributed Sensor Networks.

Eltoweissy, M., Moharrum, M., & Mukkamala, R. (2006). Dynamic Key Management in Sensor Networks. *IEEE Communications Magazine*, *44*(4), 122–130. doi:10.1109/MCOM.2006.1632659

Eschenauer, L., & Gligor, V. (2002). A Key Management Scheme for Distributed Sensor Networks. In *Proceedings of the ACM Conference on Computer and Communication Security*, (pp. 41-47). ACM. doi:10.1145/586110.586117

Giannetsos, T., & Dimitriou, T. (2012). LDAC: A Localized and Decentralized Algorithm for Efficiently Countering Wormholes in Mobile Wireless Networks. *Journal of Computer and System Sciences*, *80*(3), 618–643. doi:10.1016/j.jcss.2013.06.015

Gong, S., Zuo, X., Mei, L., & Zhao, R. (2014). Enhanced Key Management Scheme Based on Random Key Pre-distribution for Wireless Sensor Networks. In Future Information Technology (LNEE), (vol. 309, pp. 343-348). Springer. doi:10.1007/978-3-642-55038-6_52

Guo, S., Leung, V., & Qian, Z. (2010). A Permutation-Based Multi-Polynomial Scheme for Pair wise Key Establishment in Sensor Networks. In *Proceedings of IEEE International Conference on Communications*, (pp. 1-5). IEEE.

He, X., Niedermeier, M., & Meer, H. (2013). Dynamic Key Management in Wireless Sensor Networks: A Survey. *Journal of Network and Computer Applications*, *36*(2), 611–622. doi:10.1016/j.jnca.2012.12.010

Lazos, L., & Poovendran, R. (2004). SeRLoc: Secure Range-Independent Localization for Wireless Sensor Networks. In *Proceedings of the Third ACM workshop on Wireless security*, (pp. 21-30). ACM.

Lazos, L., & Poovendran, R. (2006). HiRLoc: High-resolution Robust Localization for Wireless Sensor Networks. IEEE Journal on Selected Areas in Communications, 24(2), 233-246.

Lazos, L., Poovendran, R., & Capkun, S. (2005). Rope: Robust Position Estimation in Wireless Sensor Networks. In *Proceedings of the Fourth International Conference on Information Processing in Sensor Networks*, (pp. 324-331). Academic Press.

Liu, D., & Ning, P. (2003). Establishing Pair wise Keys in Distributed Sensor Networks. In *Proceedings of the ACM Conference on Computer and Communications Security*, (pp. 52-61). ACM.

Liu, D., Ning, P., Liu, A., Wang, C., & Du, W. (2008). Attack Resistant Location Estimation in Wireless Sensor Networks. *ACM Transactions on Information and System Security*, *11*(4), 1–39. doi:10.1145/1380564.1380570

Mi, Q., Stankovic, J., & Stoleru, R. (2012). Practical and Secure Localization and Key Distribution for Wireless Sensor Networks. *Ad Hoc Networks*, *10*(6), 946–961. doi:10.1016/j.adhoc.2011.12.008

Niu, Y., Gao, D., & Gao, S. (2012). A Robust Localization in Wireless Sensor Networks against Wormhole Attack. Networks, 7(1), 187-194. doi:10.4304/jnw.7.1.187-194

Rasmussen, K., Capkun, K., & Cagalj, M. (2007). SecNav: Secure Broadcast Localization and Time Synchronization in Wireless Networks. In *Proceedings of the ACM International Conference on Mobile Computing and Networking*, (pp. 310-313). ACM.

Savarese, C., Rabaey, J., & Beutel, J. (2001). Locationing in Distributed Ad-hoc Wireless Sensor Networks. In *Proceedings of IEEE International Conference on Acoustics, Speech, and Signal Processing*, (pp. 2037-2040). IEEE.

Srinivasan, A., & Wu, J. (2007). A Survey on Secure Localization in Wireless Sensor Networks. In The Encyclopedia of Wireless and Mobile Communications. CRC Press.

Xiao, Y., Rayi, V., Sun, B., Du, X., Hu, F., & Galloway, M. (2007). A Survey of Key Management Schemes in Wireless Sensor Networks. Computer Communications, 30, 2314-2341. doi:10.1016/j.comcom.2007.04.009

Zhang, T., He, J., & Yu, H. (2012). Secure Localization in Wireless Sensor Networks with Mobile Beacons. International Journal of Distributed Sensor Networks.

Zhu, S., Setia, S., & Jajodia, S. (2003). LEAP: Efficient Security Mechanisms for Large-scale Distributed Sensor Networks. In *Proceedings of the ACM Conference on Computer and Communications Security*, (pp. 62-72). ACM.

Chapter 3
Energy Efficient Congestion Control in Wireless Sensor Networks

Awais Ahmad
Kyungpook National University, South Korea

Sohail Jabbar
COMSATS Institue of Information Technology, Pakistan

Anand Paul
Kyungpook National University, South Korea

Seungmin Rho
Sungkyul University, South Korea

ABSTRACT

Avoiding from congestion and provision of reliable communication characterising the low energy consumption and high data rate is one of the momentous challenges at Media Access Control (MAC) layer. This become more difficult to achieve when there is energy constraint mixed with mobility of nodes. Same issue is addressed in this underlying paper. Here we have proposed a Time-Sharing Energy Efficient Congestion Control (TSEEC) technique for Mobile Wireless Sensor Networks. Time Division Multiple Access Protocol (TDMA) and Statistical Time Division Multiple Access Protocol (STDMA) are major constituents of this technique. These helps in conserving the energy by controlling the sleeping, waking up and listening states of sensor nodes. Load Based Allocation and Time Allocation Leister techniques further helps in conserving the network energy minimizing the network congestion. First mentioned technique is designed on the basis of STDMA Protocol and uses the sensor node information to dynamically assign the time slots while later said technique is does the job of mobility management of sensor node. This Time Allocation Leister techniques further comprises of Extricated Time Allocation (ETA), Shift Back Time Allocation (SBTA), and eScaped Time Allocation (STA) sub techniques for

DOI: 10.4018/978-1-4666-9941-0.ch003

Copyright ©2016, IGI Global. Copying or distributing in print or electronic forms without written permission of IGI Global is prohibited.

managing the joing and leaving of nodes to cluster and redundant\absence of data for communication respectively. To control the movement of mobile sensor nodes, we have also introduced mobility pattern as part of TSEEC that helps in making the protocol adaptive to traffic environment and to mobility as well. A comparitive analysis of proposed mechanism with SMAC is performed in NS2 along with mathematical anslysis by considering energy consumption, and packet deliver ratio as performance evaluation parameters. The results for the former outperforms to that of later. Moreover, comparative analysis of the proposed TSEEC with other MAC protocols is also presented.

INTRODUCTION

Wireless interconnection of autonomous sensors over some shared medium make up wireless sensor network. This emerged network has its large number of applications including temperature, pressure, vibration, sound and etc. The collected data from the target environment is passed on to base station either directly or through multihop communication fashion. The new set of mobile WSN is the result of advancement in static WSN along with distributed robotics (Bergbreiter and Pister 2003). The network architecture of MWSN is same as is of WSN, however implicit or explicit mechanism are provided to move these sensors in space over time (Zeinalipour et al. 2008). Relative means or absolute means are used to derive the coordinates of MWSN. Localization techniques is a typical example of former while Geographical Positioning System is of later (Lingxuan and David 2004). Based on movement speed, MWSN are categorized into i) highly mobile ii) mostly static and iii) hybrid. In the first case, movement is of the speed of human, cars or airplanes. Very low velocity such as that of moving robotics comes under second given case. The thrid category is the combination of both above. Such as moving cars with sensors installed therein (Zeinalipour et al. 2008). Some exceptional challenges such as energy efficiency and congestion are posed due to the perceptibly dynamic characteristics of MWSN as well as applications of traditional wireless networks.

Since the sensor nodes are battery operated and replenishment of battery is almost impossible in some scenarios especially in the applications related to MWSN. So, energy is considered as the most vital design confront in such networks. This is also important in increasing the network lifetime and ultimately comes up with better performance with respect to latency, throughput and bandwidth parameters. This is one of the key reason that current research is mainly focused on minimizing the energy consumption ratio.

On one hand, various activities of radio such as transmission, reception, Idle listening to the channel effect much on draining the node's energy. On the other hand, energy conservation factor is also affected by the congestion and its related effects. Congestion either arises due to fastly approaching the data at some node compared to its processing and further transmission to the ultimate destination or due to simultaneous sending of data from numerous nodes at sink nodes. This also comes with the effect of packet loss that again requires their retransmission. So, its become obvious to have its disastrous effect on node's energy. In such scenarios, efficient and effective MAC protocol has its indispensable role in minimizing the energy consumption ratio. It allocates the medium resources in sensible way to have its contribution in achieving the target of lesser energy draining.

Based on the above detailed discussion relating to sensor node's constraint resources, this current underlying study is on the purpose of implementing a novel mobility aware energy efficient MAC protocol for Mobile WSN. The awakening, sleeping, and listening states of nodes (scheduling algorithm) is developed for the proposed MAC has the aim to achieve relatively better performance in energy conservation. TSEEC has the assumption that Mobile Sensor Nodes have their location information through some localization technique (Zeinalipour et al. 2008). Such information is used to predict the mobility pattern via the technique presented in (Zaidi and Mark 2004)

In our proposed scheme, causes of congestion are classified into two classes. I) Node level congestion (NLC) that is caused by the overflow of buffer and ii) link level congestion (LLC) is caused by pumping of large data into channel sent by various nodes deployed in neighborhood at the same time. So, exploiting this scenario, we have designed a protocol that is based on STDMA. This has its role in performance improvement in both NLC and LLC. Thus ultimately comes up with the result of network energy efficiency. A hello packet having some defined statistical information is communicated with Cluster Head (CH) by each mobile sensor node. Based on this collected information, a modified TDMA technique is used for timeslot allocation by CH to its mobile sensor nodes in its specific cluster. This technique gives energy efficiency to network and also play its role in minimizing the congestion at sink node in Mobile WSN. Various shortcomings in the static WSN (Zorbas and Douligeris 2011) (Zorbas et al. 2012) are solved by mobile sensor nodes. This proposed technique also works in the scenarios of joing, or leaving the cluster as well as where there is redundant data or no data from environment at all to communicate to the cluster head.

BACKGROUND

In this section, we overview some of the existing schemes and their challenges associating with building congestion, delay, and energy consumption for WSN. WSN have some of the unique design level constraints because of the limited signal range, processing power, storage capabilities, as well as communication pattern. Thus, it is quite necessary to take into account such constraints while deploying wireless sensor nodes. We now briefly explain some of the existing related schemes.

Majority of the issues in WSN are end-to-end delay, congestion, and delay. Such problem arises when many neighboring nodes transmit their data to their respective cluster head. Such communication consume considerable amount of energy due to congestion as well as delay.

Based on the various protocols developed in the field of WSN, one of the protocol that is based on the randomized coordination algorithm, which improves the efficiency of the network (Zaidi and Mark 2014). The basic idea of the proposed scheme is based on the scheduling system among various nodes, which turn-on their transceiver when required. Such sensors are then responsible to play a role of coordinator node. In such scenario, where a coordinator node selection is based on the scheduling system usually depends on the hop-by-hop communication, direct communication, depends on the network capacity. In the proposed scheme, the number of the coordinators are kept as minimum in order to enhance the network lifetime. Moreover, GAF scheme is proposed based on the geographical location information (Xu et al. 2001). In such scheme, the geographical area is divided into fixed square grids. Each grid square are equal in size. The sensor nodes deployed in such region should follow the strict scheduling period of sleep and awake to minimize the battery lifetime. However, one sensor node should be awake all the time among these sensor nodes with a view to reach its destination.

One the exiting MAC protocol i.e., PCAP based on the handshaking mechanism of RTS and CTS that helps in avoiding collision in the network (Guo et al. 2006). In such scheme, RTS and CTS badges the sensor nodes in order to perform their respective tasks. The throughput of the proposed scheme is about 20%, which is much more acceptable in WSN environment. Moreover, the RTS and CTS incorporates additional delay in the network. With a view to avoid congestion in WSN environment, various other schemes are proposed, such as CSMA/CA in IEEE 802.15.4 for CD-WPAN. In this scheme, the usage of beacon enable mode is introduces (Sokullu and Donertas 2008). Moreover, the sharing of information is achieved through cross layer communication. Such scheme is not a useful remedy to avoid congestion in the WSN because of many-to-one traffic in the network.

Additionally, one of the major threat to WSN environment is congestion at sink node. Such congestion occurs because of receiving incoming packets from various sourced. Such congestion could be controlled by Robust Routing Algorithm with Fair Congestion Control (RRA-FCC) (Liu et al. 2008). Typically in WSN, there are almost one or two sink nodes, depending on the number of sensor nodes. However, various research have been conduction in which more than two sink nodes are considered.

Another alternate way to control congestion in WSN is to deploy mini sink, which is responsible for retrieving of data from various sensor nodes in the network (Khan et al. 2007). When the data is retrieved, aggregation technique is applied in order to reduce the number of carrier packets in the network. Such reduction results in minimizing the number of hops towards destination as well as conserve the energy during transmission. As it is a known fact that wireless sensor nodes are operated by batteries. If these sensor nodes are deployed in a region where changing of these battering are not possible, than there is a great chance of failing of a network. In order to cope with such problem, a protocol called TRAMA is designed that is used for energy conservation as well as avoiding congestion (Chih and Lollini 1994). The proposed scheme is based on the TDMA technique, which splits the timeslots into two sections. Among these, one part is based on random access protocol, whereas, the latter is based on scheduled access period. However, the main disadvantage of the proposed scheme is used to stop the long communication process that lead towards the excessive amount of dropping of packets.

Adhoc on Demand Distance Vector Routing protocol (AODV) based on scheme is proposed that is capable of reducing network segment error along with the control information (Ahmed and Hussain 2011). These factor could be achieved by manipulation of the path finding, and also addressed various constants of reliability, link repair with minimum overhead, and improved error control mechanism. In the proposed scheme, the sensor nodes relayed their data towards their cluster head in three sections, such as mobility factor, zone head selection, route maintenance and node mobility.

Adhoc on Demand Distance Vector Routing protocol (AODV) based on scheme is proposed that is capable of reducing network segment error along with the control information (Ahmed and Hussain 2011). These factor could be achieved by manipulation of the path finding, and also addressed various constants of reliability, link repair with minimum overhead, and improved error control mechanism. In addition, a scheme based on greedy technique is proposed, which calculates the cost function for each nodes, and then route the data (Zou et al. 2005). The phenomena through which cost is calculated is close to Euclidean distance of the shortest path from node towards their based station. Later, when the greedy routing is unable to perform, the packet uses the ghi-cost-to-low-cost rule and relay the data towards its coordinator or sink node. Usually, AODV protocol suffers from performance degradation where nodes are mobile. However, it performs well when the nodes are static (Das et al., 2000). In addition, there is no such mechanism in AODV protocol that helps in finding the cost and willingness of anoce to be a part of data transmission.

In case, if we are considering DMAC in WSN, it usually consumes extra energy (Kyoungseok et al. 2007). To deal with such constraint, improved energy efficiency of DMAC is proposed that is based on full sleep cycle. Such technique is used to reduce the congestion, whereas, enhance the energy efficiency in the network.

In WSN, the energy conservation usually depends on the distance between sensor nodes i.e., greater the distance, more quickly depletion of energy will happen, and vise-versa. Moreover, other reasons for wastage of energy is the state of the energy i.e., idle listening, collision, over hearing, central packet, and overmitting (Anastasi et al. 2009). In order to cope which such issues, various other techniques are proposed that helps in minimizing energy consumption, such as duty cycling, and data driven technique. However, the mentioned scheme still need acceptable scheduling techniques, which helps in minimizing congestion in WSN (Anastasi et al. 2009). Several other well known protocols including CCP (Xing et al. 2002), SPAN (Chen et al. 2002), HEED (Younis and Fahny 2004), ECODA (Tao and Yu 2009), and LCM (Shih et al. 2013) that helps in minimizing the said problem. These schemes do not cope with the delay incorporated in the network.

In the literature, MICA1 mote is an example whose energy consumption is given in the Table 1 (MICA2 Mote Datasheet). However, various other schemes that helps in minimizing energy consumption based on compression are discussed in (Pradhan and Ramchandran 2003) (Tang and Raghavendra 2004) (Wu, Chen 2006) and (Xiong et al. 2004).

For data collection procedure, several algorithms are proposed that efficiently collect the data in WSN environment. Such schemes are based on clustering phenomena that uses residual energy. Moreover, clusterhead rotating mechanism is also introduced for the same purpose. To address these problems, fuzzy logic based clustering scheme is developed along with the extension of energy prediction. The proposed scheme is used for enhancing the network lifetime by evenly distribution of workload among various sensor nodes (Lee and Cheng 2012).

Table 1. MICA sensor specification

State	Energy Consumption (mW)
Transmission	80
Reception/ Idle	30
Sleep	0.003

In the given scenarios of WSN, where link breakage is one of the severe issues while using mobile wireless sensor nodes as well as abrupt movement, the proposed scheme work well in such condition by incorporating mobility pattern along with the scheduling technique. In addition, the proposed scheme also helps in minimizing the time required for the sensor node to communicate with the clusterhead.

PROPOSED SOLUTION

Application Scenario

It is a challenging task to reduce the congestion at sink node. Hence, a novel design of MAC protocol is required that focuses on energy efficiency as well as congestion control. The designed technique helps in enhance the network lifetime as well as efficiency. Due to the inherent characteristics of the WSN, a sensor node could be classify into two stages, such as mobility phase as well as stagnant phase. In mobility phase, the node move rapidly, whereas, in stagnant phase, the node movement is very less. Therefore, the proposed scheme uses stagnant phase, where a node could stay at some location for a period of time. For those scenarios, where a stagnant phase does not exists, there the scheduling as well as time sharing phenomena is not possible to implement. In the designed algorithm, stagnant phase is quite long that fulfills the basic criteria for scheduling as well as time sharing. In addition, the sensor node location information is used for decision making model as shown in Figure 1.

Overview

In a given scenario, where we have n number of mobile sensor nodes, which are deployed in a coordinate plane. The nodes that are deployed are grouped together to form a respective cluster, whereas, in each cluster, a static cluster head is deployed as shown in Figure 2.

Formation of the cluster head (CH) is described in later sections. In the given cluster, the assignment of time slots to newly joined mobile sensor nodes is with the help if statistical time division multiplexing (STDMA). The STDMA is based on the statistical information of the nodes, i.e., node ID, their battery information, and location information. After getting these information, the CH evaluates each node statistical information and assign dynamic timeslots to all sensor nodes in the given cluster.

The proposed scheme is composed of two main strategies that is dealing with the delay in a network. These strategies are Load Based Allocation, and Time Allocation Leister. Figure 3 summarizes the process of proposed scheme (TSEEC).

CLUSTER HEAD FORMATION

Cluster Head (CH) is based on self-capability of organizing of sensor nodes. As mentioned earlier that all the sensor nodes are aware of their location information. Moreover, we assume that they are aware of their neighboring information in their vicinity head. The terminology vicinity head is used for neighboring of a sensor nodes, i.e., signal range. Initially, the deployment of the sensor nodes is settle on, hence at the beginning, selection of CH is on hand. However, due to the homogeneous nature of the sensor nodes, the node having maximum sensor nodes is selected as CH. Afterward, each sensor node attaches itself to CH on the basis of received signal strength. For instance, a sensor node receives request from neighboring CH, than the following algorithm is followed.

$$Sweight_i > Sweight_j \qquad \forall_j$$

where $Sweight$ is weight or strength of received signal of the invitee CH.

If $Sweight_i = Sweight_j \qquad \forall_j$

Then the selection is on the basis of

$$Eweight_i > Eweight_j \qquad \forall_j$$

where $Eweight$ is the weight of energy level of invitee CH.

If $Sweight_i = Sweight_j \qquad \forall_j$

random selection is made.

LOAD BASED ALLOCATION

In load based allocation phase, the Ch assign timeslots to each mobile sensor node in a cluster. In a given cluster, mobile sensor node send a Type-I Hello Packet to its CH. The format of Type-I Hello Packet is Type-I_Broadcast_msg {node ID, battery information, location information, propagation time}.

Figure 1. Typical working of TSEEC

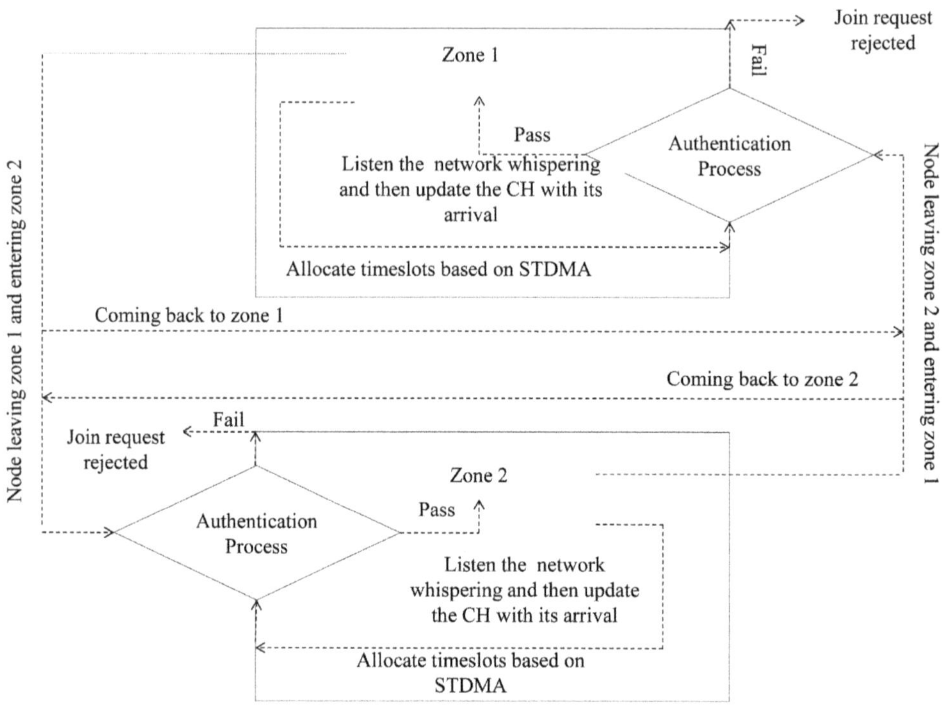

Figure 2. Deployment of sensor nodes along with cluster head

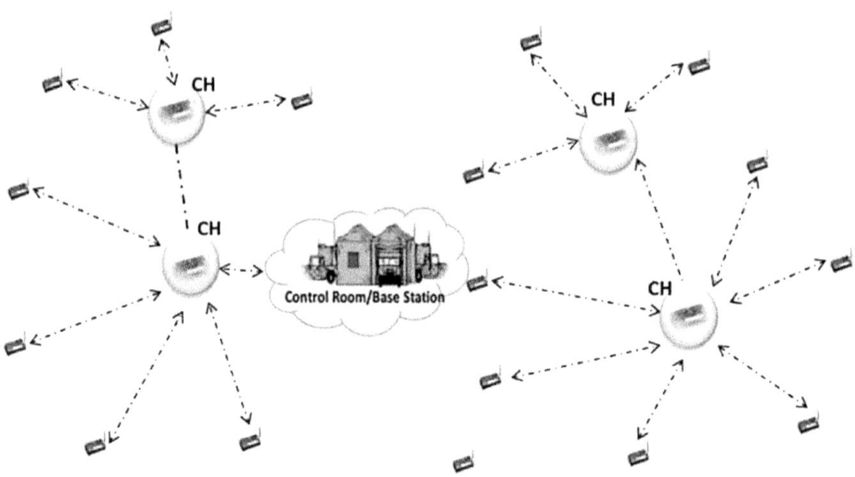

Figure 3. Working of TSEEC

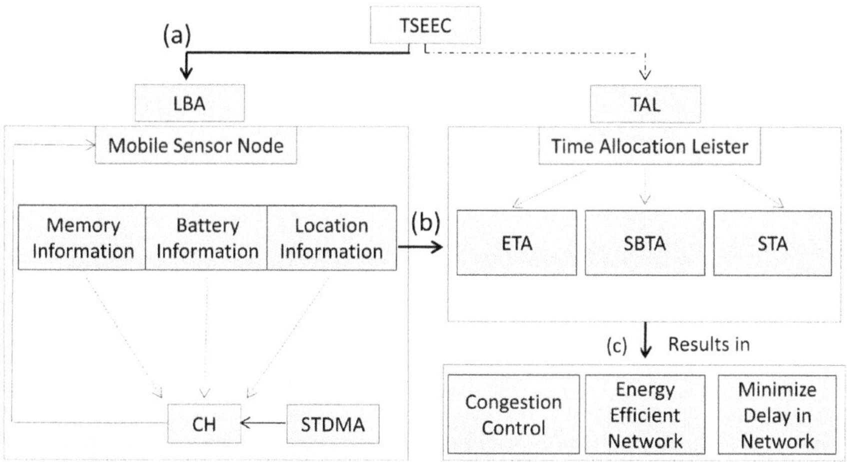

After getting these information, the CH compute the distance between each mobile sensor nodes and itself by using time of arrival technique. Afterward, the CH is then assigns weights to all of its member nodes on the basis of distance and other information available in Type-I Hello Packet. Correspondingly, the proposed scheme uses the weighted values as well as broadcast_msg as a matrix for assigning time slots to all of its member nodes. Equation 1 is used to calculate the weighted values.

$$\omega = \left(\frac{1-\left(ToA\right)}{C} \text{ x } 100\right) + \text{K} \tag{1}$$

In a given equation, ω represents the weight, K represents priority values of the candidate node, and c represents the cost, i.e., statistical information of a node. Moreover, C is calculated by considering the mean of the statistical information of the ith node.

$$\text{Cost } (C_i) = \frac{\text{battery life time of a node} + \text{Location information of a node)}}{2} \tag{2}$$

The algorithm for assignment of time slots is as follows.

When the weight is calculated for an mobile sensor node node, this information is broadcast to the cluster by Type-II_broadcast_msg. Type_II broadcast message is in the following form: Type-I_Broadcast_msg{node id's, beginning time of cycle,

Algorithm 1.

```
Mobile_sensor_node
Mobile Sensor Node (MSN) node_ID, Battery_life, Location_Info,
time_of_sending
        Hello_Packet(){
node_ID = MSN→node_ID
                Battery_life = MSN→Battery_life
                Location_info = MSN→Location_Info
                Location_info=MSN→time_of_sending}
while(!total_mobile_sensor_nodes)
        MSN Hello_ Packet() to CH
        TOA(){
        distance_MSN=mobile_sensor_node→arrival_
tim        e-MSN→time_of_sending}
        CH assign Weights to MSN
        End while
```

scheduling information, dynamic time slots}. Every node becomes aware of mobile sensor node (MSN) information exist in the same cluster after receiving Type_II.

Figure 4, elaborates statistical information based dynamic time slot allocation by considering an example scenario. A cluster with 5 MSNs (n_1, n_2, n_3, n_4, n_5) and one cluster head (CH) are considered in the example scenario. It is assumed that all MSNs broads Type-I message to CH_1. CH_1, then, finds the weight by using two kinds of information i.e. i) The information, taken from the header of the broadcast message of all, and ii) the measured distance by using the ToA. Table 2 describes the metrics that are assumed after computing Type-I broadcast message information.

Table 2 contains the information, which are available to broadcast as a Type-II broadcast message by CH1 in the cluster. After receiving Type-II broadcast message from five MSN nodes, the information about the starting time of the data communication cycle and scheduling information and also information about other nodes are received by the MSN nodes in return. This provides time synchronization efficiency among all the MSNs i.e., the node 4 (n_4) while communicating with the CH1, and the others MSNs (n_2 n_5 n_3 n_1) turn their transceiver to listening state. While in listening state, the MSNs cannot communicate with other nodes or CH1, as it is busy in listening channel. After completing its timeslots by n4, n2 (which has the second highest weighted value) turns on its radio and start communicating with the CH, whereas, n4 turns its radio to listening state. This mechanism continues till the last node of the cluster accomplishes its communication. The mentioned mechanism repeated cycle by cycle.

Figure 4. Scenario illustrate the Type-I and Type-II broadcast messages to calculate weight for each MSN and assign dynamic timeslots to all mobile sensor nodes

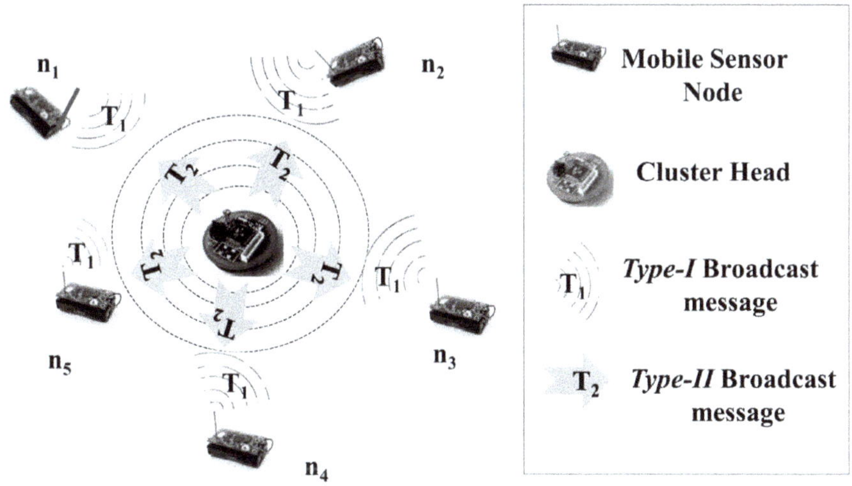

At acquiring statistical information from various MSNs, strategic LBA, which is based on STDMA, helps and allocates timeslots by evaluating statistical information. However, mobility aware TAL technique exists that exploits free timeslots, done when node joins or leaves the cluster or absense of data is in the region of activity. Mainly, TAL is composed of three basic loops, such as joining, leaving, and absence of data in the following scenario: Extricated Time Allocation (ETA), Shift Back Time Allocation (SBTA), and eScaped Time Allocation (STA) as shown in Figure 5.

Table 2. Example scenario of calculated weighted value

Node ID	ToA (µsec)	Battery Information (J)	Location Information (x,y,z, Coordinate Plane)	Weighted Value	Assigned Timeslots (Sec)
n4	150	20J	i	25	20
n2	135	18J	j	21	15
n5	100	14J	k	18	13
n3	85	12J	l	14	10
n1	70	10J	m	13	5

Figure 5. Sharing of time slots among different mobile nodes

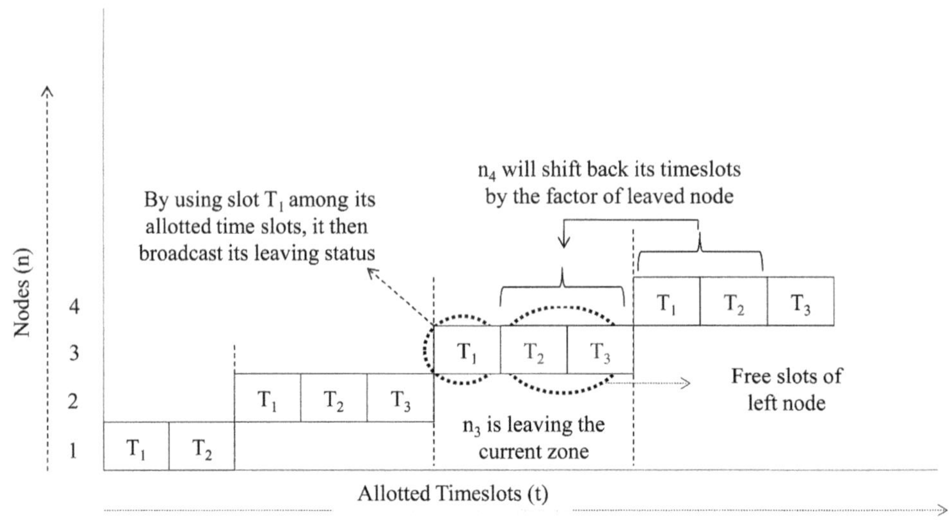

TIME ALLOCATION LEISTER

The fast changing in the position of sensor node results in variation in network topology. Therefore MWSN's scheduling processes influenced by the abrupt change in topology and location of various sensor nodes and also this change affects the timeslots allocation in the network.

The problem of fast alteration in topology as well as location (transferring sensor node from cluster to another cluster) is controlled by TAL mobility model. The three prongs of TAL are: Extricated Time Allocation (ETA), Shift Back Time Allocation (SBTA) and Escaped Time Allocation (STA).

1. **ETA:** In this scenario, the allocated timeslots is unutilized by the sensor node when there is no data is in transmission. In this technique, the MSN senses the repeated data or "redundant data" or, it does not perform any data sensing. Then, the free timeslot message to neighboring MSNs are transmitted by that MSN to shift back their timeslots as per free time timeslots kept over. Moreover, while entering MSN into a new cluster, free timeslots is allocated to new enrolled MSN. In order to provide efficiency to the mobile WSN, for further utilization of timeslot, it must mull over two scenarios, as shown as in Figure 5.
    a. In first scenario, free timeslot is allocated by CH to all recently added MSNs into the cluster.
    b. Secondly, allocation of free timeslots to the neighboring MSN of the cluster, which is done by CH (neighboring node will shift back its timeslot).

2. **SBTA:** In this case, we assume that MSN leaves from its cluster. At the time of leaving cluster, it broadcasts its timeslot and leaving information. The neighboring nodes start shifting back their timeslots as free timeslots is empty, which is left by the departed MSN.
3. **STA:** ETA and SBTA, in combine, makes a STA. By this technique, MSN has no data to sense or has redundant data and other MSN for leaving the cluster.

MOBILITY MODEL FOR TAL

The performance of the Mobile sensor node is predicted by TSEEC by using location information. Localization technique is well deliberated in WSN (Zaidi and Mark 2004). However, most of the authors in WSN assume the known physical location of a sensor node. Therefore, by TSEEC, it is assumed that each MSN is aware of his own location information by using (Ali et al. 2005).

Let, $\Gamma(\alpha + \Gamma_1) \rightarrow$ current mean (x,y) of α in F_i,

where x = x coordinates and, y = y coordinates.

In the proposed scheme, AR-1 model (Zaidi and Mark 2004) is used to predict MSN mobility pattern. A column vector S_t is defined to represent the state of the MSN at time t.

$$S_t = [x_t, \dot{x}_t, \ddot{x}_t, y_t, \dot{y}_t, \tilde{y}_t]^{\tilde{\ }} \quad (3)$$

where, St is the state of the mobility, position of a node is defined by (x_t, y_t), $(\dot{x}_t, \dot{y}_t)$ defines the velocity of a node, and symbol ($\tilde{\ }$) denotes the metric transpose operator. Whereas, $(\ddot{x}_t, \tilde{y}_t)$ is the acceleration in x and y direction for a node, respectively.

Hence, AR-1 model (Zaidi and Mark 2004) offers,

$$S_{t+1} = A_{st} + \omega_t \quad (4)$$

where, A is a 6 x 6 transformation metric, ωt is the vector i.e., 6x1 discrete-time zero mean with White-Gaussian process having autocorrelation function $R_\omega(k) = \delta k\, Q$, where $\delta_0 = 1$ and $\delta_k = 0$ when $k \neq 0$. Since the Q metric is the covariance metric of ω_t. A and Q values are based on training data by using Yule-walker Equation (Lin and Oppenheim 1987). Detailed explanation is given in (Zaidi and Mark 2004) (Zaidi et al. 2004) (Lim et al. 1987).

Moreover, at given time t, mobility state information of a MSN $\dot{\check{s}}_t$ at time t+1 is predicted by using $\dot{\check{s}}_t$ at time t. However, the optimized MSN's predictable state $\dot{\check{s}}$t+1, while considering minimum mean square error sense is calculated by:

$$\dot{\check{s}}_{t+1} = A^i\dot{\check{s}}_t \quad (5)$$

When mobility of a MSN is estimated, the mobility information is broadcasted in the cluster. In this procedure, MAC header is modified by adding the information about mobility of a MSN. Let F_i represents frame, the expected mean (x, y) of ith in a frame F_{i+1} is measured by node α as $\Gamma(\alpha + F_1) = \text{average}\ (\dot{\check{s}}_{t+0}.\ \dot{\check{s}}t{+}1, \dot{\check{s}}_{t+2} \ldots \dot{\check{s}}_{t+max})$.

When the value of $\Gamma\ (\alpha + F_1)$ is measured, the header of every frame of ith node is supplied by that information. Data packet format after adding information is: Format $\{\Gamma\ (\alpha + F_1)$, Header, Data$\}$. Rest of the cluster nodes keep in listening state for receiving of $\Gamma(\alpha + F_1)$. Same procedure is followed by each MSN in cluster and then MSN update information when they receive any data packet. The working of TSEEC is pseudo coded in Algorithm 2.

EXPERIMENTAL RESULTS AND DISCUSSIONS

NS-2 simulator is used to evaluate the TSEEC protocol performance by comparing results with existing protocols and scheduling techniques.

SIMULATION ENVIRONMENT

Simulation is performed by taken various number of mobile wireless sensor nodes from 50 to 100 by deploying over an area of 1000m X 1000m.

The distance between the nodes varies from time to time, hence totally depends on the mobility pattern of each MSN in the given cluster. While considering 50 nodes, we use two clusters, 25 nodes and a static CH in each cluster. However, for 100 nodes, three clusters are taken with same ratio of one static CH in each cluster. Initially each nodes are given 15 Joules energy level. IEEE 802.11 MAC type is used to minimize network delay. Whereas, CH uses IEEE 802.11n by availing frame aggregation technique (Feng et al. 2011) for delivering data packets at surface station. Constant bit rate (CBR) is used by each MSN to generate data packets. Hello packet is transmitted only once at the start of the communication.

Algorithm 2. Represent working of TSEEC

```
N = node
Number of nodes = 50
Z = Zone (z1, z2)
Cluster Head = CH1, CH2
        Loop i← 0 to 24
                zI = node [i]
                Z1 = z1 + CH1
        Loop i = ← 25 to 50
                z2 = node [i]
                Z2 = z2 + CH2
        Loop j = 0 to 24
                Node [i] ← t[j]
        Loop j = 25 to 50
                Node[j] ← t[i]
Proc TAL
Loop i← 0 to 24
        If (payload_node = = 0) // condition 1
                CH1← message← node[i]
                        then
                                New_arr_node[i] ←
t[i]        ← CH1
else        // condition 2
        CH1←leave_msg← node[i]
                Node [i+1] ← t[i]
        CH1→shiftback T of other nodes by a factor
of                ti+1, ti+2
else
combine (condition1 + condition2)
Repeat procTAL for zone2
```

MESSAGE INTERVAL TIME (MIT)

Message interval time results of TSEEC, while using with periodic or without periodic active/listening scheduling techniques, are depicted in Figures 6. Figure 6 (a) shows results in 50 MSNs with two clusters with 1 CH in each cluster i.e., CH1 and CH2, 25 MSNs in each CH. Results show that periodic active/listening scheduling technique is more efficient in energy consumption as compared to the non-periodic

Figure 6. Inter arrival time in seconds under low traffic load

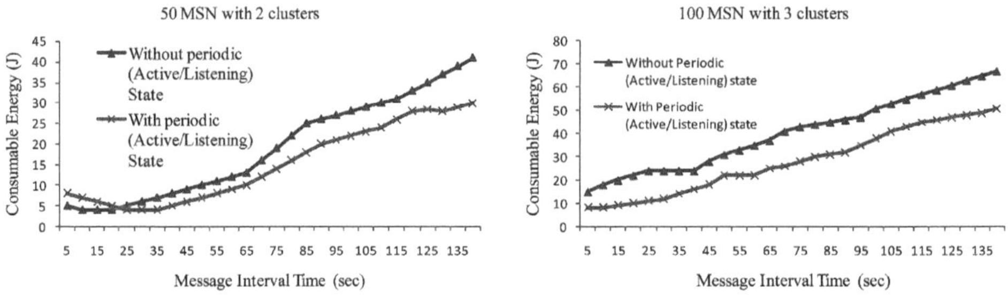

(without periodic) active/listening scheduling technique. In periodic active/listening scheduling, MSNs, with waiting state to start communication with the respective CH, switch their radios on to a listening mode. Very least energy i.e., 35 Joules is consumed by this technique to listen without any data packet transmission.

However, all MSNs turn their radios to full active mode to sense the environment in without periodic active/listening scheduling technique. For illustration, 68 Joule of energy are consumed in 50 nodes setup while considering TSEEC without periodic active/listening technique. Therefore TSEEC with periodic active/listening scheduling technique is recommended as it does not have the load of extra sensing of an MSN. Hence only 35 Joule of energy is consumed in the active/listening scheduling technique throughout the communication period. On the other hand in 100 MSNs setup, as shown in Figure 6(b), we can see the increase in the amount of energy consumed by MSN in without periodic active/listening i.e. 118 Joule of energy consumption. This is due to the extra packet loss by poor allotment of times-lots to each MSN. However, results show that TSEEC performed better when used with periodic active/listening scheduling technique if larger number of MSNs are considered i.e., 43 Joules of energy consumption.

ENERGY LEVEL OF MOBILE SENSOR NODE

In Figure 7, the comparison of both protocols, i.e., SMAC and proposed TSEEC is shown. The implementation of SMAC in a given scenario results in long but same sleep/active period, which results in dropping of data packets. Such packets are dropped during sleep state. Thus, we can say that the protocol like SMAC is not suitable for mobile sensor network since it also affect the scheduling process. In such technique, when a mobile node advertise its information to its neighboring nodes, the amount of received data packets could be dropped due to sleep state in a given cluster. As a result, the mobile will be lacking-off its neighboring informa-

Figure 7. Calculating energy level at every mobile sensor node

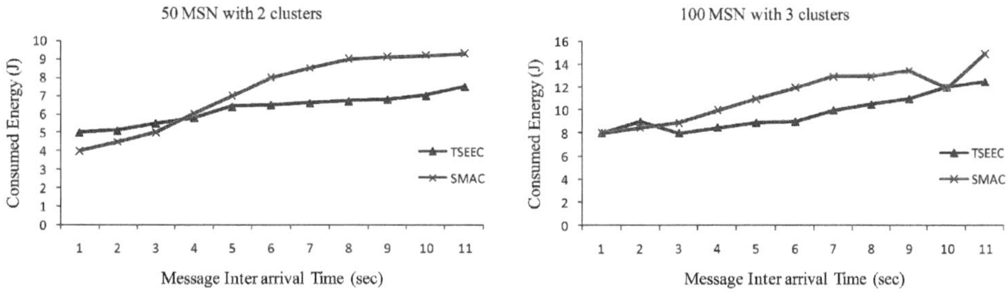

tion because of the sleep state. Therefore, in a given figure, the SMAC consumes 9 Joules of energy in one complete cycle. Whereas, the proposed TSEEC consumes quite lower amount of energy since it uses aperiodic sleep/awake schedule due to STDMA. In case, if we increase the number of mobile wireless sensor nodes as shown in Figure 7 (b), the performance of the SMAC is quite poor, i.e., it consumed 14.5 Joules as compared with the proposed TSEEC scheme, i.e., 11.5 Joules.

DELIVERY RATIO

Figure 8 (a) (b) shown the calculation of packet delivery ratio of both protocols, i.e., SMAC and the proposed protocol. In SMAC, greater amount of energy is saved due to long sleeping period. However, it results in excessive amount of packet drop ratio due to mobility factor as well as sleep schedule. Therefore, the mobile sensor nodes does not receive data packets, which results in scarce information about the cluster environment. Hence, it effects the overall data communication i.e., maximum amount of data packets are dropped due to the lack of cluster information. If the speed of the mobile sensor node increases, the overall scheduling process would be effected, which results in 0.5% of deliver ratio. On the other hand, the proposed scheme

Figure 8. Calculated packet delivery ratio under Low Traffic Load (LTR)

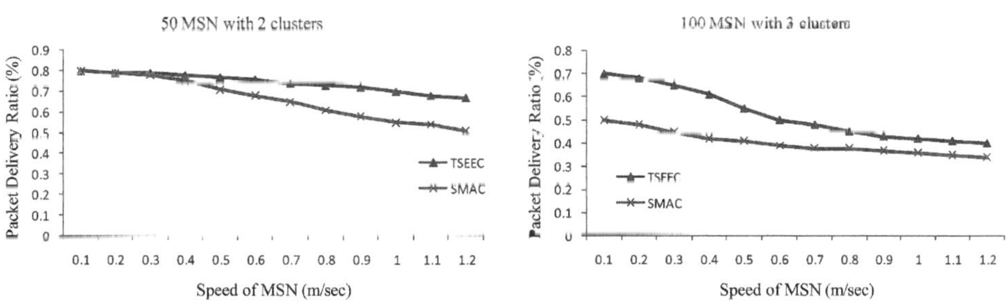

achieved 0.69% of delivery ratio. Moreover, if we increase the number of mobile sensor nodes to 100 as shown in Figure 8. (b), the packet delivery ratio decrease to 0.4%. In such condition, the proposed TSEEC system packet drop ratio is 0.3%.

802.11n ENERGY LEVEL

The implementation of 802.11n at CH is delineated in Figure 9. Various tests have been performed by the employment of scheduling techniques, such as idle listening, transmitting state, listening state and receiving state. In the given figure, less amount of energy has been consumed in receiving state, i.e., 2 Joules in one communication cycle. Whereas, CH in transmitting state consumes 4.2 Joules because of the aggregation feature (see (Feng et al. 20110 for more details). In listening state, the CH consumes almost negligible amount of energy. In the beginning of the communication, the CH listens its member nodes in order to get statistical information. On the basis of these statistical information, the CH assign timeslots. As a result, the CH cannot keep itself in idle state, therefore, idle state does not consume excessive amount of energy, i.e., 0.2 Joules.

The total energy consumption can be calculated as:

$$C_e = T_E - T_S \tag{6}$$

Figure 9. Energy status of IEEE 802.11n

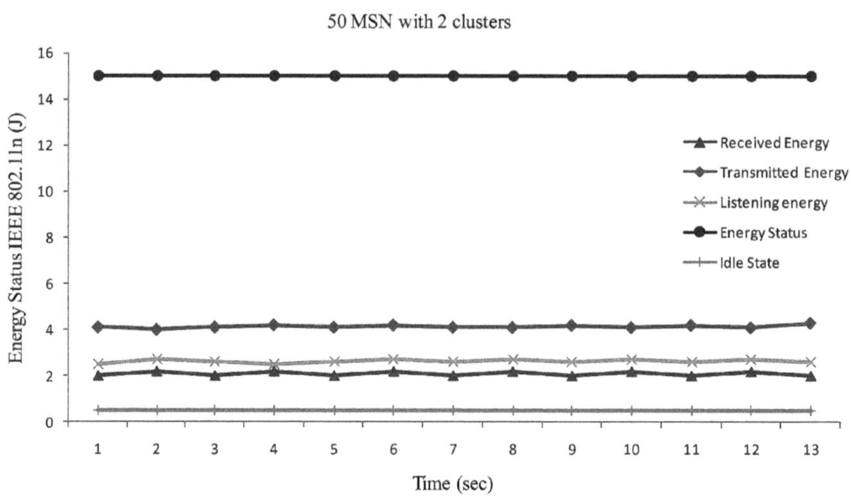

COMPARATIVE ANALYSIS WITH EXISTING MAC PROTOCOLS

In Table 3, the comparison has been made with rest to syschronization, communication pattern, type of the protocol, and adaptively changes to atmosphere. In Table 3, the two SMAC protocols, namely; SMAC and DMAC, have the same features. However, the cross layer MAC is not considered in this comparison. On the other hand, WiseMAC and TRAMA have diverse characteristics. There protocols are based on TDMA that increase the usage of TDMA in the context of energy efficiency. The mentioned protocol uses dynamic allocation of time to different nodes, whereas, uses the high ratio of sleep schedule. Such technique results in avoiding collision in the network vis-à-vis the CSMA protocol. On the other hand, the WiseMAC is used for downlink infrastructural WSN that is based on the synchronized preamble sampling focusing on low traffic.

Moreover, SIFT protocol is used in event driven scenarios, i.e., when an event is sensed, the sensor will turn on its transceiver. In case, if the comparison is made between 802.11 MAC protocol and SIFT, the SIFT takes the edge on 802.11 since it effectively decreases the latency. Furthermore, DMAC is an energy efficient protocol that achieves low latency. Such low latency is achieved because of the diverse timeslots to various leaf nodes. Such technique helps in successive data transmission. Comparatively, the newly proposed technique 'TSEEC' does not incorporates various challenging features, such as synchronization, battery power, and timeslots. The main advantage of the proposed scheme is the welcoming of new nodes and timeslot sharing among various nodes in order to achieve better throughput.

Table 3. Comparison of TSEEC with existing MAC protocols

	Synchronization of Time	Communication Pattern	Type	Adaptively to Changes
S-MAC T-MAC DSMAC	No	All	CSMA	Average
WiseMAC	No	All	Np-CSMA	Average
TRAMA	Yes	All	TDMA/CSMA	Average
SIFT	No	All	CSMA/CA	Average
DMAC	Yes	ConvergeCast	TDMA/Slotted Aloha	Average
TSEEC	Yes	All	TDMA/STDMA	Good

CONCLUSION AND FUTURE RESEARCH DIRECTIONS

Various technique are studied so far for delay, energy wastage, and congestion control in WSN. The proposed work is also implemented that helps in minimizing congestion control, providing energy efficiency, whereas, minimizing delay for message interval and inter arrival time in WSN. Moreover, such constraints are handles by incorporation of STDMA along with sharing of timeslots in the existing TDMA technique. The mentioned techniques and tested and implemented on mobile wireless sensor nodes with multiple clusters along with the static CH. The results show that the proposed scheme has proved itself the suited solution for congestion control, energy reservation, minimizing delay and providing efficiency to the network.

In our future work, we will work on finding the neighboring mobile sensor nodes. Since mobile sensor nodes move freely that causes rapid changes in topology of the WSN network. Rapid change in topology influence TDMA scheduling that causes delay in the network. Moreover further studies reveal that the scheme can be a useful in various ubiquitous networks (Tsai et al. 2007; Paul et al. 2012; Paul 2013, Tsai et al. 2013). Hence a mobility model is required that overwhelm the absence of the neighboring node.

REFERENCES

Ahmed, U., & Hussain, F. B. (2011). Energy efficient Routing Protocol for Zone Based Mobile Sensor Networks. *Wireless Communications and Mobile Computing Conference (IWCMC), 2011 7th International*. doi:10.1109/IWCMC.2011.5982691

Ali, M. (2005). MMAC: A Mobility-Adaptive, Collision-Free MAC protocol for wireless sensor network. In *Proceedings of 24th International Conference on Performance, Computing and Communication Conference*. IEEE IPCC.

Anand Paul, Y.C. (2012). Parallel Reconfigurable Computing-Based Mapping Algorithm for Motion Estimation in Advance Video Coding. *ACM Transactions on Embedded Computing Systems*, *11*(S2), 40.

Anastasi, G., Conti, M., Di Francesco, M., & Passarekka, A. (2009). Energy conservation in WSN: A Survey. *Ad Hoc Networks*, 537–568. doi:10.1016/j.adhoc.2008.06.003

Bergbreiter, S., & Pister, K. S. J. (2003). CotsBots: An Off-the-Shelf Platform for Distributed Robotics. In *Proceeding of the IEEE International Conference on Intelligent Robots and Systems*. Las Vegas, NV: IEEE. doi:10.1109/IROS.2003.1248878

Chen, B., Jamieson, K., Balakrishnan, H., & Morris, R. (2002). SPAN: An energy efficient coordination algorithm for topology maintenance in ad hoc wireless network. *ACM Wireless Network*, *8*(5), 481–494. doi:10.1023/A:1016542229220

Chih-Lin, I., & Pollini, G. P. (1994). The tree-search resources auction multiple access (TRAMA) protocol for wireless personal communications. *Vehicular Technology Conference, 1994 IEEE 44*th. doi:10.1109/VETEC.1994.345276

Das, S. R., Perkins, C. E., & Royer, E. M. (2000). Performance comparison of two on-demand routing protocols for ad hoc networks. *INFOCOM 2000. Nineteenth Annual Joint Conference of the IEEE Computer and Communications Societies. Proceedings*. IEEE. doi:10.1109/INFCOM.2000.832168

Hu & Evans. (2004). *Localization for Mobile Sensor Networks.* Tenth Annual International Conference on Mobile Computing and Networking (ACM MobiCom 2004).

Khan, M., Gansterer, W., & Haring, G. (2007). Congestion Avoidance and Energy Efficient Routing Protocol for Wireless Sensor Networks with a Mobile Sink. *Journal of Networks*, *2*(6), 42–49. doi:10.4304/jnw.2.6.42-49

Lee, J.-S., & Cheng, W.-L. (2012). Fuzzy-Logic-Based Clustering Approach for Wireless Sensor Networks Using Energy Predication. *IEEE Sensors Journal*, *12*(9), 2891–2897. doi:10.1109/JSEN.2012.2204737

Lim, J. S., & Oppenheim, A. V. (1987). Advanced Topics in Signal Processing. Englewood Cliffs, NJ: Prentice Hall.

Liu, Y., & Liu, Y. (2008). A robust routing algorithm with fair congestion control in WSN. *Computer Communication and Networks, 2008, ICCCN '08'. Proceedings of 17th International Conference.* doi:10.1109/ICCCN.2008.ECP.167

MICA2 Mote Datasheet. (n.d.). Retrieved from http://www.xbow.com/Products/Product_pdf_files/Wireless_pdf/6020-0042-01_A_MICA2.pdf

Oh, K., & Woo, S. (2007). Improved energy efficiency of DMAC with periodic full sleep cycle for WSN with heavy traffics. *Intelligent Sensors, Sensors Networks and Information, 2007, ISSNIP 2007, 3rd International Conference.* .4496824 doi:10.1109/ISSNIP.2007

Paul, A. (2013). Dynamic Power Management for Ubiquitous Network Devices. *Advance Science Letters, 19*(7), 2046–2049. doi:10.1166/asl.2013.4634

Pradhan, S. S., & Ramchandran, K. (2003). Distributed source coding using syndromes (DISCUS): Design and construction. *IEEE Transactions on Information Theory, 49*(3), 626–643. doi:10.1109/TIT.2002.808103

Sokullu, & Donertas. (2008). *Combined effects of mobility, congestion and contention on network performance for IEEE 802.15.4 Based Networks.* Computer and Information Science, 2008, ISCIS '08', 23rd International Symposium on Digital Object Identifier.

Tang, C., & Raghavendra, C. S. (2004). *Compression Techniques for Wireless Sensor Networks.* Book Wireless Sensor Networks Kluwer Academic.

Tao, L., & Yu, F. (2009). ECODA- Enhanced congestion detection and avoidance for multiple class of traffic in sensor network. *Communication, 2009, APCC 2009, 15th Asia-Pacific Conference on.* doi:10.1109/APCC.2009.5375498

Tsai, A. C., Paul, A., Wang, J. C., & Wang, J. F. (2007). Efficient Intra Prediction in H.264 based on Intensity Gradient approach. *IEEE International Symposium on Circuits and Systems* (ISCAS 2007), (pp. 3952-3955). doi:10.1109/ISCAS.2007.378665

Tsai, H. C., Chen, B. W., Wang, J. F., & Paul, A. (2013). Enhanced long-rang Personal Identification based Multimodal Information of Human Features. *Multimedia Tools and Applications.*

Wang, S. S., & Chen, Z. P. (2013). LCM: A Link-Aware Clustering Mechanism for Energy-Efficient Routing in Wireless Sensor Networks. *IEEE Sensors Journal*, *13*(2), 728–736. doi:10.1109/JSEN.2012.2225423

Wu, M., & Chen, C. W. (2006). Multiple Bit Stream Image Transmission over Wireless Sensor Networks. IEEE & Wiley Interscience. doi:10.1002/9780471784173.ch13

XiaoxingGuo., M. R., & Frater, M. J. R. (2006). A Propagation-delay-tolerant Collision Avoidance Protocol for Underwater Acoustic Sensor Networks. In *Proc. MTS/IEEE OCEANS*.

Xing, G., Wang, X., Zhang, Y., Lu, C., Pless, R., & Gill, C. (2002). Integrated coverage and connectivity configuration of energy conservation in sensor network. *ACM Transactions on Sensor Networks*, *1*(1), 36–72. doi:10.1145/1077391.1077394

Xiong, Z., Liveris, A. D., & Cheng, S. (2004). Distributed source coding for sensor networks. *IEEE Signal Processing Magazine*, *21*(5), 80–94. doi:10.1109/MSP.2004.1328091

Xu, Y., Hiedemann, J., & Estrin, D. (2001). Geography informed Energy Conservation for Ad Hoc Routing. In *Proceeding of the Seventh annual ACM lst IEEE International Conference on Mobile computing and Networking* (MobiCom). doi:10.1145/381677.381685

Younis, O., & Fahmy, S. (2004). HEED: A hybrid energy-efficient, distributed clustering approach for ad hoc wireless networks. *IEEE Trans. On Mobile Computing*, *3*(4), 660–669. doi:10.1109/TMC.2004.41

Zaidi, Z. R., & Mark, B. L. (2004). Mobility Estimation for Wireless Networks Based on an Autoregressive Model. In *Proc. IEEE Globecom 2004*. doi:10.1109/GLOCOM.2004.1378999

Zaidi, Z. R., Mark, B. L., & Roshan, K. (2004). A Two-tier Representation of Node Mobility in Ad Hoc Networks. In *Proc. IEEE Sensor and Ad Hoc Communications and Networks (SECON)*. Santa Clara, CA: IEEE. doi:10.1109/SAHCN.2004.1381913

Zeinalipour-Yazti, D., Papadakis, H., Georgiou, C., & Dikaiakos, M. D. (2008). *Mobile Sensor Network Data Management. Parallel Processing Letters Journal.*

Zorbas, D., & Douligeris, C. (2011). Connected coverage in wsns based on critical targets. *Computer Networks*, *55*(6), 1412–1425. doi:10.1016/j.comnet.2010.12.023

Zorbas, D., Razafindralambo, T., & Lilli, I. (2012). *Wireless Sensor Network Re-deployment under Target Coverage Constraint.* New Technology, Mobility and Security (NTMS), 2012 5th International Conference on.

Zou, L., Lu, M., & Xiong, Z. (2005). A Distributed Algorithm for the Dead End Problem of Location Based Routing in Sensor Networks. *Vehicular Technology, IEEE Transactions on*, *54*(4), 1509 – 1522.

Chapter 4
Energy Efficient Image Compression and Transmission in WSN

Wajeeha Aslam
National University of Sciences and Technology (NUST), Pakistan

M. Usman Akram
National University of Sciences and Technology (NUST), Pakistan

Muazzam A. Khan
National University of Sciences and Technology (NUST), Pakistan

Nazar Abbas Saqib
National University of Sciences and Technology (NUST), Pakistan

Seungmin Rho
Sungkyul University, South Korea

ABSTRACT

Wireless sensor networks are greatly habituated in widespread applications but still yet step behind human intelligence and vision. The main reason is constraints of processing, energy consumptions and communication of image data over the sensor nodes. Wireless sensor network is a cooperative network of nodes called motes. Image compression and transmission over a wide ranged sensor network is an emerging challenge with respect to battery, life time constraints. It reduces communication latency and makes sensor network efficient with respect to energy consumption. In this paper we will have an analysis and comparative look on different image compression techniques in order to reduce computational load, memory requirements and enhance coding speed and image quality. Along with compression, different transmission methods will be discussed and analyzed with respect to energy consumption for better performance in wireless sensor networks.

DOI: 10.4018/978-1-4666-9941-0.ch004

Copyright ©2016, IGI Global. Copying or distributing in print or electronic forms without written permission of IGI Global is prohibited.

1. INTRODUCTION

Wireless sensor network (WSN) comprises dozens to several thousands of inexpensive, resource constrained and autonomous devices called sensor nodes. Nodes are interlinked with the high energy base stations also called sinks shown in Figure 1. All of the data is transmitted, aggregated and filtered at sinks. Typically a sensor node is small device having low power battery, microcontroller, and sensor, and transceiver, hardware architecture as depicted in Figure 2. Low cost and low complexity of nature demands lead to compromise on transmission rate, reliability, efficiency and power consumption. Although WSN is widely used in military, civilian, emergency and different environmental situations, yet now it's a challenge for us to transmit human sensed data like image through wireless network with low power consumption within available bandwidth.

Figure 1. Sensor networks

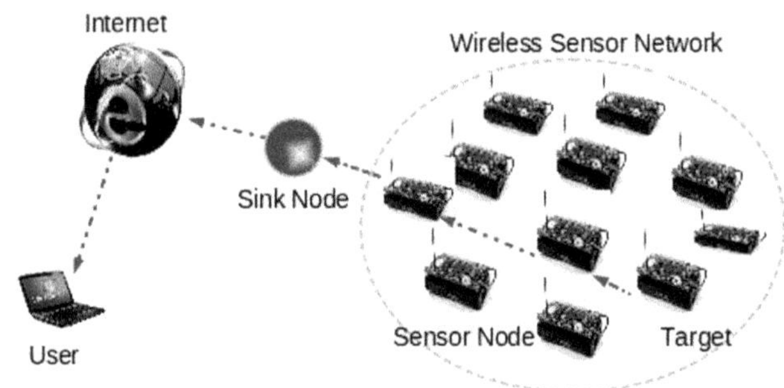

Figure 2. Sensor node's hardware architecture

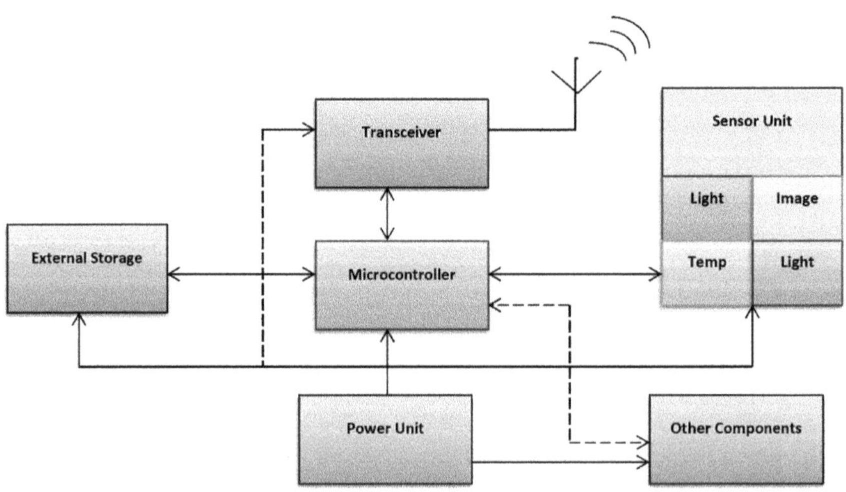

In distributed network a single node has not enough computational power to compress large amount of data that's why computation of compression is divided among nodes evenly. Conservation of energy in sensor network is directly proportional to the number of bits transmitted. In this paper work we will analyze and compare different image compression and transmission techniques regarding energy efficient in wireless sensor network.

2. LITERATURE REVIEW

2.1 Image Compression

Transmission of image requires large memory, high bandwidth and high energy consumption. Multimedia transmission is an error tolerant protocol in network. Image compression techniques are utilized to conserve energy, reduce latency and improve quality of transmission rate. Typically compression algorithms are classified into Discrete Cosine Transmission (DCT) based and Discrete Wavelet Transmission (DWT) compression.

JPEG is most popular and greatly used algorithm for image compression. Its discrete cosine transmission based compression. Before discussing in detail it should be noted that black and white image pixels range from 0 to 255. Pure white is presented by 0 and pure black by 255. Pixels from 0 to 255 show different shades of grey but in colored image pixels ranges from -128 to 127. Raid et al. (2014) presented JPEG process steps for color images [1]. Those steps are as following

- Image color space is commuted as RGB to YCbCr and partitioned into 8 * 8 blocks.
- As DCT enforced from top to bottom and left to right so each block is shifted from [0-255] to [-128 to 127] pixels.
- Each blocked is quantized and then undergone through entropy encoding.
- Compressed image is decoded through inverse discrete cosine transformation.

Kaddachi et al. (2011) discussed FPGA based on discrete cosine transformation derived in zonal form [2]. Zonal DCT reduce the coefficients in image block to be computed. Hence reduced DCT coefficients quantized and encoded cause reduction in cost of whole production chain. Hardware is implemented in such a way that compression settings can be changed according to image distortion and energy consumption. It enhances processing time and reduces the energy consumption.

Figure 3. Image compression scheme: Compression technique step and decompression technique step [1]

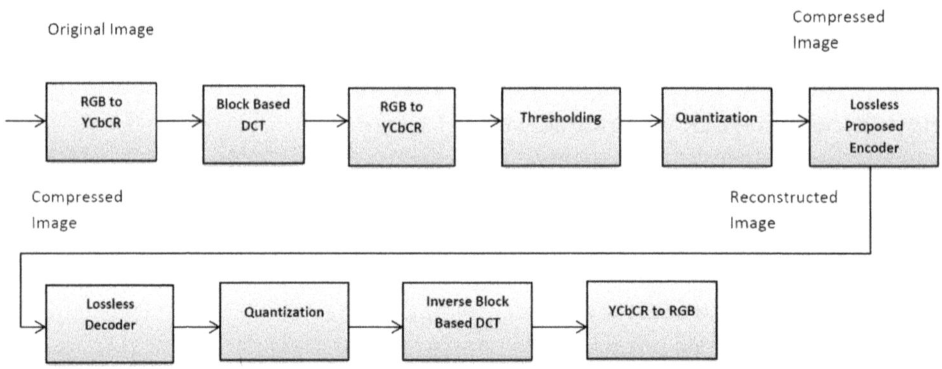

Figure 4. FPGA based block diagram for image compression [2]

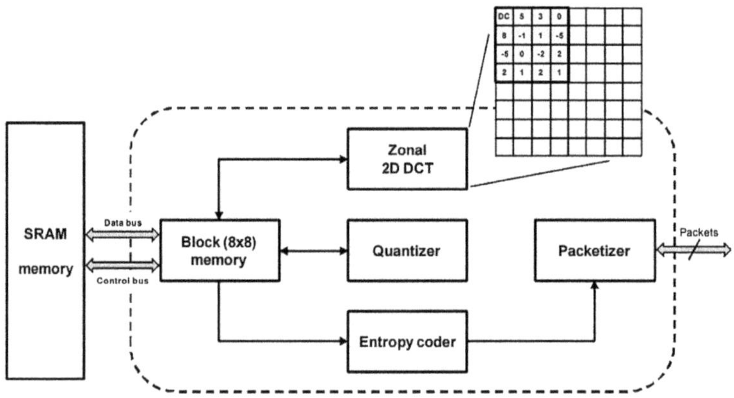

An approach grounded on wavelet transmission called SHPS (Skipped High Pass Sub Band) is used. In this approach high pass coefficients do not compute that save energy consumed for computational consumption. It reduces the number of execution steps and assists in managing time and energy in sensor networks. Nasri et.al (2010) presented work on SHPS [3]. According to which SHPS skips high sub band in vertical direction in each transformation step. Resulted low sub band of horizontal direction is further decamped into further vertical and horizontal sub bands. Computational load is reduced by bypassing two sub bands out of four sub bands at each transformation step. Figure 5 depicts SHPS decomposition transformation result.

Figure 5. SHPS technique compression [3]

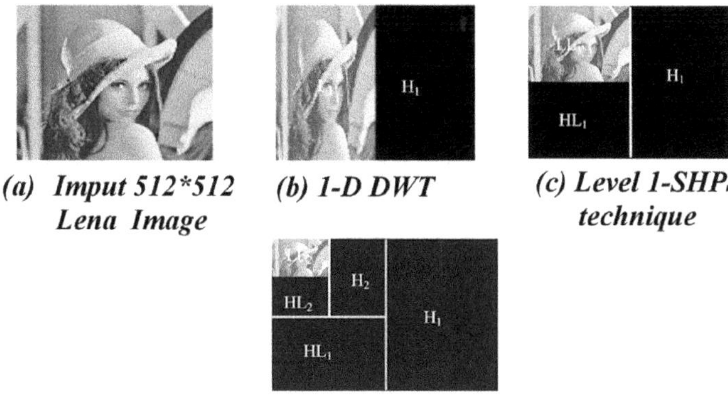

(a) Imput 512*512 Lena Image ***(b) 1-D DWT*** ***(c) Level 1-SHPS technique***

(d) Level 2-SHPS technique

Main idea behind SHPS in WSN is to exchange data through broadcast. Routing algorithm is assumed to run in self organizing nodes. Cluster head receives the query from source node and then select the set of nodes those run wavelet transformation. Each cluster head selects low pass band and forwards remaining part image to next cluster. This image compression and broadcasting of image continues till it reaches its final destination sink node. Figure 6 depicts flow distributed image compression work.

Karthikeyan et al. (2013) presented work on JPEG2000 [4]. It is based on discrete wavelet transform. After DWT image sub bands are quantized to reduce precision and then converted into sign magnitude before entropy encoding. Code block result of sub band division is independently coded that generate embedded bit stream. Bit stream is of two types one is sign bit plane and other is magnitude type bit plane. EBCOT is used by two tier entropy coder. Tier one is context based arithmetic encoder. To optimize life time of network and increase speed of processing, data is broadcasted in all of nodes. Image transmission is done in cluster network in which source node divides the image into tiles and then transmits in a package of nodes

Figure 6. SHPS technique compression in distributed WSN [3]

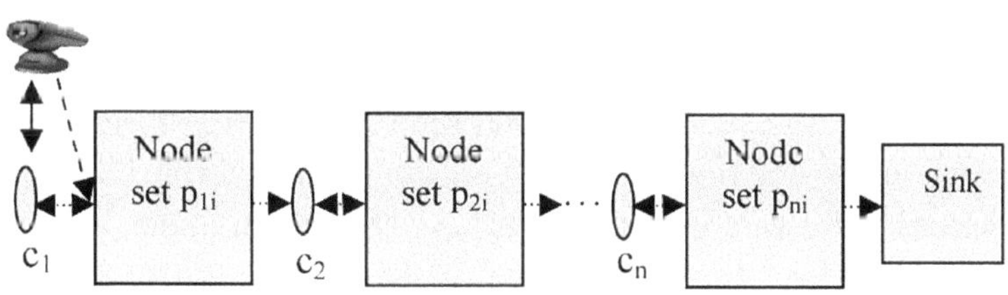

Figure 7. Up to 3 level multi hop network decomposition for data exchange in distributed network [4]

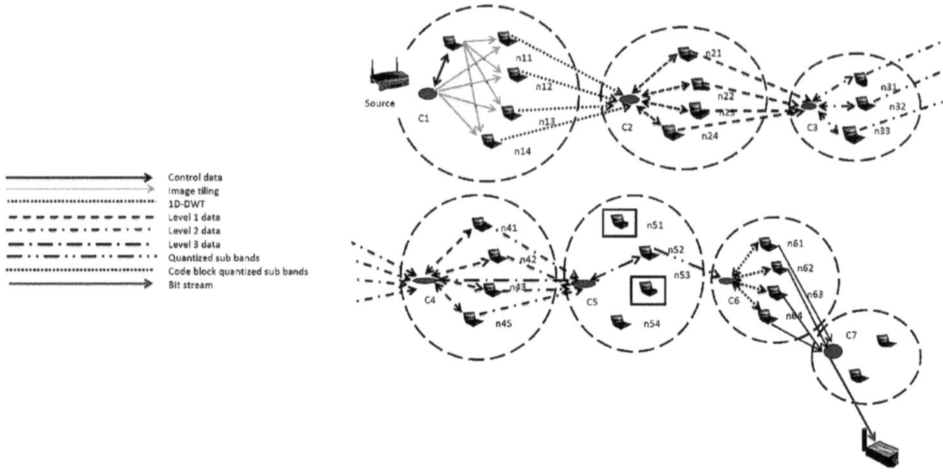

in cluster. Those nodes run 1D wavelet transform on tiles and send intermediate result to the cluster head. After two 1D wavelet transform data is further forwarded from one cluster head to another cluster head for same process. Distribution of work optimizes the network speed and life time.

Dawood et al. (2012) presented work on image compression in wireless sensor network [5]. In JPEG2000 work load is reduced by dividing the task among nodes. Two methods are followed one is dividing by rows and columns and second is tiling of image. This process of compression speeds up processing time and reduce overhead of computation at one node and involve in energy consumption to some extent.

2.1 Image Transmission

Radio Transceiver in node architecture consumes most part of energy while transmitting and receiving signals. Although compression of image reduces size and overhead on network but for efficient energy consumption transmission criteria's are also kept under consideration.

Lecuire et al. (2007) presented work on efficient image transmission in sensor network [6]. Two image transmission schemes 1) open loop image transmission and 2) closed loop image transmission were presented in their work. Both methods are grounded on image wavelet transform and semi reliable transform. Semi reliable transmission is end to end transmission. According to it, intermediate nodes decide either to discard or forward packets regarding to battery of node.

Figure 8. Energy transmission for uncompressed image [6]

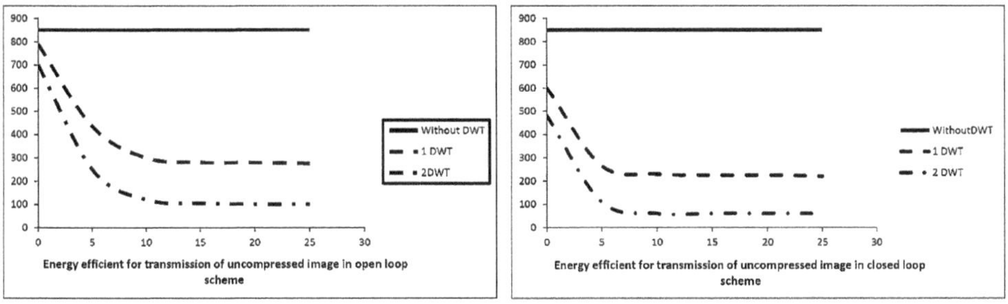

In Open loop scheme Intermediate nodes are independent of available energy while making decision. These do not need feedback. When packets are received by node then it need only two piece of information. One is priority of the packet and second is total priority level. In closed loop scheme intermediate nodes are totally dependent of available energy while making decision. Feedback of intermediate nodes performs basic role. The delay in feedback shows direct relationship with the distance between concerned nodes. Closed loop path is not much more complex than Open loop path. But it maintains path from source to destination in a regular way. In fact if end to end path is consistent and regularly maintained then closed loop path is best scheme else open loop path scheme is best.

Image compression reduces the overhead of computation of large sized data. It reduces the amount of energy consumed for transmission. (Lecuire, 2007) [6] Transmission of compressed image dissipated 57% less energy as compared to un-compressed image. As depicted in figure 8 and 9. In both of open loop and closed loop scheme percentage difference of energy is same. It shows that transmission behavior is not changed by use of any compression algorithm.

(PPGIa,E.B, CPGEI, CPGEI, & PPGIa, 2012) [7]. presented selective decode and forward SDF scheme based on DWT. It shows energy efficiency with incremental decode and forward IDF. SDF is the most cooperative employed protocol.

Figure 9. Energy transmission for compressed image [6]

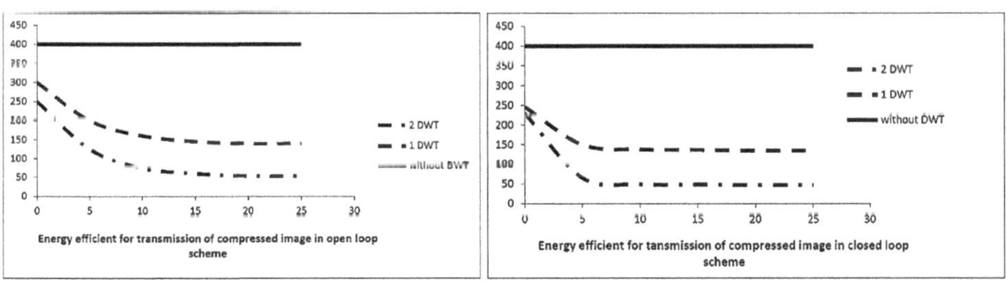

Figure 10. Energy transmission SDF with low image quality loss [7]

(a) Original image, MSSIM = 1 (b) LL_1, MSSIM = 0.9301
(c) LL_2, MSSIM = 0.8324 (d) LL_3, MSSIM = 0.6054

In this only relay node forward those messages which can be decoded correctly at receiver side. IDF is most likely SDF but in this relay node only send on the messages only whenever demanded by the destination.

The cooperative mode of communication helps to save transmission energy. It uses a cooperative communication scheme which comes about in two time slot chunks at other end non cooperative communication scheme takes place in one time slot chunk. Both of cooperative and non-cooperative communication schemes keep under consideration two things 1st packet loss rate and 2nd end to end throughput. IDF is the most energy efficient scheme that requires feedback of the channels. Lower resolution part of image in SDF relay is sent from source to destination node. If destination in SDF is not able to receive image from source then it will be able to get from relay. This technique is more concerned with overall reduction of energy consumption with minimum loss of image quality. Instead of forwarding entire missed image only lost low resolution band of images are forwarded. This procedure is performed with expense of image quality at low level. Figure 10 shows the effect of SDF on image where MSSIM is mean of structural similarity of index method and LL present low sub band result of subsequent DWT.

Nasri et al. (2011) presented adaptive wavelet image compression and organization algorithm (AWICOA) [8]. It ensures low computational load and efficient energy consumption. It is basically concerned with resource constrained multi hop network. It undergoes through wavelet transform viva lifted scheme (LS 9/7).

AWICOA is consisting of two techniques SHPS and EHPS (PPGIa, 2012) [7]. SHPS (skipped high pass sub band) preserve energy by bypassing least significant sub band. EHPS (eliminating high pass sub band) only applies low pass filter for

Figure 11. Energy consumption with compression [8]

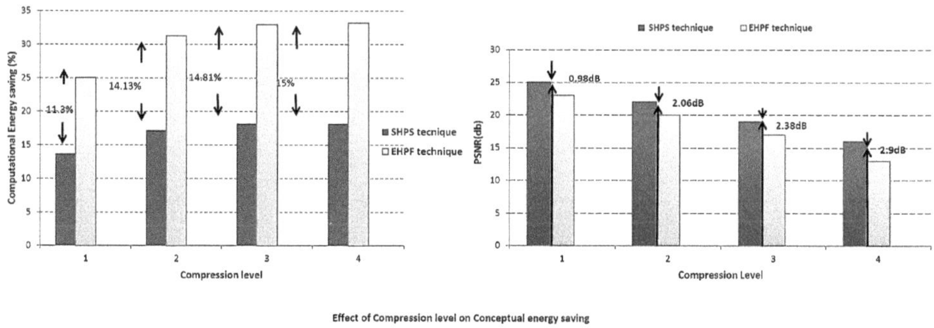

decomposition in vertical order at each step of compression. SHPS avoid high pass sub band generation in column transform. In this proposed system multiple of packets are shared in relay of intermediate clusters. Relay of clusters ensure avoidance of redundancy of packets. It involves intermediate nodes in transmission of packet with multi hop to preserve memory. When EHPS is applied then energy saving is 13.7% less as compared to LS 9/7 at 1st compression level. At 2nd compression level energy saving is 17% at expense of loss of image quality. SHPS save energy up to 25% at 1st compression level. At 2nd compression level it reduces energy consumption up to 31% at expense of image quality loss. This is depicted in figure 11.

3. ANALYSIS AND COMPARISON

In general best compression technique for image transmission in wireless sensor network can be decided based on these criteria listed in block list box.

In session A we discussed compression techniques. Each technique carries its own worth. Certain criteria are considered to decide worth of scheme. JPEG is good with energy consumption. It computes with high speed, operates on small sized

Figure 12. List circles for best image compression and transmission criteria

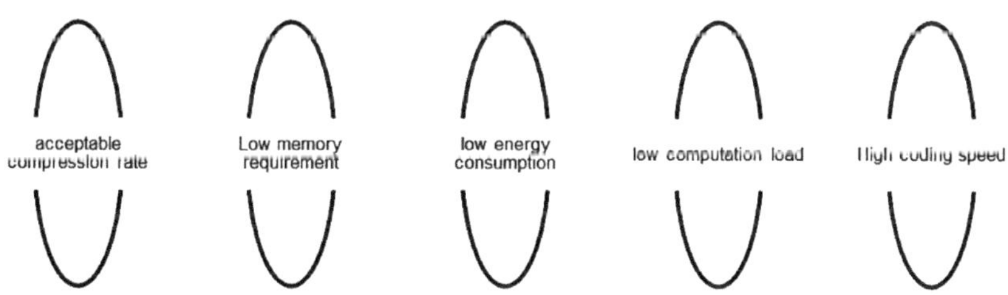

Table 1. Analysis overview for image compression

Name of Technique	Transformation Method	Image Size	Computational Load	Memory Requirement	Coding Speed	Compression Quality
JPEG	DCT	Small	Low	Low	High	Low
FPGA Based Zonal DCT	DCT	Flexible according to requirement	Bit High	Flexible according to the requirement	Medium	Acceptable
SHPS	DWT	Flexible according to requirement	Low	Low	High	High
JPEG 2000	DWT	Small	High	High	Average	High

image and results image quality loss. FPGA based Zonal DCT is flexible with respect to image size. It accommodates different image size with adaptation of its service. It provides acceptable compression rate and flexible with respect to memory. JPEG2000 is good for small sized image but require more energy due to high demand of memory and computational load. SHPS is a very improved technique. It provides soft corner for varying size of image, need low memory, low computational load and high coding speed. It provides a good quality of image with very small loss.

In session B we discussed transmission of image with respect to energy conservation. Different of schemes are discussed with a bird eye view. Transmission schemes are of their worth according to provided network and energy constraints. If path is not persistent and no need of feedback then open loop is a good scheme. If path is consistent and need feedback then closed loop is goon one. In a cooperative network SDF technique is good. In cooperative and persistent with feedback IDF is a good technique. In multi hop network AWICOA is a good technique. AWICOA is considered best if it applies SHPS compression scheme for image quality at receiver end.

Overall comparison work shows that compression techniques put impact on quality of image but do not at transmission method. Transmission method remains same before and after the compression of image. Compression just gives a flora of reduction in computation load. SHPS is an improved image compression technique. It reduces computational load and provides a good quality product with very small loss. It is need of demand that transmission method is decided what to be used.

Table 2. Analysis overview for image transmission

Name of Technique	Type of Network	Computational Load	Energy Consumption	Compression Technique	Quality of Image
Open loop scheme	Maintain at time of need. No need of feedback	Low	Less than 57% with compressed image	DWT/semi reliable transmission	Very minimum loss of image quality
Closed loop scheme	Maintained always. Need of feedback	High	Less than 57% with compressed image	DWT/semi reliable transmission	Very minimum loss of image quality
SDF	Cooperative	Low	Low	DWT	Minimum loss of image quality
IDF	Cooperative Need of feedback	Medium	Low	DWT	Minimum loss of image quality
AWICOA	Multi hop	Low	Less than 13.7% at 1st DWT and 17% at 2nd DWT energy saved	SHPS/DWT	Low quality at 2nd DWT compression
AWICOA	Multi hop	Low	Up to 25% at 1st DWT and 31% & at 2nd DWT energy saved	EHPS/DWT	Low quality at 2nd DWT compression

4. CONCLUSION

Multimedia communication over wireless sensor network is emerging challenge with respect to energy consumption. Transmission of huge sized data requires high memory requirement that consumes lot of energy. Sensors are energy constraint in nature. An impressive compression scheme can reduce overhead of computation and memory requirements. Comparative study shows SHPS is an effective image compression technique with varying image size and provide an effective quality product. Transmission behavior is never altered by change of any compression scheme. Compression can enhance quality product but can never change behavior of transmission method. It is nature of network and demands that decide which transmission scheme is best.

ACKNOWLEDGMENT

Special thanks to Dr. Muazzam A. Khan for directing in a good way and providing fruitful and remarkable guidance in field of Sensor Networks.

REFERENCES

Dawood, M. S., Ahila, L., Sadasivam, S., & Athisha, G. (2012). Image Compression in Wireless sensor networks- A survey. *International Journal of Applied Information Systems*, *1*(9), 11–15. doi:10.5120/ijais12-450227

Kaddachi, M. L., Makkaoui, L., Soudani, A., Lecuire, V., & Moureaux, J. M. (2011, December). *FPGA-based image compression for low-power Wireless camera Sensor Networks.* Paper Presented at Next Generation Networks and Services (NGNS), 2011 3rd International Conference on. doi:10.1109/NGNS.2011.6142537

Kapthikeyan, A., Shankar, T., Srividhya, V., Sarkar, S., & Gupte, A. (2013). Energy Efficient Distributed Image Compression Using JPEG2000 in Wireless Sensor Networks (WSNS). *Journal of Theoretical and Applied Information Technology*, *47*(2).

Lecuire, V., Duran-Faundez, C., & Krommenacker, N. (2008). Energy-efficient image transmission in sensor networks. *International Journal of Sensor Networks*, *4*(1-2), 37–47. doi:10.1504/IJSNET.2008.019250

Manhas, E. B., Jr., Brante, G., Souza, R. D., & Pellenz, M. E. (2012, April). *Energy-Efficient Cooperative Image Transmission over Wireless Sensor Networks.* Presented at of Wireless Communication and Networking Conference (WCNC), 2012 IEEE, Shanghai, China. doi:10.1109/WCNC.2012.6214121

Nasri, M., Helali, A., Sghaier, H., & Maaref, H. (2011, December). *Energy conservation for image transmission over wireless sensor networks.* Presented at Microelectronics (ICM), 2011 International Conference on. doi:10.1109/ICM.2011.6177376

Nasri, M., Helali, A., Sghaier, H., & Maared, H. (2010, October). *Energy-Efficient Wavelet Image Compression in Wireless Sensor Network.* Paper Presented In Communication in wireless environments and ubiquitous systems: New challenges (ICWUS), 2010 international conference on, Sousse, Tunisia.

Raid, A. M., Khedr, W. M., El-dosuky, M. A., & Ahmed, W. (2014). *Jpeg Image Compression Using Discrete Cosine Transform- A Survey.* arXiv preprint arXiv:1405.6147

Chapter 5
Novel Energy Aware Algorithm to Design Multilayer Architecture for Dense Wireless Sensor Networks

Naveen Chilamkurti
La Trobe University, Australia

Sohail Jabbar
COMSATS Institute of Information Technology, Pakistan

Abid Ali Minhas
Al-Yamamah University, Saudi Arabia

ABSTRACT

Network layer functionalists are of core importance in the communication process and so the routing with energy aware trait is indispensable for improved network performance and increased network lifetime. Designing of protocol at this under discussion layer must consider the aforementioned factors especially for energy aware routing process. In wireless sensor networks there may be hundreds or thousands of sensor nodes communicating with each other and with the base station, which consumes more energy in exchanging data and information with the additive issues of unbalanced load and intolerable faults. Two main types of network architectures for sensed data dissemination from source to destination exist in the literature; Flat network architecture, clustered network architecture. In flat architecture based networks, uniformity can be seen since all the network nodes work in a same mode and generally do not have any distinguished role.

DOI: 10.4018/978-1-4666-9941-0.ch005

Copyright ©2016, IGI Global. Copying or distributing in print or electronic forms without written permission of IGI Global is prohibited.

1. INTRODUCTION

Network layer functionalities are of core importance in the communication process and so the routing with energy aware trait is indispensable for improved network performance and increased network lifetime. Designing of protocol at this under discussion layer must consider the aforementioned factors especially for energy aware routing process. In wireless sensor networks there may be hundreds or thousands of sensor nodes communicating with each other and with the base station, which consumes more energy in exchanging data and information with the additive issues of unbalanced load and intolerable faults. Two main types of network architectures for sensed data dissemination from source to destination exist in the literature; Flat network architecture, clustered network architecture. In flat architecture based networks, uniformity can be seen since all the network nodes work in a same mode and generally don't have any distinguished role. So usually no conservation of energy is supported by itself from its architectural setup. In cluster based network architecture, autonomous nodes are grouped in a centralized way to designate one node as head and other nodes as members in called clustering. Group is named cluster. Designated head is termed as cluster head and its member nodes are called as cluster members. Figure 1 depicts a complete scenario of complete network architecture.

Cluster head (CH) acquires the sensed data from cluster member nodes, aggregates and forwards it to other cluster heads directly or via some transit node (multi-hop) or directly (direct-hop) to the base station. From these two communication models, direct hop fits better for small scale networks where the nodes communicate directly to the base station. The approachable network size is a function of maximum communication range of node. For large scale networks, multi-hop communication model provides scalability through the transit nodes' assistance to destine the data to far distantly placed Base Station (BS). Simulation based comparative analysis of this direct hop and multi-hop communication to the base station is given in detail by Jabbar, Minhas, Paul, & Rho (2014). Hence, in clustered network architecture, nodes are designated different roles. Some head the cluster, some become member of cluster, some are ready to be set as cluster head, some assist the cluster head in performing its job, and some acts as communication bridge for the a cluster node to communicate with other network node. Figure 2 shows the possible roles of a node during its life time depending upon the underlying clustering algorithm. The role of the node that are underlined in the figure are the part of every clustering algorithm. For better understanding of the role, short description is also given with each one.

Literature is rich in cluster based routing protocols that mostly encompass, i) cluster designing, ii) route establishment and iii) cluster head rotation processes. Cluster designing consists of cluster head selection, cluster member affiliation to

Figure 1. Complete scenario of clustered network architecture

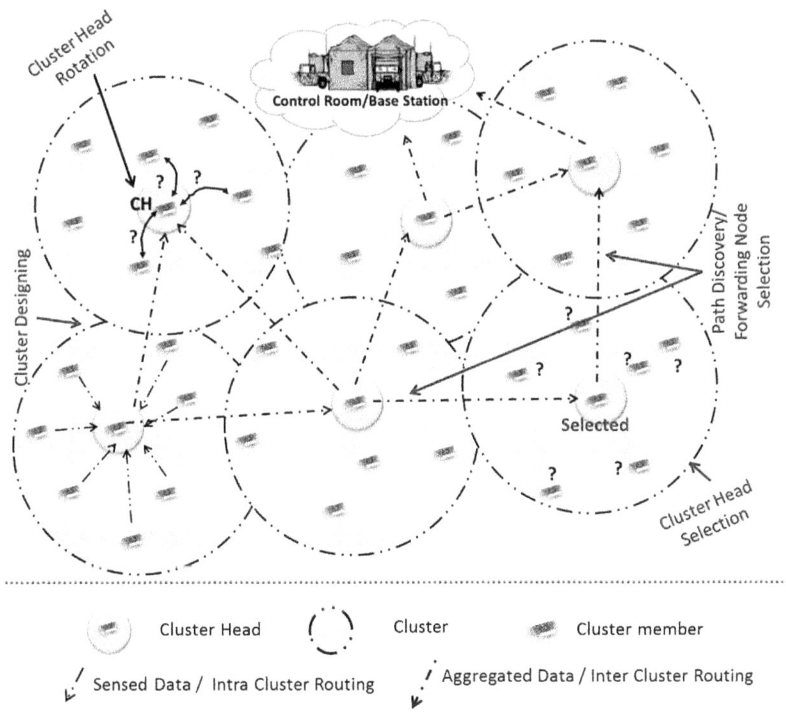

Figure 2. Various roles of node during its life time in clustered network architecture

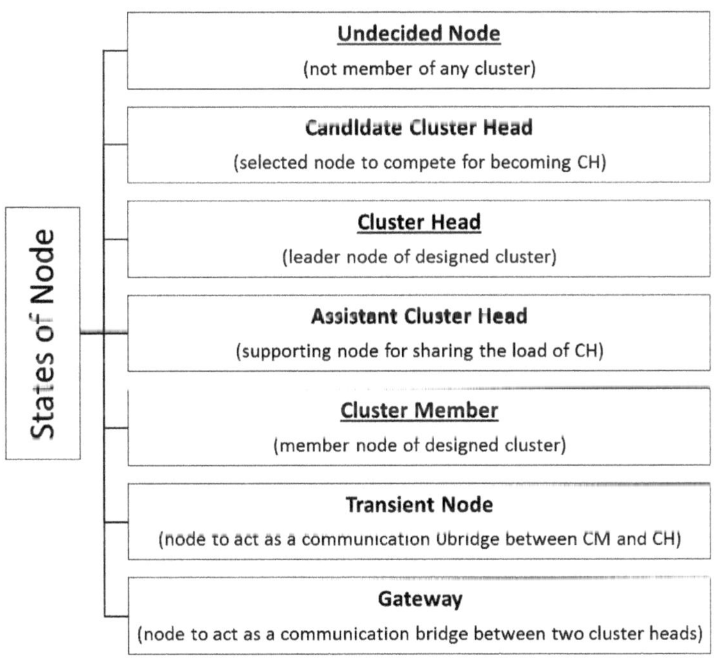

cluster head and time slot assignment for communicating the sensed data to CH. All the proposed algorithms for the same either accomplish these processes with central control of BS (Muruganathan, Ma, Bhasin, & Fapojuwo, 2005) or locally controlled distributed style (Yang & Lee, 2012). Both the techniques have their pros and cons. In Centralized Cluster Designing (CCD), the deployed nodes communicate their information to the Base Station (BS). This information can be node energy level, node degree (number of neighbors of a node that is also called as node density), geographical location, and output of some decision metrics' calculation depending upon the underlying CCD algorithm (Muruganathan, Ma, Bhasin, & Fapojuwo, 2005). This information communication can be direct or multi-hop using transient nodes depending upon the network scale. Base station applies some clustering algorithm; K-Nearest Neighbor (Samet, 2008), K-Means (Cheung, Li, Ng, & Huang, 2008), Centroid (Mahmud, Rahman, & Akhtar, 2012), or Affinity Propagation (Guan, Shi, Marchese, Yang, & Liang., 2011) etc., for the selection of cluster head and communicates the decision with the deployed network nodes. CH then broadcasts its status. Listener nodes affiliate themselves to the most suitable CH.

Proximity to CH, distance of CH to BS, energy, load or the combination of these can be the affiliation decision parameters. Figure 3 gives a typical scenario of centralized cluster designing approach. In Clustesr 'R', node *i* communicates its required decision parameter's value to BS in three hops.

In Distributed Cluster Designing (DCD), nodes communicate the values of selection metric(s) to each of their neighboring nodes. The node that has the best value of selection metric(s) is designated as the Cluster Head (CH). CH advertises its status which helps the nodes to join it as cluster members depending upon Received Signal Strength Indicator (RSSI) or any other assignment metrics (Jabbar, Butt, Najm-us-Sehr, & Minhas, 2011) that is almost the same as stated before in case of CCD. The data communication process in this designed network architecture is split into intra-cluster (within cluster – cluster member nodes to cluster head) and inter-cluster (between clusters – from CH of one cluster to CH of other cluster) data transmission. This data communication style leads to significant energy saving. The more optimized algorithms for designating the network nodes as cluster heads and cluster members, their inter-communication and CH role rotation have proved the priority of clustered network over the flat architecture based networks. Clustered network is considered to be the most energy efficient architecture due to its ease in route discovery, fault tolerance, data aggregation and shortest possible end-to-end delay nature. According to Soroush et al. (Naeimi, Ghafghazi, Chow, & Ishii, 2012) clustering schemes show significant advantages over flat strategies. They enumerated the following advantages of clustering schemes that introduce them as the most compatible protocols with WSNs attributes:

Figure 3. A typical centralized cluster designing (CCD) approach

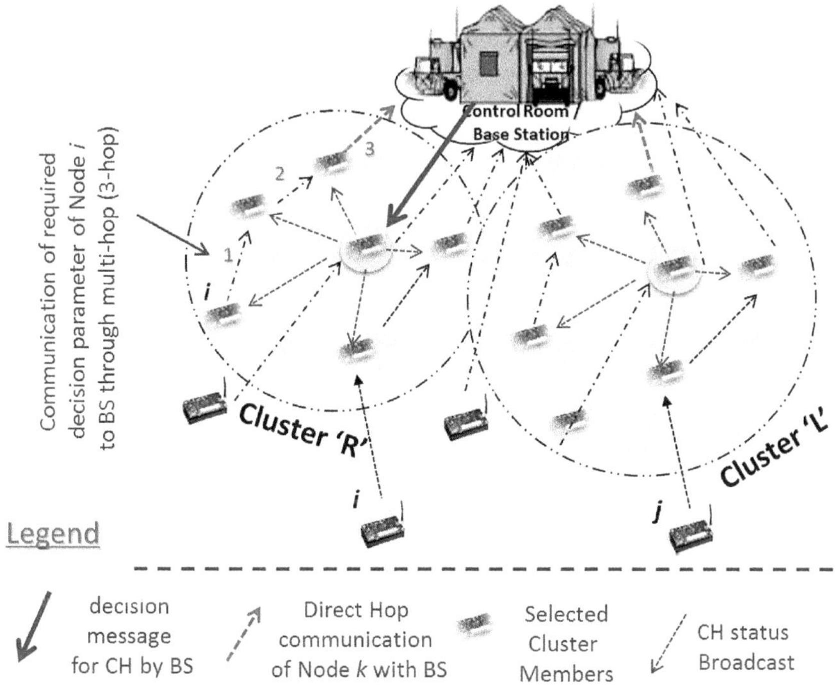

1. Overall transmission power is minimized.
2. The imbalanced load over the deployed nodes is almost balanced with the equal size clusters and centralized management.
3. Bandwidth constraint issue is resolved by efficiently reducing the bandwidth demand.
4. The overhead for routing and topology maintenance is reduced due to centralized management.
5. Data aggregation can be effectively used to eliminate the redundant and highly correlated data.
6. In the inter-cluster and Intra-cluster communication, the collision and interference can be reduced by the use of multi-power levels.
7. The manageability and scalability of the network is improved.
8. Routing table becomes too small due to localizing the route setup within the cluster boundaries.

In the subsequent section, a brief comparative study of existing well known and state of the art ideas relating to cluster designing is given in broader perspective. More-over the compensation to their deficiencies is addressed from the solutions provided in sub section 4.2.

2. COMPARISON OF EXISTING RELATED TECHNIQUES

The general trend in cluster designing scene is either centralized decision at base station for cluster head selection and its members called centralized cluster designing (Muruganathan, Ma, Bhasin, & Fapojuwo, 2005) or distributed decision by exchanging information between neighboring nodes until cluster head and its members are selected called distributed cluster designing (Lee & Lee, 2012). Both the techniques drastically create mess in energy consumption due to too much broadcasting especially in large networks as well as message exchange until some final decision is made. Broadcasting is one of the major factors that squeezes the node's energy. This needs to be considered and should be minimized at each functional level i.e. forwarding node selection to neighbor table or routing table, from cluster designing to cluster head selection and cluster head rotation and from aggregation process to routing process. Though the available literature is rich with high quality articles on energy aware cluster designing, yet they seem to be lacking in covering key aspects to handle the various factors of energy efficient cluster designing. P. T. Yang et al. (Yang & Lee, 2012) require either GPS equipped sensors or some non-GPS technique to complete the working of their proposed solution. Former method increases the cost while in later case, algorithms to correctly identify the node's location are to be implied. That is an extra overhead to face. Same is the case with the idea proposed by C. Li et al. (Li, Ye, Chen, & Wu, 2005), J. Yu et al. (Yu, Qi, Wang, Guo, & Gu, 2011), and Z. Yu et al. (Yu, Liu, & Cai, 2008). Some proposed solutions have formula with parameters like neighbor nodes' distance to BS, their energy, their nodal density, etc. that need intensive message exchange for collecting related information. The proposed solutions by M. Aslam et al. (Aslam, et al., June 2012), G. Chen et al. (Chen, Li, Ye, & JieWu, 2007), Z. Yu et al. (Yu, Liu, & Cai, 2008) and C. Li et al. (Li, Ye, Chen, & Wu, 2005) follow the same fashion to a complete cluster designing.

K. Lee et al. (Lee & Lee, 2012) adopts distributed cluster designing approach in their proposed solution; A Self-Organized and Smart-Adaptive Clustering and Routing Approach for Wireless Sensor Networks (SOSAC). The involved fitness function in the solution requires information of node degree and node energy locally (one hop distance) and out of its local area. Rich message exchange is the inseparable part of this information collection process. Another node's energy draining factor is centralized calculation and decision making. The proposed solutions by Li, Ye, Chen, & Wu (2005), and Chen, Li, Ye, & Jie Wu (2007) are its typical examples. The proposed solution; An Energy-Efficient Unequal Clustering Mechanism for Wireless Sensor Networks (EEUC) by Li, Ye, Chen, & Wu (2005) requires that each node must have the value for maximum distance and minimum distance of node from BS among all other nodes. These values are the pre-requisites to complete the calculation for identifying the maximum competition radius. On considering the worst case of this scenario, it has all the drawbacks related to the centralized cluster designing in large scale networks i.e. re-broadcasting of received broadcast from the far distant node to communicate their info to the BS. BS then broadcasts the extracted info of maximum distance and minimum distance from the received information.

Hence based on the available literature, it can be concluded that to have the need of an energy efficient cluster designing algorithm. This energy efficient solution must address all the aforementioned shortcomings in existing related ideas. More-over, we believe that a hybrid solution exploiting the pros of distributed and centralized cluster designing approaches and escaping the cons of these both techniques is highly appreciable to ameliorate the over network performance. It also must eradicate the cause of relatively high energy consumption during the cluster designing process.

Our novel layer based hybrid algorithm for cluster head and cluster member selection come up to novel communication architecture. Since its substantial constituent is cluster designing, so it is named as MCDA. Proposed design not only has its effect in lessening the blind broadcasting but also decreasing the message exchange. Furthermore, decisions are made locally, so minimize the message exchange. This energy saving endeavor is further backed by non-location aware idea and localized parameter encompassing calculations for cluster head and cluster member decision making. Proposed solution also encapsulates the beauty of efficient centralized decision-making i.e. one hop for various aspects of cluster designing and energy aware distributed cluster head selection and Cluster member allocation process. Novel algorithms for time slot allocation, minimizing the cluster head completion candidates, and node affiliation to cluster the head play underpinning roles to achieve the target. These incorporations in MCDA result in minimizing transmissions, suppressing unfavorable response of transmissions and near equal size and equal load clusters.

3. ASSUMPTIONS AND DEFINITIONS

In this section, the assumptions that are made during the design of network and simulation model are presented. Some of the scenario-related definitions are also given. Guidelines for few of our assumptions are taken from (Muruganathan, Ma, Bhasin, & Fapojuwo, 2005).

Assumption 1 (Homogenous Nodes): All the sensor nodes have the same configuration; sensing and communication range, i.e. 100m and have the same energy level, i.e. 15J.

Assumption 2 (Communication Radius Model): The communication range of a sensor node $'A'$ has the radius 'R' that is centered at 'c'. It can be defined as

$$CR(c,R) = \{A, q \in S :| D(A-q) \leq R_A\}.$$

CR represents communication radius, *S* represents the set of deployed nodes and $D(A-q)$ is the distance between nodes A and q in the deployment area. R_A is radius of sensor node $'A'$.

Assumption 3 (Reliable Communication Link): We have assumed AWGN channel and have adjusted Signal to Noise Ratio (SNR) in such a way that signal reaches the destination node with an energy that the overall detection probability is more than 0.5. In our case, the threshold value of detection probability is 0.4 for a readable received signal.

Assumption 4 (Reliable Network Nodes): The deployed network nodes are reliable and secure and will not be malfunctioning or be hacked or die suddenly. The deployed nodes are considered dead when their energy approaches a defined threshold, say 0.5J in our case.

Definition 1 (Communicating Neighbor Set): All the nodes (q_i) which fulfill the following criteria are considered in the neighborhood of node $p.\left|D(A-q)\right| \leq R_p$, $\forall_i$ Where $i = 1,2,3,...,n$ and R_p is the communication radius of node p.

4. ENERGY AWARE ALGORITHM FOR DESIGNING MULTILAYER ARCHITECTURE

This section comprises of two subsections. First is the summarized presentation of research work given in (Jabbar, Minhas, Paul, & Rho, 2014). The given research work is a novel energy aware algorithm for designing multilayer architecture for dense wireless sensor networks. Proposed design technique is innovative in its idea. Secondly, a constructive ramification of first work is given to ameliorate the performance in network life time. Novel algorithms for time slot allocation, minimizing the cluster head completion candidates, and cluster member selection\node affiliation to cluster head play underpinning roles to achieve the target. These incorporations in MCDA result in minimizing transmissions, suppressing unfavorable response of transmissions and near equal size and equal load clusters. We have done extensive simulations in NS2 and evaluate the performance of E-MCDA in energy consumption at various aspects of energy, packets transmission, number of designed clusters, number of nodes per cluster and un-clustered nodes. It is found that the proposed mechanism optimistically outperforms the competition with MCD and EADUC.

4.1 Multilayer Cluster Designing Algorithm

Multilayer Cluster Designing Algorithm for Lifetime Improvement of wireless sensor network; MCDA by Jabbar et al. (Jabbar, Minhas, Paul, & Rho, 2014) is a hybrid approach in its communication architecture perspective and architectural design perspective. MCDA uses multilayer approach comprising of first flat layer in the footprint of BS and the subsequent clustered layers. Designing of former layer is initiated centrally whilst distributed fashion is applied in the designing of later. The deployed nodes in flat layer are termed as first layer nodes. Authors' started the network clustering from second layer up to network boundary. The cluster heads in the second layer are selected by the elected decision maker nodes of first layer. Neighbor Counter, Decision Maker Nodes and Packet Sequence ID with Postfix Counter are the key factors in designing the clusters. Where Neighbor Counter is used for preferring one node over others for selection of various positions i.e. decision maker, cluster head at various steps, Decision Maker Nodes are used for selection of cluster heads from the subsequent layer, and Packet Sequence ID with Postfix Counter is a packet id that is used for grouping of nodes and for preferring one node over others from that group for becoming CH. Second layer nodes elect the node with highest node density as their decision maker node. Second layer nodes communicate their nodal density on their turn to their decision maker node to take part in competition of becoming CH. Time slots are assigned to these nodes based on Time Division Multiple Access (TDMA) technique. When the first node of second layer communicates its nodal density to decision maker node, it assigns sequence number with postfix counter '0' to this packets. All the recipient nodes of second layer nodes saves this packet sequence number and become the part of same group. All the nodes having packet with same packet sequence number are included in the same group. Only those nodes of a group communicate their nodal density to decision maker nodes which have highest nodal density than their previous nodes. These nodes increment the postfix counter that gives two fold advantages; i) to let the other member nodes of same group to know about their nodal density, ii) to let the non-member nodes to know that neither they should continue this postfix counting nor they should save any info of other group's member nodes. This postfix counter assists the packet sequence number in separating the members of one group from other. The node right after the last member of first group communicates to its decision maker node and assigns the new packet sequence number with postfix counter '0'. After collecting the nodal density of second layer's selected nodes, the decision maker nodes elect the CH having highest nodal density among its second layer's addressed nodes. The elected cluster heads broadcast "Join Request" packets. This is to inform other sensor nodes of its availability as CH. Recipient nodes send their consent message in the form of "join accept" message to become the cluster members. If "join Request" message is received from more than one CHs then decision of membership is based on the current load on CH i.e. CH having less number of member nodes is preferred to be attached with it.

4.2 Extended- Multilayer Cluster Designing Algorithm

In a continuation of proposed idea given by Jabbar, Minhas, Paul, & Rho (2014) and summarized in previous section, a constructive ramification of this idea is proposed with the name Extended MCDA (E-MCDA) to ameliorate the performance in network life time. Novel algorithms for time slot allocation, minimizing the CH competition candidates, and cluster member selection to CH play underpinning roles to achieve the target. These incorporations in MCDA result in minimizing transmissions, suppressing unneeded response of transmissions and near equal size and equal load clusters. We have done extensive simulations in NS2 and evaluate the performance of E-MCDA in energy consumption at various aspects of energy, packets transmission, number of designed clusters, number of nodes per cluster and un-clustered nodes. It is found that the proposed mechanism optimistically outperforms the competition with MCDA and EADUC. More-over, the assumptions made for homogeneity of nodes, communication radius model, reliability of the communication link, communication neighbor set and network nodes are same as are explained in section 3. The figures from Figure 4 to Figure 12 and tables from Table 1 to Table 5 show the working of the extended version. Time slot allocation at network setup phase is introduced to make the cluster designing process more energy efficient in Section 4.2.1, energy efficient cluster head selection idea is explained in Section 0, 'Required Node Degree' based cluster member selection for near equal size clusters is given in Section 4.2.3. Section 0 discusses the comparative analysis of Extended MCDA with MCDA and another state-of-the-art cluster designing algorithm.

4.2.1 Time Slot Allocation at Network Setup Phase

The proposed improvement of introducing the time slot allocation at network setup phase minimizes the message broadcast during selection of $Node_{(dm)}$. It also has its role in the election of cluster heads by suppressing the turns of nodes with equal and lower degree to take part in said processes since the decision for the same is based on the highest node degree that has already been done. All network nodes

Figure 4. Formation of first layer

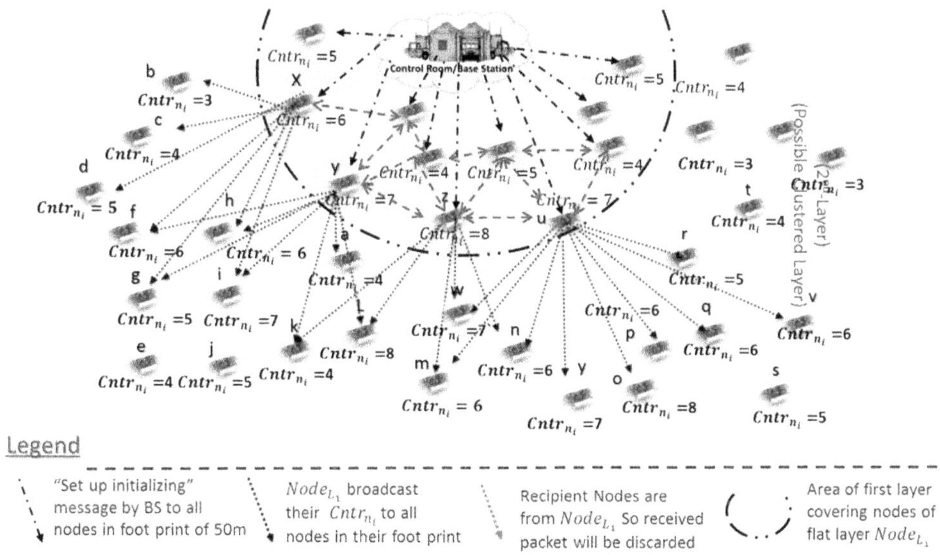

broadcast beacon message $(Msg(Beacon))$ that is piggybacked with the information of their number of neighbors. With this all recipient nodes including BS set up their neighbor table containing Node id and No. of neighbors (Line 1 to Line 4 of Algorithms 1). BS broadcasts initialization message $(Msg_{(setup_{init})})$ to initiate the cluster designing process. Recipient nodes are designated as first layer nodes $(Node_{L_1})$ (Line 5 to Line 7 of Algorithm 1). $Node_{L_1}$ send their number of neighbors exclud-

Table 1. Representation of $Msg_{(CL_{init})}$ messages by $Node_{L_1}$ recipients nodes from $Node_{L_2}$

Selected Nodes from $Node_{L_1}$	Waiting Time for Group of Recipient Nodes of $Msg_{(CL_{init})}$ Message $T_{wait(t_i)}$	$Msg_{(CL_{init})}$ Message Recipient Nodes' Group
Y	$1200\mu s$	f, g, i, k, l, a, w
Z	$7500\mu s$	a, k, l, w, m, n
U	$450\mu s$	$\boldsymbol{w}, \boldsymbol{m}, \boldsymbol{n}$, y, p, o, r, q, v

ing their neighbors that are not from $Node_{L_1}$ i.e. $Nodes_{(rem)}$ s(Line 8 to Line 13 of Algorithm 1). Now the BS assigns the node Ids to $Node_{L_1}$ according to their $Nodes_{(rem)}$ value (Line 14 to Line 16 of Algorithm 1). Total time slots allocated to the $Node_{L_1}$ i.e. $(Total_{TS})$ is calculated by BS as given in Equation 3, Equation 4 and Equation 5 (Line 17 to Line 20 of Algorithm 1).

$$\Delta t_i = \Delta t_{i-1} + \Delta t \tag{3}$$

$$\Delta t_i = (T_{est_{tx}} + T_{GI}) \tag{4}$$

$$Total_{TS} = \sum_{i=1}^{n} \Delta t_n + \Delta t \tag{5}$$

Now ultimately, BS broadcasts flat layer initialization message $(Msg_{(FL_{init})})$ to formally initiate the designing of clustered network. $(Msg_{(FL_{init})})$ contains $Total_{TS}$, Δt and newly allocated $Node_{ids}$ (Line 21 and Line 22 of Algorithm 1). Recipient nodes of $(Msg_{(FL_{init})})$ calculate $T_{wait(t_i)}$ as in Equation 6 (Line 23 to Line 25 of Algorithm 1).

$$T_{wait(t_i)} = Total_{TS} - \Delta t_i \quad or \qquad Total_{TS} - (\Delta t + i) \tag{6}$$

To initiate the clustering process in the second layer nodes $(Node_{L_2})$, clustering layer initialization message $(Msg_{(CL_{init})})$ that has the information of $T_{wait(t_i)}$ and Cnt_{n_i} is broadcasted. Only $(Node_{L_2})$ receive the packet and all other nodes that are of $Node_{L_1}$ discard the packet (Line 26 and Line 32 of Algorithm 1).

Refer to the Figure 4, there are 11 nodes in footprint of BS to receive this message and are comprising of the first layer. $Msg_{(FL_{init})}$ message consists of 128 bits and transmission time for a packet is $128\mu s$ per packet (message). If Δt is the time slot allocated to a node which consists of estimated transmission time $T_{est_{tx}}$ and the safe gap in the transmission of next node called guard interval T_{GI} then first node from $Node_{L_1}$ is allocated Δt_1 that is equal to Δt as in Equation 7.

$$\Delta t = (T_{est} + T_{G1}) = (128 + 28)\mu s = 153\mu s \tag{7}$$

The next allocated time starts right after the completion of this Δt_1 and so on. Calculation for the same is given as follows based on $\Delta t_i = \Delta t_{i-1} + \Delta t$:

Let $i = 1$ and $\Delta t_O = 0$,

So, $\Delta t_1 = \Delta t_0 + \Delta t = 153\mu s$,

If $i = 2$ and $\Delta t_1 = 153\mu s$,

Then $\Delta t_2 = \Delta t_1 + \Delta t = (153 + 153)\mu s = 306\mu s$

If $n = i$ and $\Delta t_1 = 306\mu s$,

Then $\Delta t_3 = \Delta t_2 + \Delta t = (306 + 153)\mu s = 459\mu s$

Similarly,

If $i = 10$ and $\Delta t_1 = 1377\mu s$,

Then $\Delta t_{10} = \Delta t_9 + \Delta t = (1377 + 153)\mu s = 1530\mu s$

and

If $i = 11$ and $\Delta t_1 = 1530\mu s$,

Then $\Delta t_{11} = \Delta t_{10} + \Delta t = (1530 + 153)\mu s = 1683\mu s$

Hence based on the above calculations, total transmission time $Total_{TS}$ of first layer nodes having 11 nodes in number is $1683\mu s$ i.e. $(128\mu s + 25\mu s) \times 11$.

$$Total_{TS} = (128\mu s + 25\mu s) \times 11 = 1683\mu s$$

Since BS knows the number of nodes in the first layer as is discussed previously. So this all calculation of $Total_{TS}$ of first layer nodes is performed by BS and is piggybacked in $Msg_{(FL_{init})}$ message to be sent to nodes of first layer. We are in a plan that all the $Node_{L_2}$ must not participate in the competition of being CH in order to lessen the broadcasting since the CH announcement of highest degree node suppresses the broadcasting of node with equal and lower degree node. Moreover there is not any possibility that lower degree node may announce its candidacy of being CH before the higher degree nodes and both these nodes are in the communication range of each other.

For the said purpose, each node calculates $T_{wait(t_i)}$ that is equal to $Total_{TS} - \Delta t_i$ Sample calculation for the same is also performed in subsequent paragraphs.

Let $Total_{TS} = 1683$ and $i = 1$

If $\Delta t_1 = 153\mu s$ Then $T_{wait(t_1)} = 1683\mu s - 153\mu s = 1530\mu s$

If $\Delta t_2 = 306\mu s$ Then $T_{wait(t_2)} = 1683\mu s - 306\mu s = 1377\mu s$

If $\Delta t_3 = 459\mu s$ Then $T_{wait(t_3)} = 1683\mu s - 459\mu s = 1224\mu s$

Similarly,

If $\Delta t_{11} = 1683\mu s$ Then $T_{wait(t_{11})} = 1683\mu s - 1683\mu s = 0\mu s$

We can have a look over the calculated values of Δt_n , and $T_{wait(t_i)}$ from the succeeding process. We notice that the output is in the increasing order in case of Δt_n and in decreasing order in case of $T_{wait(t_i)}$. This is also depicted in part 'a' and part 'b' of Figure 6.

$Node_{L_1}$ broadcast this calculated value of $T_{wait(t_i)}$ along with Cnt_{n_i} to $Node_{L_2}$ in $Msg_{(CL_{init})}$ for further calculation of $T_{net_{wait}}$ for assigning the turns of second layer nodes to take part in selection of decision maker nodes $\left(Node_{(dm)}\right)$ and later on in CH election. Hence based on the above mentioned discussion, all the $Node_{L_2}$ are at same level to be in fair competition of being CH. The subsequent layers are

Figure 5. E-MCDA Function for Flat Layer

ALGORITHM 1. Function of E-MCDA for Flat Layer

START

$Node_{all}$ broadcasts $Msg(Beacon)$
$Node_{all}$ Setup $Table_{(nbr)}$
BS Setup $Table_{(nbr)}$
$R_{(B_{T_x})}$ = 100m
$Msg_{(setup_{init})}$ Contains N_Cnt_{BS}
BS broadcast $Msg_{(setup_{init})}$
Set $Node_{(R_x)}$ = $Node_{L_1}$
$Node_{(R_x)}$ Calculates $Nodes_{(rem)}$
If $Node_{id}[BS(Table_{(nbr)})] = Node_{id}[Nodes(Table_{(nbr)})]$ Then
    $Cnt_{(node)}++$
End if
$Nodes_{(rem)}$ = $[Nodes(Table_{(nbr)}) - Cnt_{(node)}]$
$Node_{(R_x)}$ Sends $Nodes_{(rem)}$ to BS
BS Generates $List(Tx_{seq})$ Based on $Nodes_{(rem)}$
Sort $List(Tx_{seq})$ in ascending order
Allocate $Node_{ids}$ to nodes based on their values in sorted $List(Tx_{seq})$
BS calculates $Total_{TS}$
$\Delta t = (T_{est_{tx}} + T_{GI})$
$\Delta t_n = \Delta t_{i-1} + \Delta t$
$Total_{TS} = \Sigma_{i=1}^{n} \Delta t_n + \Delta t$
$Msg(FL_{init})$ contains $Total_{TS}$, Δt and newly allocated $Node_{ids}$
BS broadcasts $Msg(FL_{init})$
$Node_{(R_x)}$ calculate $T_{wait(t_i)}$
$\Delta t_i = (\Delta t \times i)$
$T_{wait(t_i)}$ = $Total_{TS} - \Delta t_i$
$Msg_{(CL_{init})}$ = $T_{wait(t_i)} + Cnt_{n_i}$
$Node_{L_1}$ Broadcast $Msg_{(CL_{init})}$
Set Transceiver of $Node_{l_1}$ = Sleep mode for specific time interval
If $recipients[Msg_{(CL_{init})}] \notin Node_{L_i}$ Then
    Receive the packet
    Set $Node_{(R_x)}$ = $Node_{L_2}$
End If

End

Figure 6. Depiction of calculation of Δt_n *and* $T_{wait(t_i)}$

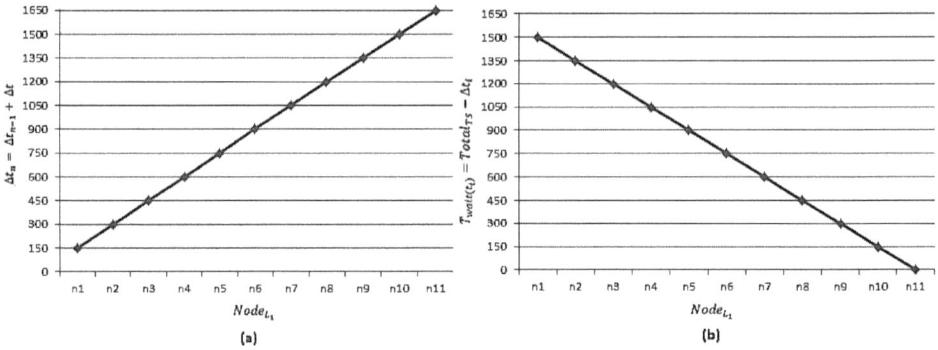

defined on the basis of transmission range of previous layer nodes (first layer nodes in this case till now). Figure 4 shows this layered representation of the deployed nodes and summarized description in the form of legend on formation of first layer. Figure 5 gives pseudo-code for the function of E-MCDA at flat layer.

We started the network clustering from the second layer up to the network boundary. The cluster heads in the second layer are selected by the elected decision maker nodes of the first layer. This saves energy that other cluster based routing algorithms consume for frequent cluster designing in the neighborhood of the base station (as in DSS algorithms) or re-clustering of the whole network (as in centralized cluster based networks). This also decreases the load on the neighbor nodes that ultimately does the load balancing and hence increases the life time of network. After setting up of first layer, $Node_{L_1}$ play their role for the designing of second layer; clustered layer. Extended MCDA designates decision maker nodes from $Node_{L_1}$ to break the conflict of claim of nodes for heading the nodes in its neighbors.

$Node_{L_1}$ broadcast $Msg_{(CL_{init})}$ message with the fields $Cntr_{n_i}$ (their node degree) and $T_{wait(t_i)}$ (time for which the $Node_{(Rx)_i}$ of $Msg_{(CL_{init})}$ message must wait before it takes any action) values for the initialization of cluster designing process at second layer. The suggested improvements in designing of second layer and carrying it on to the next layers is discussed in detail in subsequent sub section along with brief shortcomings of other two competing cluster designing methods.

4.2.2 Energy Efficient Cluster Head Selection

Extended MCDA is introducing a new ever technique for designing the clusters by sensibly managing the announcement of being CH in a distributed way. Neighbor

Counter and Decision Maker Node are the key factors in designing the clusters. Where neighbor counter is for preferring one node over others for selection of various positions (decision maker, cluster head) at various steps, and Decision Maker Node is for the selection of cluster head from the subsequent layer (especially in case of tie). The cluster designing process at second layer is initialized with the $Msg_{(CL_{init})}$ message by $Node_{L1}$. Its recipients that are not from $Node_{L_1}$ comes in second layer and are called second layer nodes $(Node_{L2})$. These nodes set up their array of neighbor count of first layer nodes and compared them. Among those, the node which has the highest neighbor count $(Max\ Cnt_{ni})$ is elected as decision maker node $(Node_{(dm)})$ (Line 1 to Line 4 of Algorithm 2). Recipients of $No\ \ Node_{L_2}$ start their timers as soon as they receive this message and set it $T_{wait(t_i)}$ whose value is set by the $Msg_{(CL_{init})}$ message sent from $Node_{L_1}$. A node may receive $Msg_{(CL_{init})}$ message from more than one $Node_{L_1}$. In this case only the first $T_{wait(t_i)}$ value is considered and all others' are ignored. $T_{wait(t_i)}$ value is different for different $Msg_{(CL_{init})}$ messages recipient groups (Line 5 to Line 10 of Algorithm 2). Line 11 to Line 13 of Algorithm 2 give an abstract level calculation of $T_{net_{wait}}$ .For having more practical exposure for the aforementioned proposed solution, following example is demonstrated in detail with its depiction in Figure 7.

Figure 7. Election of first cluster head in 2nd Layer

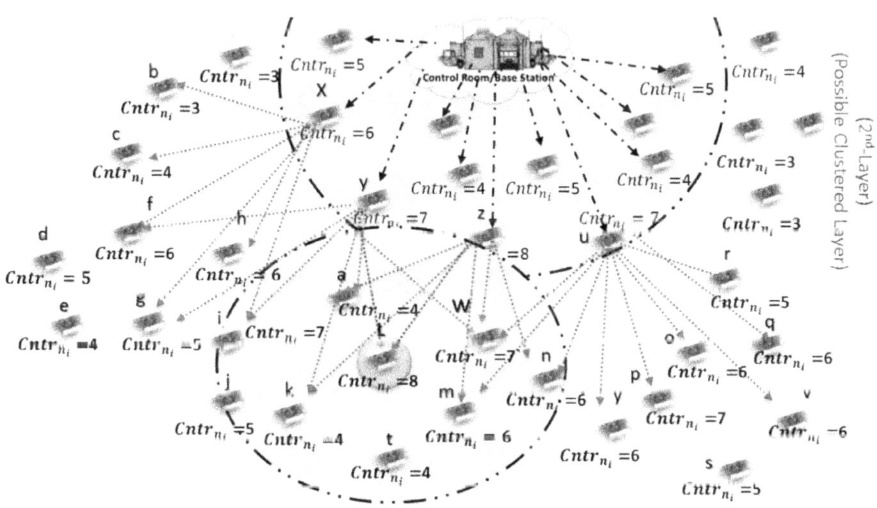

Table 2. Representation of $Node_{L_2}$ that receive multiple $Rqst_{DM_{Elec}}$ messages from $Node_{L_1}$

Nodes Receiving $Msg_{(CL_{init})}$ Messages From	
Y *and* Z	Z *and* U
$k, l, a,\ w$	$w,\ m,\ n$

Let in a scenario, nodes $Y, Z, and\ U$ are selective nodes from first layer with $T_{wait(t_y)} = 1200$, $T_{wait(t_z)} = 750$, and $T_{wait(t_u)} = 450$. $Msg_{(CL_{init})_y}$. message from node **Y** is received by f, g, i, k, l, a, w and the a, k, l, w, m, n nodes receive $Msg_{(CL_{init})_z}$ message from node **Z** . While the recipients of $Msg_{(CL_{init})_u}$ message from node $U\ are\ n, y, p, o, r, q, v$. Referring to Table 1, Node ids in bold are the nodes those are receiving $Msg_{(CL_{init})}$ messages from more than one first layer nodes. The node Y among nodes $Y, Z, and\ U$ is the first node to broadcast its $Msg_{(CL_{init})}$ message followed by node Z. and then node U . A the recipients of $Msg_{(CL_{init})_y}$ message from node Y have their $T_{wait(t_y)}$ equal to $1200\mu s$ and of node **Z** have their $T_{wait(t_z)}$ equal to $750\mu s$.

In case of recipients of $Msg_{(CL_{init})}$ messages from node **U**, $T_{wait(t_u)}$ is equal to $450\mu s$. Node receiving $Msg_{(CL_{init})}$ messages from more than one first layer nodes have their $T_{wait(t_i)}$ equal to that from which it receives $Msg_{(CL_{init})}$ messages in first. Tabular representation of this scenario is demonstrated in Table 1, Table 2 demonstrates second layer nodes in separate in given scenario that is also depicted in Figure 7 which are at the intersection of communication range of $Msg_{(CL_{init})}$ messages from more than one first layer nodes.

So, each node from $Node_{L_2}$ waits for $T_{wait(t_i)}$ time received in $Msg_{(CL_{init})}$ messages.This equalizes the waiting time of all $Node_{L_2}$ to take any action ahead to cluster designing process. So, after finishing first layer nodes' transmission i.e. completion of $1683\mu s$ as is calculated in previous example, the first node that broadcasts $Msg\left(Rply_{DM_{Elec}}\right)$ message in response of $Msg_{(CL_{init})}$ message is the node having its $Net_{Cntn(i)}$ value highest among $Node_{L_2}$ i.e. also has the lowest $T_{net_{wait}}$ value (Line 14 to Line 15 of Algorithm 2).

$Node_{L_2}$ In continuation of the above mentioned scenario, among all the listeners of $Msg(Rply_{DM_{Elec}})$ messages from node (f, g, i, k, l, a, w), $Z(a, k, l, w, m, n)$ and $U(w, m, n, y, p, o, r, q, v)$, nodes $f, l, and\ o$. have the same Cnt_{n_i} value but greater than all other nodes. their cluster head election waiting time $(T_{ch_{elec}})$ is less than all other nodes of same level i.e If. Calculation o $(T_{ch_{elec}})$. that is given in Table 3 for selected nodes of $Node_{L_2}$. To break the tie among nodes $f, l, and\ o$ in accessing the channel, CSMA/CA algorithm is tested to be the best choice (Line 16 to Line 23 of Algorithm 2).

Table 3. Calculation for Net Waiting Time $T_{net_{wait}}$ for $Node_{L_2}$ for energy aware designing of clusters at 2nd Layer (the node having the lowest net node degree $Net\_Cnt_{n(i)}$ has the highest net waiting time $T_{net_{wait}}$ value and vice versa)

$Node\_ID$	i^{th} Node Degree (Cnt_{n_i})	Neighbor from $Node_{L_1}$	$Net_{Cntn(i)}$	$T_{ch_{elec}} = 1 - Log_{10}(Net_{Cntn(i)})$	$T_{net_{wait}} = T_{ch_{elec}} + T_{wait}$
f	6	2	4	0.397940009	1650.39794
g	5	1	4	0.397940009	1650.39794
h	6	2	4	0.397940009	1650.39794
i	7	2	5	0.301029996	1650.30103
j	5	1	4	0.397940009	1650.39794
k	4	2	2	0.698970004	1650.69897
l	8	2	6	0.22184875	1650.221849
m	6	2	4	0.397940009	1650.39794
w	7	2	5	0.301029996	1650.30103
n	6	2	4	0.397940009	1650.39794
y	7	1	6	0.22184875	1650.221849
p	6	1	5	0.301029996	1650.30103
o	8	1	7	0.15490196	1650.154902

$* \; T_{wait} = T_{ch_{elec}} + (T_{est_{L1}} + T_{GI}) \times Cnt(Node_{L_1}) = T_{ch_{elec}} + (128\,\%s + 25\,\%s) \times 11$

Let node l. is given the chance to access the network. It's all listeners $(i,\ j,\ k,\ a,\ t,\ m.\ w,\ n)$ suppress their turn of broadcasting $Msg(Rply_{DM_{Elec}})$ at their allocated time slot since their $Net_{Cntr_{ni}}$ is lower (Line 24 to Line 26 of Algorithm 2). Now the node with next highest $Net_{Cnt_{n(i)}}$ value among the remaining $Node_{L_2}$ senses the medium using CSMA algorithm and on finding it free, it transmits $Msg(Rply_{DM_{Elec}})$ say this node is node o. Its listeners $(\ n,\ y,\ p,\ s,\ v,\ q,\ r)$ have lesser $Net_{Cnt_{n(i)}}$ value.

So they suppress their turn and accept the highest degree transmitting node, node o in their footprint as their CH. This process continues until all second layer nodes are engaged into clusters. Figure 8 shows the completion of clustering process in 2nd layer having one cluster highlighted with node 'f' as CH. If receives more than one $Msg(Rply_{DM_{Elec}})$ messages then $Msg(\mathrm{Re}\,ply_{Node_{(dm)}})$ message is sent to only that node which have Max Cnt_{n_i} (Line 27 to Line 33 of Algorithm 2). Figure 9 shows the pseudo code for the function of E-MCDA for designing of clustered layer.

Figure 8. Step 1: Completion of the clustering process in 2nd Layer having one cluster high $Node_{(dm)}$ lighted with node 'f' as cluster head

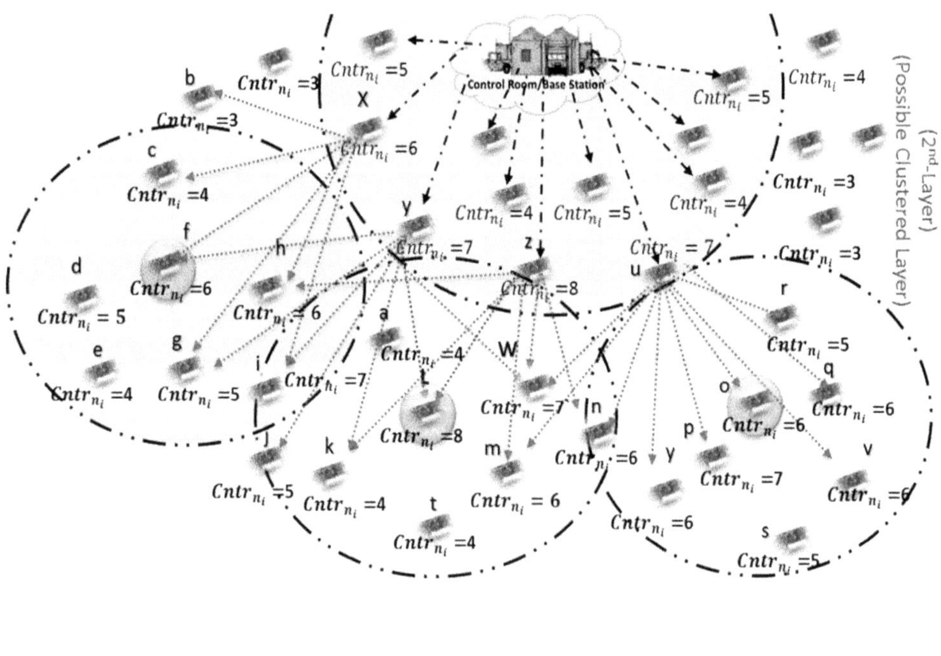

Figure 9. E-MCDA function for clustered layer

ALGORITHM 2. Function of E-MCDA for Clustered Layer

START

If Node $\in Node_{L_2}$ Then

  Set_up array of $Cnt_{ni} \in Node_{L_1}$

  Compare $Cnt_{ni} \in Node_{L_1}$

  Elect $Node_{(dm)}$ having Max Cnt_{ni} among $Node_{L_1}$ in array

  Set timer = $T_{wait_{ti}}$

    If Cnt [Received ($Msg_{(CL_{init})}$)] >1 Then

    Consider only 1st

    Ignore all later

    End if

  Each recipient [[$Msg_{(CL_{init})}$] calculates $T_{net_{wait}}$

  $Net\_Cntr_{n(i)} = Cntr_{n(i)} - (Node(s)_{nbr}\ of\ Node_{L_1})$

  $T_{ch_{elec}} = 1 - Log_{10}(Net_{Cntr\,n(i)})$

  $T_{net_{wait}} = T_{ch_{elec}} + T_{wait}$

  $Msg\left(Rply_{DM_{Elec}}\right) = Node_{(dm_{id})} + Cnt_{n_i}$

  $Msg\left(Rply_{DM_{Elec}}\right)$ has unique Id

  If $\left(T_{net_{wait}}\right)_i < \left(T_{net_{wait}}\right)_j$ Then $\forall_j\ where\ j = 0\ to\ n\ \ and\ \ j \neq i$

    If $Cntr\left(T_{net_{wait}}\right)_i > 1$ Then

      Randomly select one based on CSMA/CA algorithm

    End if

    If $T_{net_{wait}} = true$ Then

    Selected $Node_i$ from $Node_{L_2}$ Broadcasts $Msg\left(Rply_{DM_{Elec}}\right)$ to elected $Node_{(dm)}$

    End if

  End if

  If $Node_{(Rx)} = Node_{L_2}$ then

    Suppress their Turn to Broadcast $Msg\left(Rply_{DM_{Elec}}\right)$

  End if

  If Cnt [Received($sg\left(Rply_{DM_{Elec}}\right)$ at $Node_{(dm)} > 1$ then

    $Node_{(dm)}$ setup array of Cnt_{n_i}

    Compare $Cnt_{n_i} \in Node_{L_1}$

    Elect $Node$ among $Node_{L_1}$ having Max Cnt_{n_i} in array

    Designate the elected node as CH

    Send $Msg\left(Reply_{Node(dm)}\right)$ $Msg\left(Reply_{Node(dm)}\right)$ to elected CH

  End if

  End if

End

Figure 10. Step-1: Nodes at intersection area (highlighted in dark gray)

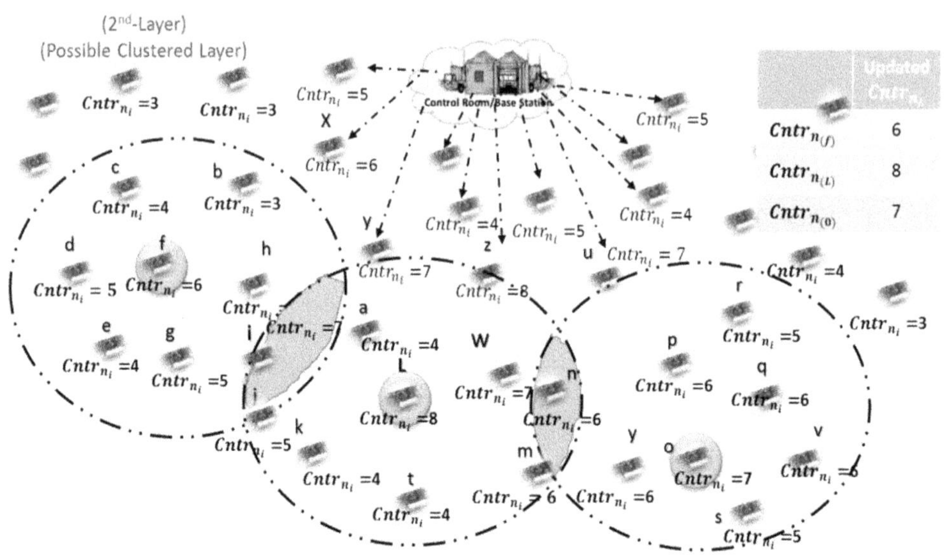

4.2.3 'Required Node Degree' Based Cluster Member Selection for Near Equal Size Clusters

This improved version 'required node degree' based node affiliation algorithm' helps much in designing near equal size clusters. Figure 10 is showing the said typical scenario in the Extended MCDA architecture where nodes i ., and n are at intersection area (communication range of more than more CHs is overlapping each other) as is highlighted in grey.

In a continuation of this,ure 11 depicts the completion of the clustering process at layer 2 with resolving of 'nodes at intersection' issue. To understand how the affiliation of nodes at intersection area to the CHs helps in designing near equal size clusters, we are proposing an algorim ich is expined with an example in subsequent paragraphs.

Let in a scenario having 3 clusters f , l ., and o . with $Cntr_{n(f)} = 6$, $Cntr_{n(l)} = 8$, and $Cntr_{n(o)} = 7$ and cluster head $CH(f)$, $CH(l)$, and $CH(o)$. Two nodes i . and j are at the intersection of cluster f and cluster l . and two other nodes m . and n . are at the intersection of clusters o .nd l . . Node i . sends $Msg_{(join_{acpt})}$ message to $CH(f)$. with piggybacking the $Cntr_{n(f)}$ value of its competitor CH i.e. CH_l . The competitor listener CH decrements its $Cntr_{n(l)}$ value counter by 1. Since

$Cntr_{n(f)} < Cntr_{n(l)}$ ds $Msg_{\left(join_{acpt(ack)i}\right)}$ message to node i with piggybacking its updated $Cntr_{n(f)}$ value to jo it as its cluster member.

Step 2 in Table 4 shows the update status of clusters at this stage. As $CH\left(l\right)$ has lost its one potential member node, so it sends its latest value of $Cntr_{n(l)}$ to update its neighbors. Now the turn is of node m . that is placed at the intersection of communication range of cluster l . and cluster o . Since $Cntr_{n(l)} = Cntr_{n(o)}$, so tie rises in selection between these two. On random choice let $CH(l)$ is selected. Tn node ses $(Msg_{\left(join_{acpt}\right)})$ message to $CH\left(l\right)$ in the same fashion as mentioned in case of Node i . and receives $(Msg_{\left(join_{acpt}\right)})$ message by $CH\left(l\right)$ with piggybacking its updated $Cntr_{n(l)}$ value. Here $CH\left(o\right)$ decrements its $Cntr_{n(o)}$ value counter by 1 due to loss of one of its potential member node and sends its latest $Cntr_{n(o)}$ value to update its neighbors. Step 3 in Table 4 shows the update status of clusters at this stage. Similar process continues until remaining two nodes j and n . in the given scenario are also affiliated to the clusters having lowest current $Cntr_{n(i)}$ value among their inviter. Step 5 in Table 4 shows the final status of clusters f ., l ., and o after breaking the overlap conflict. Now it seems to be near equal sized clusters.

If we do a theoretical comparison of proposed solution with the generic method of nodes' affiliation that are placed at the intersection of communication range of more than one CHs then we come up with the following Table 5. In generic meth-

Table 4. Function of proposed scheme for affiliation of nodes at intersection of communication range of more than one CH

Steps	$Cntr_{n(f)}$	$Cntr_{n(f)}$	$Cntr_{n(o)}$	Nodes at intersection of	
				f ., and l .	l , and o .
1	6	8	7	i ., j .	m, n .
2	6	7	7	j .	
3	6	7	6	j .	
		6	6	0	
5	6	5	6	0	0

Table 5. Comparison of generic and proposed technique for decision parameter for affiliation of nodes at intersection of communication range of more than one CHs

	CH Affiliation Decision Parameter	
	Generic	**Improved/Proposed**
$Cntr_{n(f)}$	6	6
$Cntr_{n(l)}$	4	5
$Cntr_{n(o)}$	7	6
Sample Variance	2.33	0.33
Performance efficiency of the proposed method over generic method		86%

od, node is affiliated to the CH with the lowest degree for the sake of equalizing the cluster size. It shows that the sample variance of the generic method is far greater than that of proposed solution.

Figure 11. Step 2: Completion of clustering process at layer 2 with resolving of 'nodes at intersection' issue

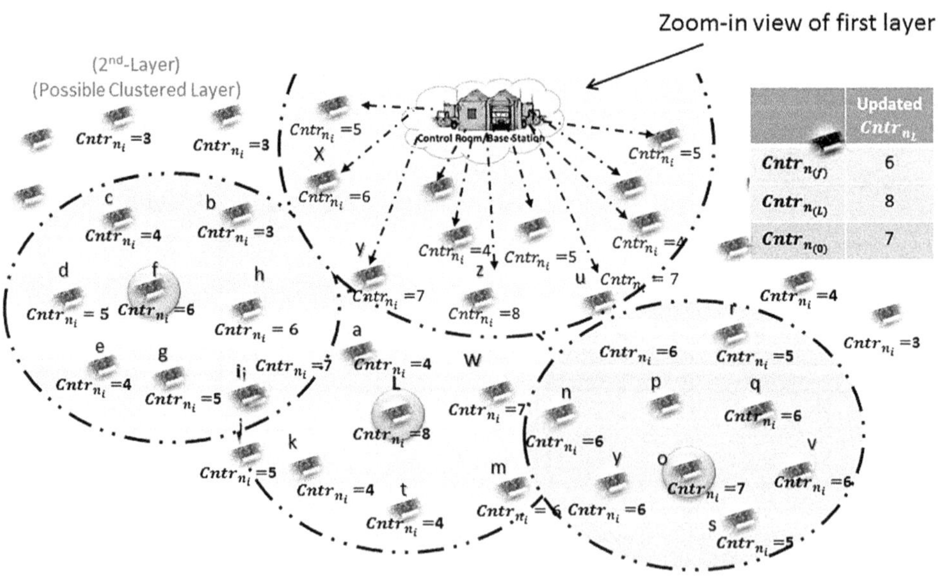

Figure 12. Sample variance of competing algorithms on equality of cluster size

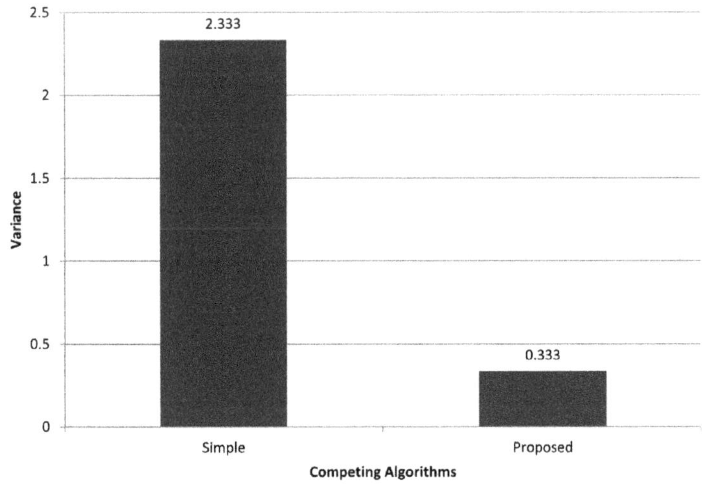

Same is also drawn in Figure 12. Derived from this result, performance efficiency (ρ) of proposed method over simple method is 86% using following formula.

$$\rho = 1 - \left(\frac{Sample\ \ Variance\ \ of\ \ Proposed\ \ Scheme}{Sample\ \ Variance\ \ of\ \ Simple\ \ Schme} \right)$$

This improved performance results in near equal sized clusters that help in a balanced load over cluster heads that results in less frequent cluster head rotation and hence better network performance and improved network life time.

4.3 Comparative Analysis of E-MCDA with State-of-the-Art Algorithms

In this Section, a comprehensive discussion is made on the comparative analysis of E-MCDA with the state-of-the-art related algorithms. Table 6 shows the simulation parameters taken during our experimentations.

Transmission of packets in the network, layer wise and per node, energy consumption in and other than packet transmission, cumulative, total and average energy consumption of network, per node, and layer wise, number of clusters, number of member nodes per cluster and number of un-clustered nodes are taken into account as performance evaluation parameters for the comparison of aforesaid algo-

Table 6. Simulation parameters

Parameter	Description
Routing Protocols	EADUC, MCDA, E-MCDA (Proposed)
Simulation Area	500 m x 500 m and 1000 m x 1000 m
Simulator	NS 2.31
Data Rate	4 Packets/Sec
TCP/IP Layer	Network Layer
Node to Node Distance	Random
Node Type	Homogenous
No. of Nodes	500
Propagation Model	Two ray ground
Initial Energy of Node	3 Joule

rithms. Since the experimentation is performed with same number of deployed nodes i.e. 500 in two different areas i.e. 500m x 500m and 1000m x 1000m, so the representation of results is given accordingly.

In this subsection, the presented brief discussion covers two aspects. First is the results taken from the simulation scenario whose parameters are same as are given in Table 6 with deployment area of 1000m x 1000m. Second is the performance efficiency of proposed solution over other two competitive solutions at the deployment areas of 500m x 500m and 1000m x 1000m. The depiction for the same is given from Figure 13 to Figure 20. Since the reasoning of better comparative performance is the same, so here we are just highlighting the impact of increased deployment area over the extracted results.

Figure 13. Total energy consumption in cluster designing in 1000m x 1000m area

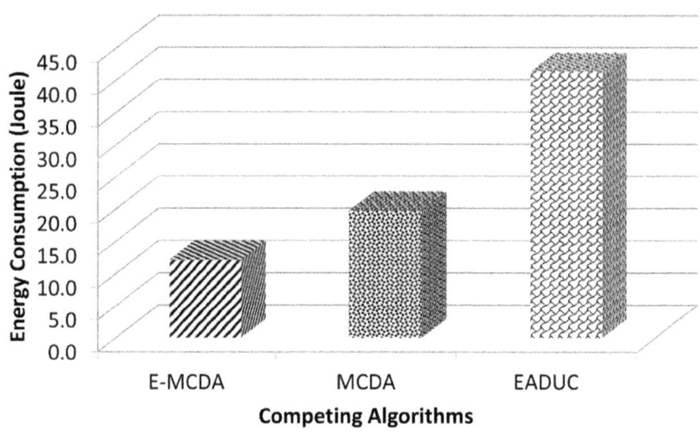

On increasing the deployment area by keeping the number of nodes and all other factors constant, on the average, the node density per unit area decreases. This new scenario may have following effect on various aspects of network and communication thereon as compared to the previously given scenario. These effects may be more number of clusters, less number of cluster members, more number of forwarding nodes and hence more end to end delay, less number of recipient nodes of same typical broadcast and hence less messages in reply, more time in completion of network set up phase, less memory and processing load due to less number of neighbor nodes and less nodes at the intersection point of neighboring nodes' (CHs) communication range. Impact of controlled broadcasting, selective candidate nodes for being CHs, and hybrid approach is also clearly reflected from the comparative results given in Figure 13. Total energy consumption in E-MCDA is still far better compared to MCDA and EADUC.

The overall consumption is decreased in underlying sparse network. Among the aforementioned effects, the major one is less number of recipient nodes of some typical broadcast and hence less message in reply. Hence, in Figure 14, performance efficiency of E-MCDA over MCDA is a bit higher in current deployment area compared to previous scenario. In MCDA, number of designed clusters is 70% more in 1000m x 1000m area compared to node deployment in 500m x 500m area while in E-MCDA, this value is 56% as is intuited from Figure 15. This aspect is stronger in EADUC compared to other two. Since in EADUC, clusters size is unequal. Farther the cluster position from the BS, bigger is its size. Therefore, this value of higher number of clusters in 1000m x 1000m area is about 45% compared to other scenario. But on the other end, the strategy adapted by EADUC in designing these unequal clusters is very expensive due to the information collection for constituent parameter. This increases the total energy consumption bar in Figure 13 and in Figure 14 as well. This style also increases the exchange of messages. Hence, number of broadcast in EADUC is 40% higher in under discussion deployment area compared to densely node deployment in 500m x 500m while in MCDA and E-MCDA, this value is 11% and 10% respectively.

Figure 16 and Figure 17 also extend this comparison in representing number of packets broadcast during cluster designing and the performance efficiency of E-MCDA in number of packets broadcast during cluster designing process over competing algorithms.

In continuation to it, its impact on energy consumption aspect is depicted in Figure 18. Extending to it, the comparison of performance efficiency of E-MCDA in energy consumption during broadcast transmission process in deployment areas; 500m x 500m and 1000m x 1000m and that of MCDA and EADUC is depicted in Figure 19.

Figure 14. Performance efficiency of E-MCDA in total energy consumption during cluster designing over competitor algorithms

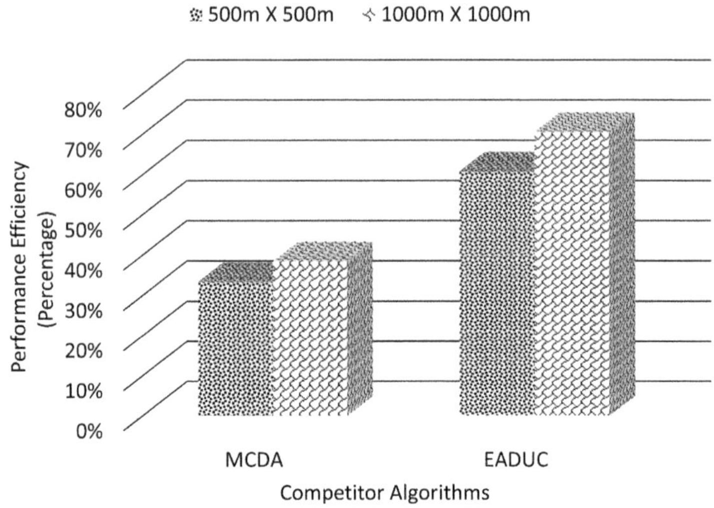

Figure 15. No. of clusters designed in 1000m x 1000m area

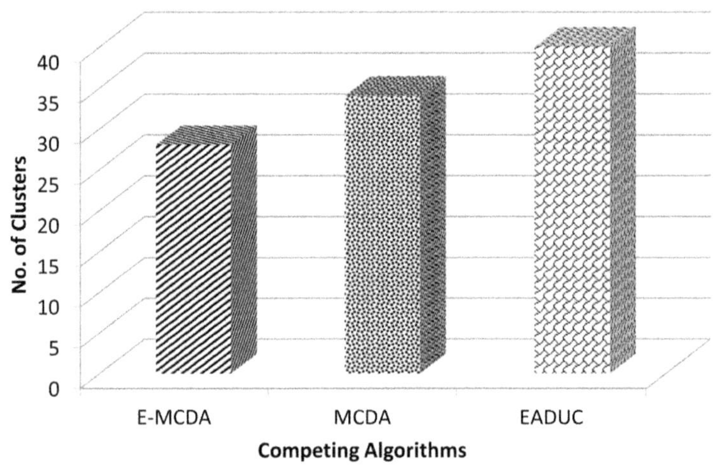

Another important effect of increasing the deployment area by keeping the other aspects same as that are of comparatively smaller area is the decrease in node density in cluster i.e. lesser CMs. This has two fold effects. One is more number of clusters in the network and the other is less nodes at the intersection point of neighboring cluster heads communication range. Sufficient detail has already been dedicated to first point in previous paragraphs. So far as decision for affiliation of

Figure 16. Number of packets broadcast during cluster designing process in 1000m x 1000m area

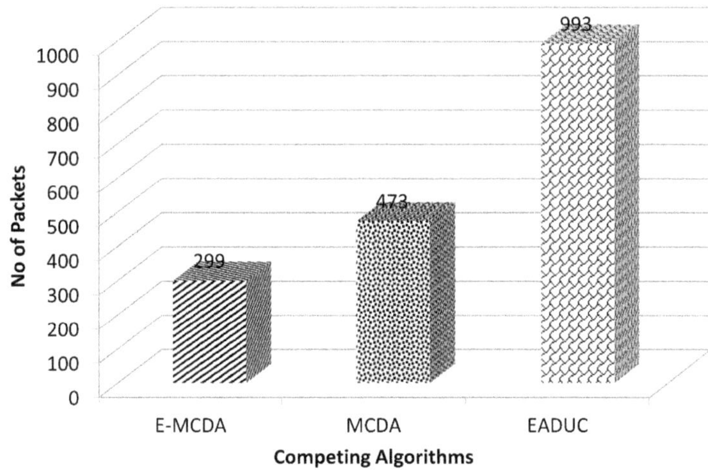

Figure 17. Performance efficiency of E-MCDA in number of packets broadcast during cluster designing process over competitor algorithms

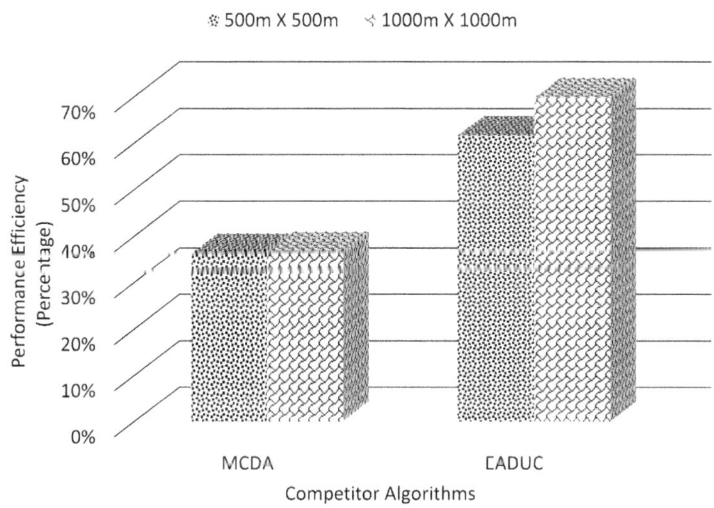

nodes at intersection points is concerned. It shapes the cluster size and ultimately has its impact on CH rotation or cluster re-designing strategy. Refer to Figure 20, a big value of standard deviation in case of EADUC shows higher difference in clusters size compared to that of in MCDA and E-MCDA. This burdens the CHs with more memory and processing load in larger clusters and special intra-cluster routing strategy as well. This further opens the numerous issues to handle.

Figure 18. Energy consumption in broadcast transmission in 1000m x 1000m area

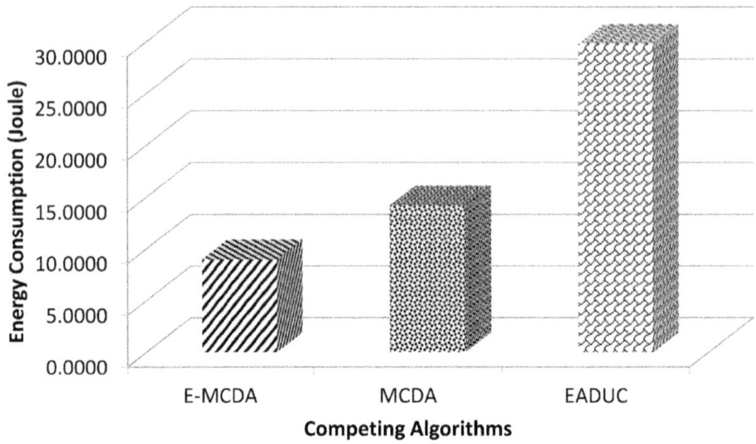

Figure 19. Performance efficiency of E-MCDA in energy consumption during broadcast transmission process over competitor algorithms

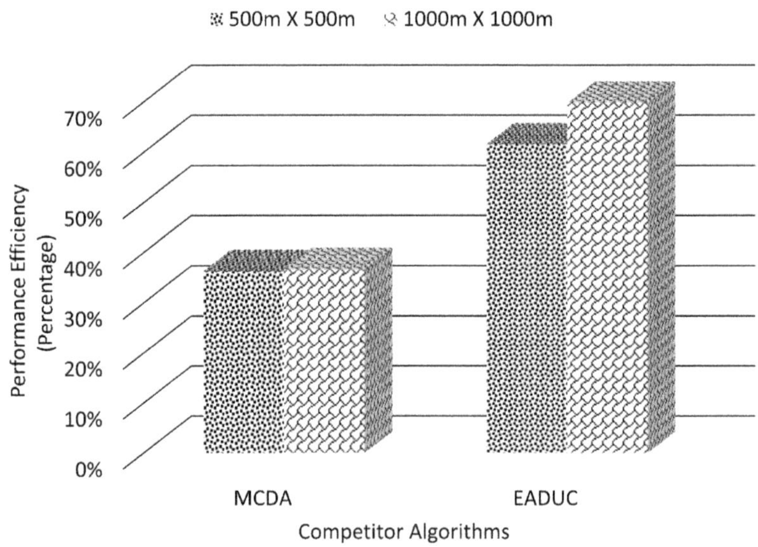

5. CONCLUSION

Idea of MCDA was improved with novel algorithms for time slot allocation at network setup phase to make the cluster designing process more energy efficient, energy efficient cluster head selection, and 'Required Node Degree' based cluster member selection for near equal size clusters. We named it Extended MCDA. These

Figure 20. Standard deviation of number of members per cluster in 1000m x 1000m

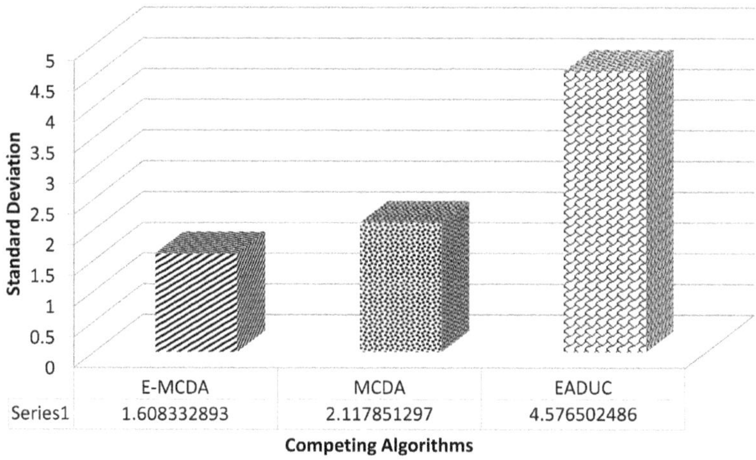

ramifications play vital roles to achieve the target. Results of our experiments have shown that E-MCDA has outperformed the two competitive mechanisms. The key performance evaluation parameters taken for this comparison were energy consumption in various operational parametric values at per node level and at layer level, no. of packets broadcast, no. of clusters designed, number of members per clusters and number of un-clustered nodes. Acceptance of E-MCDA over MCDA and EADUC in aforesaid performance evaluation parameters has been proved from the simulation based experiments. Since total energy conservation level of E-MCDA is 68% higher compared to EADUC. While in comparison to MCDA, the outperform results of MCDA regarding the same is 17%. During the clustered designing process, the messages exchange is one of the big sources of maximum energy consumption. Regarding the comparison in case of no. of packets transmitted as a performance evaluation parameter, proficiency of E-MCDA is better than EADUC with the percentage of 69. Whereas competing with MCDA, E-MCDA has shown better performance with the priority value of 19%. We have also done a comparison of energy consumption in cumulative. As in E-MCDA and MCDA nodes grouping is made into layers i.e. flat layer N_{L_1}, clustering layer 1 N_{L_2}, clustering layer 2 N_{L_3} etc. So, in accordance with this scenario of layering of nodes, cumulative energy consumption in EADUC is calculated in the same way. Hence, it is demonstrated from the experimentation and detailed analytical discussion that E-MCDA is a priority choice for clustered network design. The comprehensive and detailed, analytical and empirical comparative analysis of E-MCDA with its parent idea; MCDA and another state-of-the-art technique; EADUC has also evidenced the achievement of our research goals to propose an energy efficient algorithm for the most energy demanding part of cluster based wireless sensor networks.

6. FUTURE DIRECTIONS

Our work can be extended in many aspects starting from the proposed network architecture until energy aware routing process. Changing the selection parameter for various roles of nodes, introducing the node heterogeneity for Decision Maker Nodes, CHs, or others and the induction of localization techniques can improve the work at various aspects. Also the proposed multi-layer design can be tested for other issues like Hotspot. Keeping in view the flexibility of proposed mechanism, some of the extension in this work are suggested as follows.

The performance of first flat layer of E-MCDA can be compared to unequal clustering algorithms which have smaller cluster size in the neighbor of BS to lessen the network load over these nodes. The same goal of achieving the higher network life among the two techniques motivates for such sequence of comparative studies. In case of heterogeneous network, the same idea of multi-layer network design can be tested with changing the cluster head election criteria such as ratio of residual energy of the node to the average energy of the neighbor nodes (Yu, Qi, Wang, Guo, & Gu, 2011) or energy welfare formula (Yang & Lee, 2012) etc. In our idea of Multi-layer design, although we tried to present the pure network layer algorithm, so the involvement of any localization algorithm or GPS enabled nodes is kept away. Yet another induction of idea can be defining the layer boundaries by transmitting the broadcast signal with varying defined levels as is used by Jabbar et al. (Jabbar, Aziz, Minhas, & Hussain, 2010). This will further decrease the packets broadcasting that are used for selecting the decision maker nodes at the boundary of layers. More-over, the proposed algorithms for throughput maximization can also be used in non-conventional routing algorithms (Jabbar, Minhas, Rahseed, & Rho, 2014). Many interdisciplinary ideas relating to bio-inspired computing are briefly given in (Jabbar, et al., 2013) that can also be used to further extend and improve the proposed work. We believe that the proposed multi-layer network design can optimistically handle the hotspot issue that exists usually in sparsely deployed sensor networks. This extension of our work can be a good contribution in related work. The proposed mechanism can also has its application in densely deployed WSN for energy aware big data gathering. Similar approach based on pure clustered network is proposed in (Takaishi, Nishiyama, Kato, & Miura, 2014). Same idea can also be exploited in large deployment of nodes (uniquely identifiable embedded computing devices) that are attached to things and are interconnected to each other over the existing infrastructure of internet. This concept is termed as Internet of Things (IoT) (Miorandi, Sicari, Pellegrini, & Chlamtac, 2012). IoT is comparatively new and innovative concept encompassing number of emerging technologies where WSN is core among those.

REFERENCES

Aslam, M., Shah, T., Javaid, N., Rahim, A., Rahman, Z., & Khan, Z. A. (June 2012). CEEC: Centralized energy efficient clustering a new routing protocol for WSNs. *Conference on Sensor, Mesh and Ad Hoc Communications and Networks (SECON), 2012 9th Annual IEEE Communications Society* (pp. 103 - 105). Seoul: IEEE.

Chen, G., Li, C., Ye, M., & Wu, J. (2007). An unequal cluster-based routing protocol in wireless sensor networks. *Wireless Networks*, *15*(2), 193–207. doi:10.1007/s11276-007-0035-8

Cheung, Y.-, Li, M., Ng, M., & Huang, J. (2008). Agglomerative Fuzzy K-Means Clustering Algorithm with Selection of Number of Clusters. *IEEE Transactions on Knowledge and Data Engineering*, *20*(11), 1519–1534. doi:10.1109/TKDE.2008.88

Guan, R., Shi, X., Marchese, M., Yang, C., & Liang, Y. (2011). Text Clustering with Seeds Affinity Propagation. *IEEE Transactions on Knowledge and Data Engineering*, *23*(04), 627–637. doi:10.1109/TKDE.2010.144

Handy, M., Haase, M., & Timmermann, D. (2002). Low energy adaptive clustering hierarchy with deterministic cluster-head selection. *4th International Workshop on Mobile and Wireless Communications Network* (pp. 368 - 372). IEEE. doi:10.1109/MWCN.2002.1045790

Jabbar, S., Aziz, M. Z., Minhas, A. A., & Hussain, D. (2010). *PTAL: Power Tuning Anchors Localization Algorithm For Wireless Ad-Hoc Micro Sensors Network.* Bradford, UK: IEEE.

Jabbar, S., Butt, A. E., Najm-us-Sehr, & Minhas, A. A. (2011). TLPER: Threshold Based Load Balancing Protocol For Energy Efficient Routing In WSN. *The 13th International Conference on Advanced Communication Technology (ICACT'11).* Seoul, South Korea: IEEE.

Jabbar, S., Iram, R., Minhas, A. A., Shafi, I., Khalid, S., & Ahmad, M. (2013). Intelligent Optimization Of Energy Aware Routing In Wireless Sensor Network Through Bio-Inspired Computing: Survey And Future Directions. *International Journal of Distributed Sensor Networks*, *2013*, 13. doi:10.1155/2013/421084

Jabbar, S., Minhas, A. A., Paul, A., & Rho, S. (2014). MCDA: Multilayer Cluster Designing Algorithm For Network Lifetime Improvement Of Homogenous Wireless Sensor Networks. *The Journal of Supercomputing*, 11227.

Jabbar, S., Minhas, A. A., Rahseed, T., & Rho, S. (2014, December). Heuristic Approach for Stagnation Free Energy Aware Routing in Wireless Sensor Network. *Ad Hoc Sensor Wireless Network*, *0*, 1–25.

Lee, K., & Lee, H. (2012). A Self-Organized and Smart-Adaptive Clustering and Routing Approach for Wireless Sensor Networks. *International Journal of Distributed Sensor Networks*, 13.

Li, C., Ye, M., Chen, G., & Wu, J. (2005). An Energy-Efficient Unequal Clustering Mechanism for Wireless Sensor Networks. *IEEE International Conference on Mobile Adhoc and Sensor Systems Conference*. Washington, DC: IEEE.

Mahmud, M., Rahman, M., & Akhtar, M. (2012). Improvement of K-means clustering algorithm with better initial centroids based on weighted average. *7th International Conference on Electrical & Computer Engineering* (pp. 647 - 650). Dhaka: IEEE. doi:10.1109/ICECE.2012.6471633

Miorandi, D., Sicari, S., Pellegrini, F. D., & Chlamtac, I. (2012). Internet of things: Vision, applications and research challenges. *Ad Hoc Networks*, *10*(7), 1497–1516. doi:10.1016/j.adhoc.2012.02.016

Muruganathan, S. d., Ma, D. C., Bhasin, R., & Fapojuwo, A. (2005). A Centralized Energy-Efficient Routing Protocol for Wireless Sensor Networks. *IEEE Radio Communications*.

Naeimi, S., Ghafghazi, H., Chow, C.-O., & Ishii, H. (2012). A Survey on the Taxonomy of Cluster-Based Routing Protocols for Homogeneous Wireless Sensor Networks. *Sensors (Basel, Switzerland)*, 12. PMID:22969350

Samet, H. (2008). K-Nearest Neighbor Finding Using Max Nearest Dist. *IEEE Transactions on Pattern Analysis and Machine Intelligence*, *30*(02), 243–252. doi:10.1109/TPAMI.2007.1182 PMID:18084056

Takaishi, D., Nishiyama, H., Kato, N., & Miura, R. (2014, September). Towards Energy Efficient Big Data Gathering in Densely Distributed Sensor Networks. *IEEE Transactions on Emerging Topics in Computing*, *2*(3), 388–397. doi:10.1109/TETC.2014.2318177

Yang, P.-T., & Lee, S. (2012). A Distributed Reclustering Hierarchy Routing Protocol Using SocialWelfare in Wireless Sensor Networks. *International Journal of Distributed Sensor Networks*, *2012*, 14. doi:10.1155/2012/681026

Yu, J., Qi, Y., Wang, G., Guo, Q., & Gu, X. (2011). An Energy-Aware Distributed Unequal Clustering Protocol for Wireless Sensor Networks. *International Journal of Distributed Sensor Networks*, 2011.

Yu, Z., Liu, Y., & Cai, Y. (2008). Design of an Energy-Efficient Distributed Multi-level Clustering Algorithm for Wireless Sensor Networks. *4th International Conference on Wireless Communications, Networking and Mobile Computing* (pp. 1 - 4). Dalian: IEEE. doi:10.1109/WiCom.2008.839

Chapter 6
Cost Minimization of Sensor Placement and Routing in Wireless Sensor Networks:
Placement and Routing Issues in a Random Plane

Tata Jagannadha Swamy
Gokaraju Rangaraju Institute of Engineering and Technology (GRIET), India

Garimella Rama Murthy
International Institute of Information Technology Hyderabad (IIITH), India

ABSTRACT

Wireless Sensor Nodes (WSNs) are small in size and have limited energy resources. Recent technological advances have facilitated widespread use of wireless sensor networks in many real world applications. In real life situations WSN has to cover an area or monitor a number of nodes on a plane. Sensor node's coverage range is proportional to their cost, as high cost sensor nodes have higher coverage ranges. The main goal of this paper is to minimize the node placement cost with the help of uniform and non-uniform 2D grid planes. Authors propose a new algorithm for data transformation between strongly connected sensor nodes, based on graph theory.

DOI: 10.4018/978-1-4666-9941-0.ch006

Copyright ©2016, IGI Global. Copying or distributing in print or electronic forms without written permission of IGI Global is prohibited.

INTRODUCTION

While increasing the usage of wireless sensor networks in various applications, data collection and transmission towards Base Station (BS) are important factors. Effective cost of the sensor placement is to be minimized by proper node deployment. Sensors that can accurately detect targets at longer distances have higher cost. However, use of these expensive, long-range sensors may be prohibitive in terms of total placement cost. On the other hand, if only small range sensors are used, effective surveillance can only be achieved with a large number of these sensors. Therefore, efficient sensor placement strategies are necessary to minimize cost and yet achieve mandated levels of surveillance accuracy. Amongst all the topology schemes proposed, grid-based topology management is more effective in terms of energy efficiency and extending network lifetime. Grid-based scheme can be of a uniform grid structure or a non-uniform grid structure. Uniform grid-based routing can be modified to fit in various environments. In comparison, non-uniform grid-based routing is more energy efficient than uniform grid-based routing.

Sensor placement problem is an important not only from the point of view of cost minimization but also in terms of sensor node coverage and connectivity. In this chapter, in contrast to other attempts made in the past, a more realistic non-uniform grid placement for sensor nodes is considered.

LITERATURE REVIEW

The concept of sensor networks which has been made viable by the convergence of microelectro-mechanical systems technology, wireless communications and digital electronics. First, the sensing tasks and the potential sensor networks applications are explored, and a review of factors influencing the design of sensor networks is provided(Akyildiz, Su, Sankarasubramaniam, & Cayirci, 2002). Intuitively, a denser infrastructure would lead to a more effective sensor network. It can provide higher accuracy and has a larger aggregate amount of energy available. However, if not properly managed, a denser network can (Tubaishat & Madria, 2003). The proposed approach is aimed at optimizing the number of sensors and determining their placement to support distributed sensor networks. The optimization framework is inherently probabilistic due to the uncertainty associated with sensor detections. The proposed algorithms address coverage optimization under the constraints of imprecise detections and terrain properties. These algorithms are targeted at average coverage as well as at maximizing the coverage of the most vulnerable grid points. (Dhillon & Chakrabarty, 2003). Asymptotic optimal strip-based pattern, which is optimal to achieve node coverage and connectivity proposed (Bai, Santosh, Dong, Yun, & Ten, 2006). Optimal sensor placement patterns to achieve coverage and k-connectivity ($k < 6$). Note that k- connectivity can provide fault tolerance (Bai, Xuan, Yun, Lai, & Jia, 2008).

In this paper different mobile medium based sensor deployment strategies are investigated in depth. We discuss the boundary effect, virtual force effective distance, useless move and convergency problems of the VFA for a given ROI. The problems are solved by setting the maximum boundary coordinates, introducing the effective communication distance and constraining maximum step size. (Chen, Li, Sun, 2007). In order to extend the availability of the wireless sensor network and to extract maximum possible information from the surveillance area, proper usage of the power capacity of the sensor nodes is important. Our work describes a dynamic relocation algorithm called MaxNetLife, which is mainly based on utilizing the remaining power of individual sensor nodes as well as properly relocating sensor nodes so that all sensor nodes can transmit the data they sense to the sink (Coskun, 2008). In this paper, authors address the Wireless Sensor Network (WSN) deployment issue. They assume that the observed area is characterized by the geographical irregularity of the sensed events. They considered that each point in the deployment area is associated a differentiated detection probability threshold, WSN deployment problem is formulated as a Multi-Objectives Optimization problem, which seeks to reduce the gap between the generated events detection probabilities and the required thresholds while minimizing the number of deployed sensors. (Aitsaadi, Achir, Boussetta, Pujolle, & Tabu, 2009). Authors address the issues of maintaining sensing coverage and connectivity by keeping a minimum number of sensor nodes in the active mode in wireless sensor networks. They investigate the relationship between coverage and connectivity by solving the following two subproblems. They proved relation that if the radio range is at least twice the sensing range, complete coverage of a convex area implies connectivity among the working set of nodes. They derived Second, we derive, under the ideal case in which node density is sufficiently high, a set of optimality conditions under which a subset of working sensor nodes can be chosen for complete coverage (Zhang & Hou, 2005). Connectivity of nodes with different graphs and its structures are represent are explained (O'Rourke, 1987).

Authors discussed the important aspects of wireless sensor networks, technology, node connectivity, call management and life time of the network (Sohraby, Minoli, & Znati, 2007). Coverage or monitoring entire area by minimum number of nodes with the help of a new algorithm proposed. This algorithm is good for energy management and network life span improvement aspects(MaherRebal, Matthieu Le Berre, Faice Hnaien, Hichem Snoussi, & Lyes Khoukhi, 2014), K-monitoring problem for sensor networks and compared with ILP and Max-flow for effective node placement and connectivity (Xian Chen, Yoo Ah Kim, Bing Wang, Wei, Zhijie, Jerry Shi & Yuan, Song, 2012), authors proposed a novel protocol is based on maximum area coverage by minimum number of sensors and they also analyzed this protocol on network life time(Jing, Alhussein, & Abouzeid 2006). Cost minimization is required without losing information in the sensor field this may increases the life span of the network also. From the aspects of node or network lifetime, authors developed a novel algorithm for effective routing (Wang, Hu, & Tseng, 2005). The developed approximate algorithm is on large number of random points, such that solution is obtained in considerable time (Biagioni, & Sasaki, 2003; Sasaki, 2003). Effective routing also reduces the overall system cost in-terms of life span and energy considerations are formulated and analyzed in their paper (Chong, & Kumar, 2003). Effective, a clustering-based protocol that utilizes randomized rotation of local cluster base stations (cluster-heads) to evenly distribute the energy load among the sensors in the network . the proposed protocol is able to distribute energy dissipation evenly throughout the sensors, doubling the useful system lifetime for the networks (Heinzelman, Chandrakasan, & Balakrishnan, 2000).

Since placement affects whether there are nodes that can sense and routes to the base station, placement of sensor nodes is also important. However, there has been little research on the placement of sensor nodes and all of it is based on deterministic placement, which is not realistic when many sensor nodes are placed in a large area. In such a situation, stochastic placement is needed. (Ishizuka, & Aida, 2004). Authors develop a robust and scalable algorithm to cope with the sensor placement problem for target location under constraints of the cost limitation and the complete coverage. Grid-based placement scenario is adopted and the sensor placement problem formulated as a combinatorial optimization problem for minimizing the maximum distance error in a sensor field under the constraints. The proposed algorithm is based on the simulated annealing approach. The proposed algorithm can find the optimal sensor placement under the minimum cost limitation. Moreover, it can also find a placement with minimum distance error for large sensor fields under the cost limitation (Lin & Chiu, 2005). Fu Xiao, Yang Yang, Ruchuan Wang, &Lijuan Sun,(2014) proposed a three dimensional coverage pattern and deployment scheme for minimum number of nodes deployed for sensing activity and they analyzed the relation between node coverage and connectivity. After the optimal placement is known, we have developed an algorithm for effective data transmission from one node to another node.

SYSTEM MODEL

Sensor Placement in Uniform Grid

For simplicity a 2D grid is considered, although it could be extended to 3D by algorithm1. In this chapter two types (A and B) of sensors with different costs are considered for analysis and to create simulation environment. The main aim is to minimize cost of sensors deployment in n x n grid and this algorithm is developed based on linear programming. With the help of Linear Programming (LP) Solve package, the respective phases are simulated.

Algorithm 1: Uniform Grid Based Cost Minimization to WSNs

$$C = \sum_{i=1}^{r}\sum_{j=1}^{r}\left(CostA * a_{ij} + CostB * b_{ij}\right) \quad (1)$$

where Cost A is cost of sensor type A and Cost B is cost of sensor type B.

$a_{ij} = 1$ If sensor of type A is present at point i, j

$a_{ij} = 0$ else

$b_{ij} = 1$ If sensor of type B is present at point i, j

$b_{ij} = 0$ else

Minimize C, subject to

$$C = \sum_{i1=1}^{r}\sum_{j1=1}^{r}\left(a_{ij} * COV\left(i, j, i_1, j_1, A\right) + b_{ij} * COV(i, j, i_1, j_1, B)\right) \geq E \quad (2)$$

For all i, j

Where E is minimum number of sensors required to cover each grid point in the uniform grid with

$COV\left(i, j, i_1, j_1, X\right) = 1$ if sensor at (i, j) covers point (i_1, j_1).

$a_{ij}, b_{ij} = 0 \ or \ 1 \ for \ all \ i, j$

$aij + bij \leq 1 for\ all \quad i, j$

where E is minimum number of sensors covering each point. COV (i, j, X) =1 if sensor placed at 'i' of type X sensor (either A or B) covers the point j and is 0 otherwise a_i, b_j = 0 or 1 for all i, j.

Algorithm 2: Cost Minimization to Uniform Grid Based WSNs

Cost-minimization can also be done by using greedy approach, where RxR grid can be formed using 4x(R/2 x R/2) matrix. As shown in fig, 4 x (12x12) matrix are used to make one 24x24 matrix. Here it is possible to apply the algorithm applied on larger grids to smaller grids also for cost minimization.

Sensor Placement on Random Plane

Here a uniform grid is replaced by a set of 'n' random points on a plane. It is intended to place sensors of type A and type B with range R_A and R_B and cost C_A and C_B respectively, such that all grid points are covered with say E sensors. Here only two types of sensors are considered, but can be extended to any number of sensors.

Algorithm 3: Non-Uniform Based Sensor Field

In non-uniform based wireless sensor field the grid points are non-uniform, and also the number of sensors at each grid point is not the same. The cost minimization of sensors in non-uniform plane is represented by equation.

$$C = \sum_{i=1}^{r} (CostA * a_i + CostB * b_i) \quad (3)$$

where CostA is cost of sensor type A and CostB is cost of sensor type B.

$a_i = 1$ if sensor of type A is present at point i,

$a_i = 0$ else,

$b_i = 1$ if sensor of type B is present at point i,

$b_i = 0$ else and

C has to be minimized with respect to the equation

$$C = \sum_{j=1}^{n} \left(a_j * COV\left(i, j, A\right) + b_i * COV(i, j, B) \right) \geq E \qquad (4)$$

for all i

where E is minimum number of sensors each point must be covered with. COV (i, j, X) =1 if sensor placed at i of type X (either A or B) covers the point j and is 0 otherwise a_i, b_j =0 or 1 for all i, j.

The main drawback with algorithm 3 is that as number of points increases, it takes more time to complete the process. Like greedy approach of algorithm 2, it is also not suitable for cost minimization because of its non-regularity. Divide and conquer method implemented in algorithm2 is not suitable to fit here due to non uniformity of the grid. Table 1 gives the points with time for program compilation.

From Table 1, the time taken for complete the process increases when numbers of points are increased. If the grid has 400 points, it takes a computation time of 66 sec to obtain optimized solution. One can represent the grid of 400 points in a form of four clusters each of size 100 grid points and then apply algorithm-3 on these four clusters individually. In such case, the total computational time will be 0.01x4=0.04 sec approximately. Here one can see an important thing that could be exploited, i.e. if one could some-how group all points onto smaller groups such that each smaller group executes in time, a solution is guaranteed, which may not be optimal but still would be correct.

K-Means by Clustering

K-means is one of the simplest unsupervised learning algorithms. It solves the well known clustering problem in wireless sensor networks. In this k-means clustering the procedure is to classify a given data set through a certain number of clusters (assume k clusters) fixed a priori. The main idea behind the k-means cluster is to

Table 1. Time taken on 2GHz machine simulation

S.No	Number of Points on Graph	Time Taken to Complete the Program (in Sec)
1	100	0.01
2	200	0.1
3	400	66

Figure 1. Placements of sensor on 7x7 grids with two different types

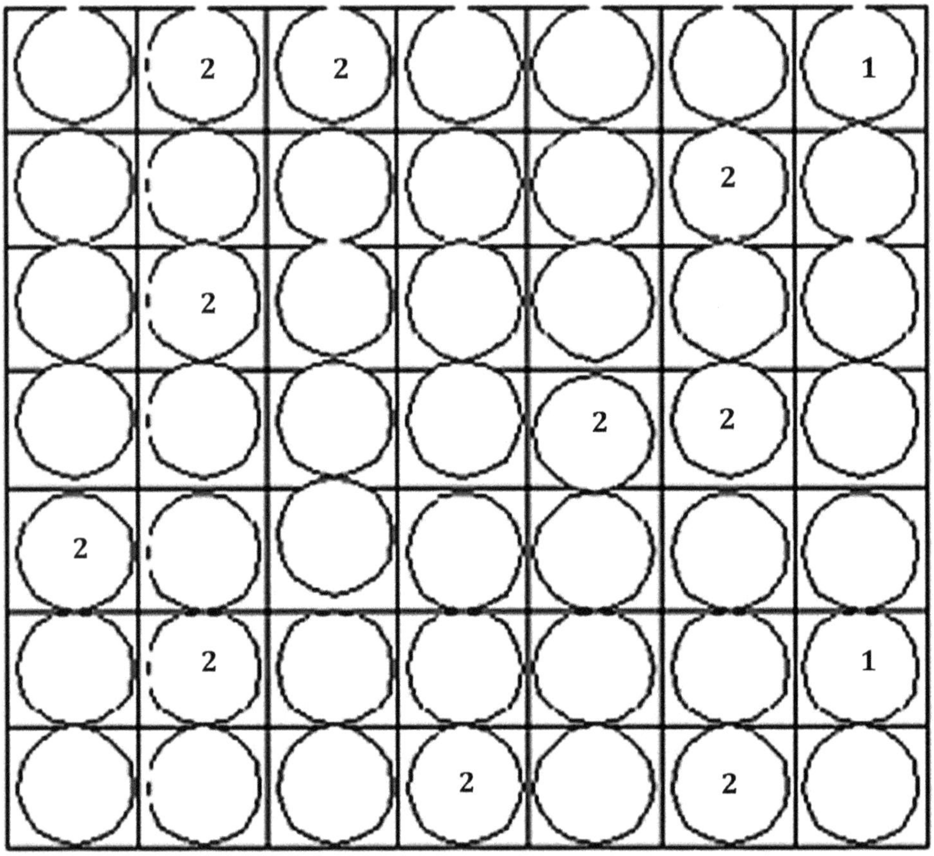

Figure 2. Division of 24X24 grids by 4x(12x12)

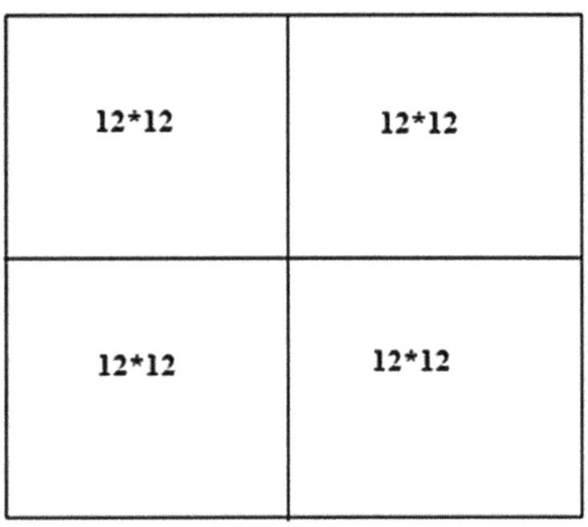

define k centroids, one for each cluster. These centroids should be placed in a cunning way because of different location causes different result and better choice is to place them as much as possible far away from each other. Consider each point belonging to a given data set and associate it to the nearest centroid.

Algorithm 4: Clustering the Points for Sensor Placement

If the algorithm 3 runs for 'm' number of points effectively, it is possible to cluster all n points into ceil (n/m) clusters and run algorithm 3 on each cluster. Here the use of k-means to cluster points is utilized. The clustering algorithm on random 1000 points on 200x200 has been shown in Figure 3. Similarly, the result of applying algorithm 3 on each cluster formed by k-means is depicted Figure 4.

Figure 3. Clustering algorithm on random 1000 points

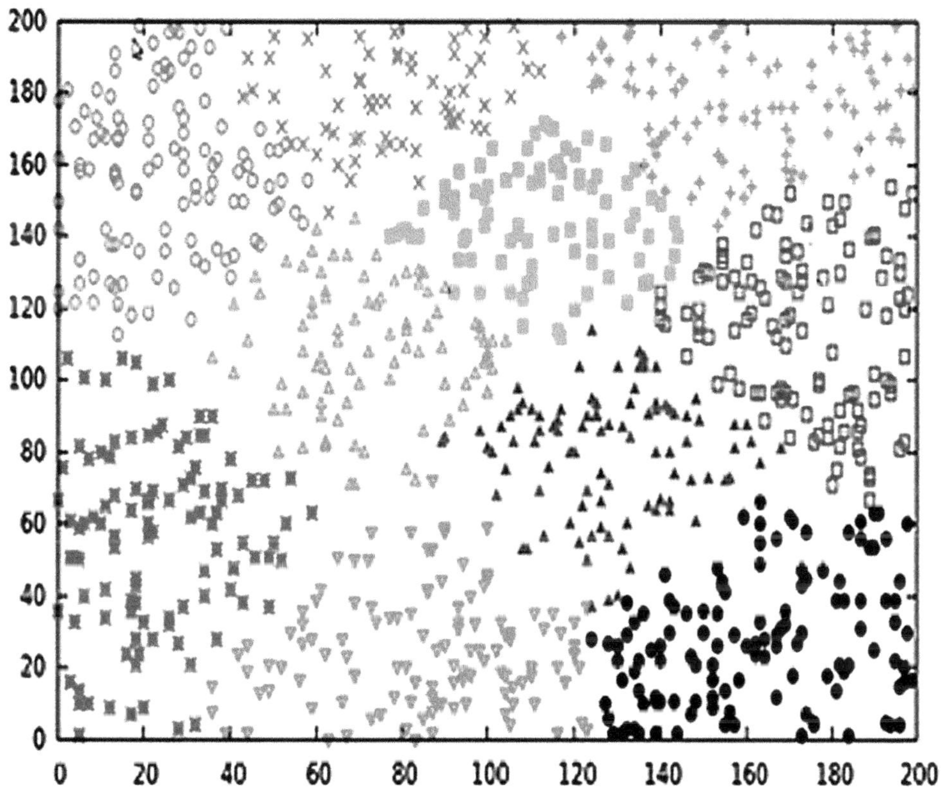

Figure 4. Algorithm 3 on each cluster of random 1000 points

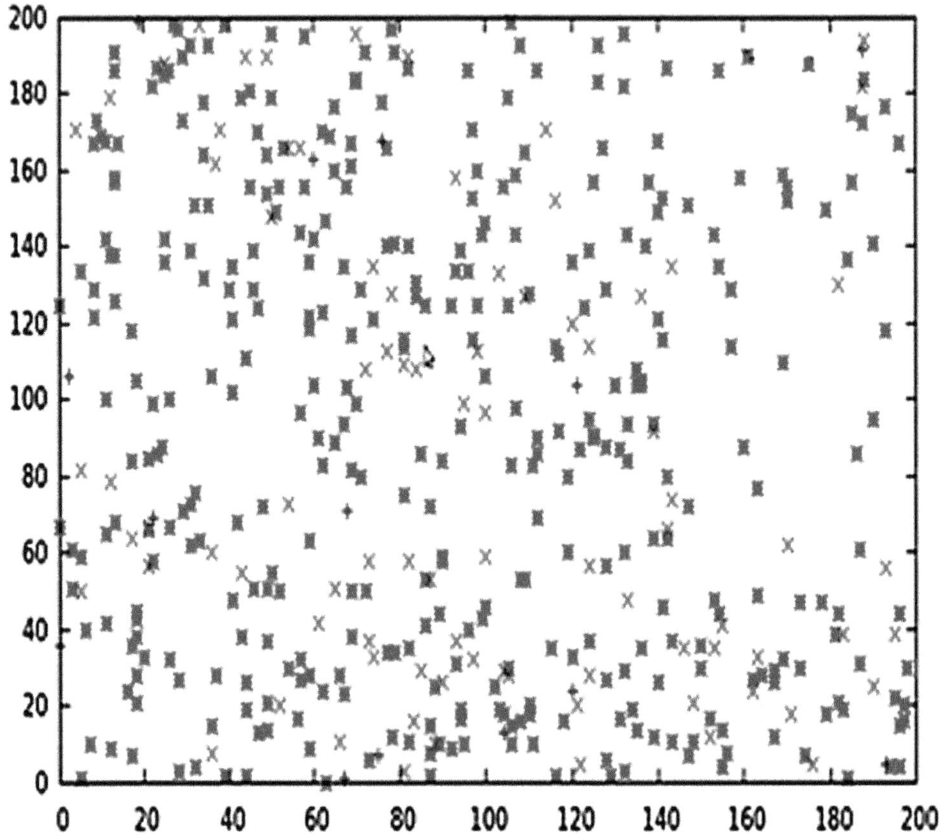

Comparison of Algorithm 2 and Algorithm 4

Consider regular NxN grid points. For algorithm2 we compute N/2xN/2 grid result and for algorithm 4 we compute ceil with [(NxN)/100] sensors, such that each cluster on an average has 100 points. Here cost of sensor A is 100 with range 1 unit and cost of sensor B is 150 with range 2 units. Table 2 represents the main advantages of using algorithm 4 with algorithm 2.

PERFORMANCE EVALUATION

Considering k-mean algorithm, one can cluster the points in such a way that the total cost is low compared to the default value. The k-means cluster takes lot of parameters and one can see its impact on cost with just two parameters: distance

Table 2. Comparison between Algorithm 2 and 4

Grid Size	Algorithm 2	Algorithm 4	Remarks
20x20	Using 4,10x10, cost 4x3100=12400	4 clusters with cost of each = 3100 Total cost is 12400	Clustering here has the same result as without clustering
24x24	Using 4 12x12, cost as 4x4300=17200	6 clusters, with each are having 100 points on average. Total cost = 17500	Larger the cluster size lower is the overall cost, higher the demand on speed of computation. computation
28x28	Using 4 14x14, cost 4x5700=22800	8 clusters 100 points on average total cost=24150	Cost difference is more due to smaller cluster size in algorithm 4 than in algorithm 2.

and initial cluster-centroid position. Using The results for 28x28 uniform grid with (RA=1, CostA=100, RB=2, CostB=150, M=2) obtained using Matlab are summarized in Table 3. The percentage of cost-difference is 0.6. Thus by changing the parameters to k-mean, one can get a better result.

Algorithm 4 on Random Graph with Modification in k-Mean Clustering

As an extension to the above work, the authors have chosen 1000 points randomly from 200x200 range and divided into 10 clusters. The results based on varying k-mean parameter obtained using Matlab are given in Table 4. It may be seen that in case of random points, changing k-mean parameters can impact cost. The number of points chosen in each cluster is also important. From Table 2, one can observe that greater the number of points in each cluster lower the cost. Thus, one must choose as large a cluster as possible. The time taken to complete the simulation process for 'g' out of 'm' maximum points in the cluster group is $t_m\ (g/m)$ where t_m is the

Table 3. Result for 28x28 uniform grid with Percentage of cost difference =150/24100=0.6%

Distance Centroid	Square Euclidean Distance square-root [(x1-x2)²+ (y1-y2)²]	Sum of absolute differences. [|x1-x2|+|y1-y2|]
Selecting K at random from X sample	24150	24100
Selecting K at uniform from range of X	24250	24200
Performance clustering on 10% of sample and the result as initial centroids	24100	24150

maximum time taken to complete the algorithm 3 to run on 'm' points. Figure 5 shows the results of clustering algorithm on 28x28 regular grid structure and Figure 6 gives the result of algorithm 3 is on each cluster group by k-means cluster concept.

Table 4. Result for 1000 random points with percentage of cost difference: 600/24100=2%

Distance / Centroid	Square Euclidean Distance square-root $[(x1-x2)^2+(y1-y2)^2]$	Sum of absolute differences. $[\|x1-x2\|+\|y1-y2\|]$
Selecting K at random from X sample	24350	24750
Selecting K at uniform from range of X	24100	24200
Performance clustering on 10% of sample and the result as initial centroids	24500	24300

Figure 5. Clustering algorithm on 28x28 regular grids

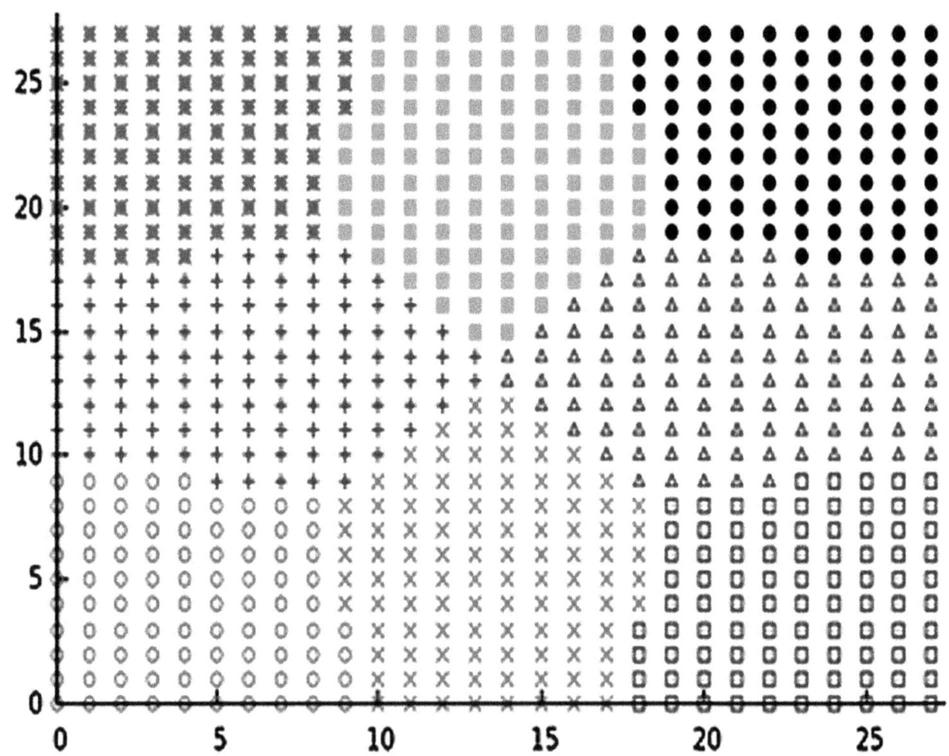

Figure 6. Algorithm 3 on each cluster by k-means

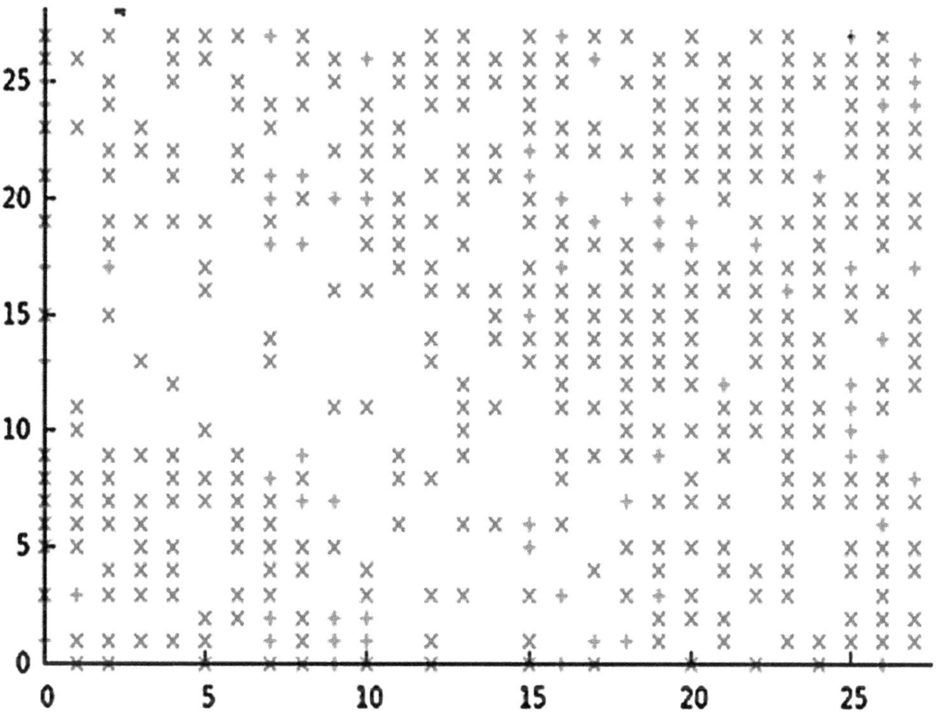

Data Transmission through Strongly Connected Nodes

These sensor nodes can collect data from all the desired nodes. But they must be able to transmit it back to other nodes or base station. First of all, a graph is prepared such that there is a directed edge from i to j, if sensor present at ith point covers point j. Once the graph is constructed, one can check whether it is connected. If it is connected only then information can flow from one node to another node. If not, strongly connected components need to be found in the graph and inter-connected.

STRONGLY CONNECTED COMPONENTS BASED ON GRAPH THEORY

In Figure 7, Sensor {B}; { C, D, E}; {F}; {B, G, F}; {C, H}; {H, J}; { K}; {G}; { I, L}; {L}; {J}; {K, L} are covered in graph by the sensors A, B, C, E, F, G, H,I, J, G, L, K. Strongly connected components are {A}, {B,E},{C,F},{D},{G,H,I,J,K,L}. Figure 8 represents the effective routing between the nodes under strongly connected components.

Figure 7. Strongly connected sensors in the network

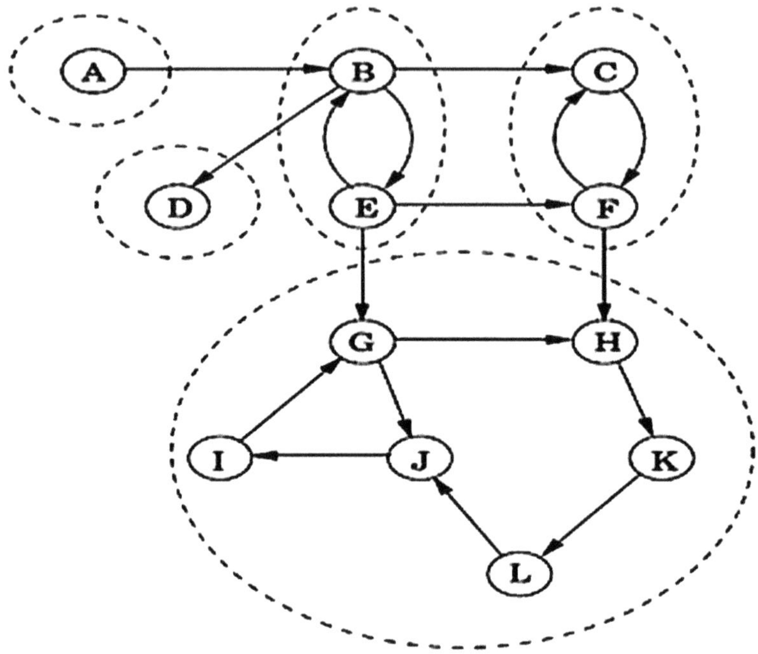

Figure 8. Strongly connected nodes and its grouping

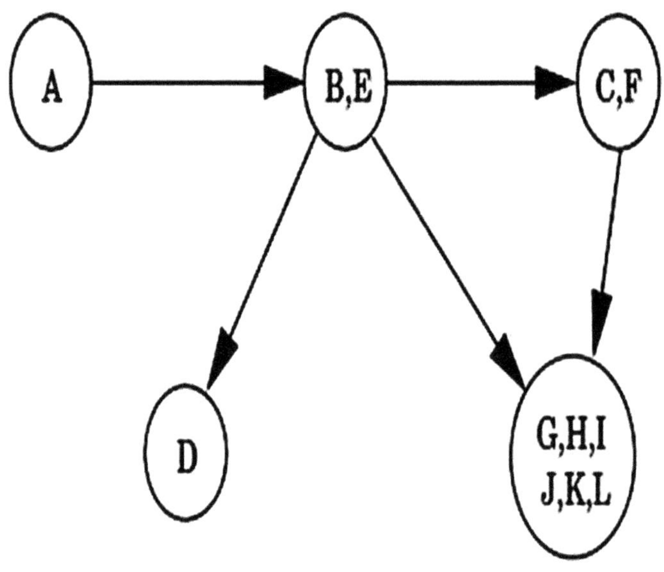

Table 5. Pre and post order numbers to sensor nodes in the network

S.No	Vertex Name	Pre-Order Number	Post-Order Number
1	A	2	1
2	B	6	3
3	C	10	7
4	D	12	11
5	E	5	4
6	F	9	8
7	G	22	11
8	H	17	16
9	I	21	12
10	J	20	13
11	K	18	14
12	L	19	15

Rules to Strongly Connected Components

1. Frame the graph with several nodes vertices and edges. Run depth-first search algorithm to node in the graph and get values of post-order & pre-order number of each vertex.
2. Orderingthemindecreasingorderofpostnumber.WegetG,I,J,L,K,H,D,C,F,B,E,A. (list Z')
3. Apply depth first search algorithm on first element of Z' and vertices other elements connected to the same element moved to r'.
4. Store r' in R' and delete element in r' from Z' and clear r'. If Z' is empty, we are done. R' has set of strongly connected components, else go to step 3.

Example:

Iteration 1:

Z' = {G, I, J, L, K, H, D, C, F, B, E, A},

r '= {G, I, J, L, K, H},

R'= {G, I, J, L, K, H},

Iteration 2:

Z’ = {{ D, C, F, B, E, A }, {D}}

r ‘ = {D},

R’ = {{G, I, J, L, K, H}, { D}}

Iteration 3:

Z’ = { { C, F, B, E, A }, {C, F}},

r’={C, F}

R’ = {{G, I, J, L, K, H}, {D}, {C, F}}

Iteration 4:

Z’ = {{B, E, A}, {B, E}},

r’= {B, E},

R’= {{G, H, L, K, H}, {D}, {C, F}, {B, E}},

Iteration 5:

Z’= {A},

r’= {A},

R’=G, H, L, K, H}, {D}, {C, F}, {B, E}, {A}}

5. Once the strongly connected components are formed, need to connect with other strongly nodes in the network. Consider two set of groups represents strongly and non strongly node connectivity i.e. *SC* and *NSC*
6. Initially take any strongly-connected component, put in *SC*, and remove it from NSC.
7. Then from all the elements in *SC*, find closest distance of all points in *NSC*. For the points in *NSC*, remove the strongly-connected component from *NSC* and put in *SC*. *R*repeat the procedure till *NSC* is empty.
8. To connect two strongly connected components, sensors need to be inserted between them.

FUTURE DIRECTION

It is possible to formulate sensor node placement problem using more than two types of sensor nodes. Thus the results in this chapter can be generalized.

The proposed clustering algorithms could be improved based on more recent algorithms such as Integer Linear Programming and others based on graph theory.

The problems of sensor placement and routing are interrelated and it is possible to form data highways on the sensor field (using proper choice of cluster head placement) and utilize them for routing with load balancing. In future one may be required to consider the placement of nodes on a non-planar grid (i.e. the sensor field is 3D).

CONCLUSION

In this chapter sensor node placement on a realistic non-uniform grid has been considered. A theoretical formulation has been made and solved. It is verified that k-mean clustering can solve the problem of deploying sensors on random points on plane, and that modification and experimentation can get us better results. Also, modifying the problem to graph theory's strongly connected components ensures transmission of data from one node to another without interruption.

REFERENCES

Aitsaadi, N., Achir, N., Boussetta, K., & Pujolle, G. A. (2009). WSN Deployment Method for Monitoring Geographically Irregular Distributed Events. *Sensors (Basel, Switzerland)*, *9*(3), 1625–1643. doi:10.3390/s90301625 PMID:22573977

Akyildiz, I. F., Su, W., Sankarasubramaniam, Y., & Cayirci, E. (2002). Wireless Sensor Networks: A Survey. *Computer Networks*, *38*(4), 393–422. doi:10.1016/S1389-1286(01)00302-4

Bai, X., Santosh, K., Dong, X., Yun, Z., & Ten, H. L. (2006). Deploying Wireless Sensors to Achieve Both Coverage and Connectivity. In *Proceedings of the 7th ACM International Symposium on Mobile ad hoc networking and computing MobiHoc'06*. doi:10.1145/1132905.1132921

Bai, X., Xuan, D., Yun, Z., Lai, T. H., & Jia, W. (2008). Complete Optimal Deployment Patterns for Full-Coverage and *k*-Connectivity (*k* · 6) Wireless Sensor Networks. In *Proceedings of the 9th ACM international symposium on Mobile ad hoc networking and computing*. doi:10.1145/1374618.1374672

Biagioni, E. S., & Sasaki, G. (2003). Wireless Sensor Placement for Reliable and Efficient Data Collection. In *Proceedings of the 36th Annual Hawaii International Conference on System Sciences*. doi:10.1109/HICSS.2003.1174290

Chakrabarty, K., Iyengar, S. S., Qi, H., & Cho, E. (2002). Grid Coverage for Surveillance and Target Location in Distributed Sensor Networks. *IEEE Transcations on Computerers*, *51*(12), 1448–1453. doi:10.1109/TC.2002.1146711

Chen, J., Li, S., & Sun, Y. (2007). Novel Deployment Schemes for Mobile Sensor Networks. *Sensors (Basel, Switzerland)*, *7*(11), 2907–2919. doi:10.3390/S7112907

Chong, C. Y., & Kumar, S. P. (2003). Sensor Networks: Evolution, Opportunities, and Challenges. *Proceedings of the IEEE*, *91*(8), 1247–1256. doi:10.1109/JPROC.2003.814918

Coskun, V. (2008). Relocating Sensor Nodes to Maximize Cumulative Connected Coverage in Wireless Sensor Networks. *Sensors (Basel, Switzerland)*, *8*(4), 2792–2817. doi:10.3390/s8042792

Cui, T., Chen, L., Ho, T., Low, S. H., & Andrew, L. L. H. (2007). Opportunistic Source Coding for Data Gathering in Wireless Sensor Networks. In *Proceedings of the IEEE Mobile Ad-hoc and Sensor Systems (MASS)*. doi:10.1109/MOBHOC.2007.4428675

Dhillon, S. S., & Chakrabarty, K. (2003). Sensor Placement for Effective Coverage and Surveillance in Distributed Sensor Networks. In *Proceedings of the IEEE Wireless Communications and Networking*. doi:10.1109/WCNC.2003.1200627

Dowell, L. J., & Bruno, M. L. (2001). Connectivity of Random Graphs and Mobile Networks: Validation of Monte Carlo Simulation Results. In *Proceedings of the 2001 ACM symposium on Applied computing (SAC '01)*. doi:10.1145/372202.372280

Fu, X., Yang, Wang, & Sun. (2014). Novel deployment scheme based on three dimensional voverage model for WSN, TSWJ. Hindawi.

Gkantsidis, C., Mihail, M., & Zegura, E. (2003). The Markov Chain Simulation Method for Generating Connected Power Law Random Graphs. In *Proceedings of the 5th SIAM Workshop on Algorithm Engineering and Experiments (ALENEX).*

Ishizuka, M., & Aida, M. (2004). Performance Study of Node Placement in Sensor Networks. In *Proceedings 24th International Conference on Distributed Computing Systems Workshops* (pp. 598–603). doi:10.1109/ICDCSW.2004.1284093

Iyengar, R., Kar, K., & Banerjee, S. (2005) Low-coordination Topologies for Redundancy in Sensor Networks. In *Proceedings of the. MobiHoc'05*. doi:10.1145/1062689.1062732

Jing, A. I., & Alhussein, A. (2006). Coverage by directional Sensors in randomly deployed wireless sensor networks. *J.Comb. Optimum, 11*, 21–41.

Khan, S. U. (2007) Approximate Optimal Sensor Placements in Grid Sensor Fields. In *Proceedings of the IEEE 65th Vehicular Technology Conference, VTC2007*. doi:10.1109/VETECS.2007.63

Lin, F. Y. S., & Chiu, P. L. (2005). A Near-Optimal Sensor Placement Algorithm to Achieve Complete Coverage/ Discrimination in Sensor Networks. *IEEE Communications Letters*, *9*, 43–45.

Maher, R., Berre, Hnaien, Snoussi, & Khoukhi. (2014). A Branch and Bound Algorithm for critical Grid coverage problem WSN. *IJDSN*, 1-9.

Miller, L. E. (2001). Distribution of Link Distances in a Wireless Network. *Journal of Research of the National Institute of Standards and Technology*, *106*(2), 401–412. doi:10.6028/jres.106.017

O'Rourke, J. (1987). *Art Gallery Theorems and Algorithms*. New York, NY: Oxford University Press.

Sohraby, K., Minoli, D., & Znati, T. (2007). *Wireless Sensor Networks: Technology, Protocols and Applications*. Hoboken, NJ: John Wiley & Sons. doi:10.1002/047011276X

Tubaishat, M., & Madria, S. (2003). Sensor Networks: An Overview. *IEEE Potential Magazine*, *22*(2), 20–23. doi:10.1109/MP.2003.1197877

Wang, Y. C., Hu, C. C., & Tseng, Y. C. (2005). Efficient Deployment Algorithms for Ensuring Coverage and Connectivity of Wireless Sensor Networks. In *Proceedings of the First International Conference on Wireless Internet*. doi:10.1109/WICON.2005.13

Xian, C. (2012). Fault-tolerant monitor placement for out of band WSN monitoring. *Ad Hoc Networks*, *10*, 62–74.

Zhang, H., & Hou, J. C. (2005). Maintaining Sensing Coverage and Connectivity in Large Sensor Networks. *Ad Hoc Sensor Wireless Networks*, *1*, 89–124.

KEY TERMS AND DEFINITIONS

Cluster Head: A node in the group which takes responsibility to data gathering from the nodes and data forwarding towards base station.

Clustering: Grouping of similar types of nodes under sensing.

Graph: A graph is a structure represents number vertices and edges.

Node Coverage: The maximum range covered by node at which it senses the information.

Strongly Connectivity: It is the process in which data is forwarded between nodes without interruption.

Uniform-Grid: Virtual grid for sensor placement.

Wireless Node: A device which can senses process and transmits and receives the similar information through independently.

Chapter 7
802.11p-Based VANET Applications Improving Road Safety and Traffic Management

Lambros Sarakis
Technological Educational Institute of Sterea Ellada, Greece

Aristotelis Papantonis
Hellenic Open University, Greece

Theofanis Orphanoudakis
Hellenic Open University, Greece

Panagiotis Karkazis
Technological Educational Institute of Sterea Ellada, Greece

Periklis Chatzimisios
Alexander Technological Educational Institute of Thessaloniki, Greece

Helen C. Leligou
Technological Educational Institute of Sterea Ellada, Greece

Theodore Zahariadis
Technological Educational Institute of Sterea Ellada, Greece

ABSTRACT

In the last few years Intelligent Transportation Systems (ITS) based on wireless vehicular networks have been attracting interest, since they can contribute towards improving road transport safety and efficiency and ameliorate environmental conditions and life quality. In order to widely spread these technologies, standardization at each layer of the networking protocol stacks has to be done. Therefore, a suite of protocols along with the architecture for the wireless environments with vehicles has been developed and standardized. Both in the US as well as in Europe the selected wireless communication protocol has been the 802.11p protocol developed by the IEEE. In this chapter, we discuss the potential impact of ITS towards achiev-

DOI: 10.4018/978-1-4666-9941-0.ch007

Copyright ©2016, IGI Global. Copying or distributing in print or electronic forms without written permission of IGI Global is prohibited.

ing the above targets of improving road safety and traffic control. We review the overall architecture and the protocol functionality and present in detail a number of applications that have been developed demonstrating selected use-cases on an 802.11p compliant system prototype. Specifically, we discuss the implementation of selected applications on the NEC's Linkbird-MX platform, which supports IEEE 802.11p based communications, showing how its functionality can be exploited to build efficient road safety and traffic management applications, and evaluate the performance of the system using an experimental testbed.

INTRODUCTION

Road safety, air pollution and traffic management are three major concerns that the residents of urban centers around the world have to face promptly given the immense growth in cities' population. Road traffic accident mortality is high among young people with transport accidents causing 8% of all losses among citizens with an age below 65 years in the EU-27, higher than any disease (Cayotte & Buchow, 2009). Across EU, transport is most dangerous in regions of Portugal, Lithuania, Latvia, Corsica, Greece and Poland. While road traffic injuries are a major cause of death and disability globally, a disproportionate number occurs in developing countries. Road traffic injuries are currently ranked ninth globally among the leading causes of disability-adjusted life years lost, and the ranking is projected to rise to third by 2020. In 1998, developing countries accounted for more than 85% of all deaths due to road traffic crashes globally and for 96% of all children killed (Murray & Lopez, 1996; Nantulya & Reich, 2002). On the other hand, air pollution is tightly related to traffic management and plays a substantial role to the climate change while burdening the commitment of Europe to decrease the CO_2 emissions. In urban areas, an increase in average speed may dramatically reduce fuel consumption, while traffic signal synchronization has the potential to increase intersection throughput for private traffic by 15%. Guiding traffic (e.g. through route advisory systems) away from problematic areas may lead to up to 8% less emissions (Murray & Lopez, 1996). Today, according to the Panorama of Transport (2009), 30% of energy is consumed for transportation of humans and goods and almost 18% of the CO_2 emissions from combustion come from road transportation (IEA, 2005). Although the broadening of the road infrastructures increases their capacity, it cannot keep up with the pace of the increase in urban populations worldwide, due to cost and time reasons, leading the city authorities to pursue "soft" measures to solve the problem (Pincus, 2011). Furthermore, there are many efforts by the scientific community to combat many critical issues that hold back the vehicular network's deployment but also support the Intelligent Transportation Systems (ITS) development by performing standardization efforts concerning vehicular communications (Kadas & Chatzimisios, 2011; Leontiadis et.al, 2011).

To tackle the previously reported issues and problems, the design and development of ITS has been pursued extensively in the last decade. These systems rely on intelligent collection and processing of information which enables decision making and information/decision dissemination to enhance the citizen's experience either through enhancing transportation efficiency or safety. They can be classified in advanced public transport systems, advanced traveler information systems, advanced Traffic Management Systems, incident management systems, electronic toll collection systems, Vehicle Information and Communications System, and Video Transmission Systems for road surveillance. ITS are expected to play a major role in enhancing road safety, transportation efficiency and improving environmental conditions, both in developed as well as in developing countries as mentioned above. Their impact will be critically affected by the adoption of standardized and low-cost technologies that can result in massive production of commodity hardware components and wide deployment of interoperable systems.

In order to achieve the objective of providing citizens using the roads with improved traffic safety, reduced traffic congestion, and environment-friendly driving, countries worldwide had to standardize communication protocols for common communication between vehicles and between vehicles and infrastructure. Vehicular networking combines wireless communication, in-vehicle sensing module, and Global Positioning System (GPS) to enable a variety of applications in road safety, traffic efficiency, and infotainment domains. In the US, the networking protocols of choice for vehicles and infrastructure associated with these patterns are included in the WAVE (Wireless Access in Vehicular Environments) family of standards formulated by the IEEE (Institute of Electrical and Electronics Engineers) and are mainly based on the IEEE 802.11p and IEEE 1609 protocols (IEEE 802.11p, 2010; IEEE 1609.0, 2013). This set of standards defines the architecture, communication model and mechanisms of high-speed short-range wireless low-latency communications.

The original IEEE 802.11, intended for WLAN, has two drawbacks within its Medium Access Control (MAC) technique; it can cause unbounded delays before channel access, as well as collisions on the channel. The IEEE 802.11p, also known as Dedicated Short-Range Communication (DSRC), is intended for vehicular ad-hoc networks (VANETs). Currently, this is the only standard with support for direct Vehicle-to-Vehicle (V2V) communication (Bilstrup, 2007; Khairnar & Kotecha, 2013). Other early systems based on mobile communications, like 3G, combined with GPS systems could only offer limited location-based services, without being able to offer broadband services as well as V2V communications with real-time response, which are critical for supporting traffic management and road safety applications over a single system. Another organization producing standards related to ITS that are active in the US is the international Society of Automotive Engineers (SAE). SAE, in collaboration with the IEEE 1609 group, contributes to the standardization of message formats that can be used by the IEEE 1609 protocol. An example of this is the standard of SAE J2735 that is intended to be used by the IEEE 1609.3 WSMP (Wave Short Message Protocol).

For the standardization of ITS in Europe, three organizations are responsible for the planning, development and adoption of European standards. The European Committee for Standardisation (CEN), which is a general standardization body and is responsible for all areas excluding the electrotechnical sector, the European Committee for Electrotechnical Standardisation (CENELEC), which is responsible for the part of electrotechnical standardization and the ETSI (European Telecommunications Standards Institute), which is responsible for standardization in the field of telecommunications. CEN standardizes the European CEN DSRC ITS technology on the 5.9 GHz radio (European Telecommunications Standards Institute, 2011). ETSI also adopts the IEEE 802.11p amendment to the IEEE 802.11 standard, which is also promoted by the Car-to-Car Communication Consortium (C2C-CC) in Europe, the major European industry consortium for vehicular communication.

In this chapter, which extends our previous work of Leligou et al. (2014), we present a practical use case based on IEEE 802.11p implementing applications targeting road safety and traffic management over wireless vehicular networking systems. We consider a generic Vehicle-to-Vehicle/-Infrastructure (V2X) communication architecture where mobile nodes (equipped with On-Board Units - OBUs) with sensing capabilities are connected to each other over a vehicular ad-hoc network (VANET) and to static nodes located at the road side (Road Side Units - RSUs) using the IEEE 802.11p standard. We discuss applications that capitalize on this architecture and are related to both the city traffic cycle and a highway case, and report on their implementation and performance evaluation using an experimental testbed. We also discuss the advantages and the feasibility of the approach, as well as we explore critical aspects for the wide penetration of such systems.

IEEE 802.11p V2X NETWORKING FRAMEWORK

The 802.11p physical layer is based on Orthogonal Frequency-Division Multiplexing (OFDM), which is an already established technology in the market; thus, providing an inexpensive solution, which has been used in other mobile environments. The 802.11-2007 standards define three different OFDM modes: 20MHz, 10MHz, and 5MHz. The community of ITS chose 10MHz bandwidth channels for mobile communications (Table 1). However, some basic modifications to the original 802.11

Table 1. 802.11p physical layer specifications

	USA	Europe
Standard / Committee	IEEE802.11p/1609.x	CEN/ETSI EN302 663
Frequency range	5850 – 5925 MHz	5855 – 5925 MHz
No. of Channels	Seven 10MHz channels (Two 20MHz channels formed by combining 10MHz channels)	Seven 10MHz channels
Modulation	OFDM	
Data rate per channel	3 – 27 Mbps	
Output power	23 – 33 dBm (EIRP)	
Communication	One direction multicasting service, One to Multi communication, Simplex communication (broadcast without ACK, multicast, unicast with ACK)	
Upper protocols	WAVE (IEEE 1609) / TCP/IP	ETSI EN 302 665 (incl. e.g. GeoNetworking) TCP/UDP/IP

OFDM were required in order to achieve robust connections and fast vehicle topology organization that led the evolution to the 802.11p protocol.

The physical layer in 802.11p is the same as the physical layer in 802.11a except for the different sampling rate. The different modes can be achieved by using reduced clock/sampling rates. 802.11a usually uses the full-clocked mode with a 20MHz bandwidth. In comparison, 802.11p usually uses the half-clocked mode with a 10MHz bandwidth. The different modes affect the following parameters:

- **Bandwidth:** In 802.11p, 10MHz is usually used, making the signal more robust against fading. The 20MHz mode is optional.
- **Carrier Spacing:** The 802.11p signal uses a carrier spacing reduced by half when compared to 802.11a.
- **Symbol Length:** The symbol length is doubled, making the signal more robust against fading.

Besides the clock rate, the Adjacent Channel Rejection (ACR) and the Spectrum Emission Mask (SEM) are also changed. Important IEEE 802.11p channel parameters are listed in Table 2.

Moving to higher communication layers, VANETs have certain requirements mainly related to mobility, high vehicle speed, rapid changes in topology and potentially large scale and density networks. These characteristics have important implications for the MAC protocol in VANETs. Based on these characteristics the

Table 2. 802.11p channel characteristics

	20MHz Bandwidth	10MHz Bandwidth	5MHz Bandwidth
Bit rate (Mbps)	6, 9, 12, 18, 24, 36, 48, 54	3, 4.5, 6, 9, 12, 18, 24, 27	1.5, 2.25, 3, 4.5, 6, 9, 12, 13.5
Modulation mode	BPSK, QPSK, 16QAM, 64QAM		
Code rate	1/2, 2/3, 3/4		
Number of subcarriers	52		
Symbol duration (μs)	4	8	16
Guard time (μs)	0.8	1.6	3.2
FFT period (μs)	3.2	6.4	12.8
Preamble duration (μs)	16	32	64
Subcarrier spacing (kHz)	312.5	156.25	78.125

802.11p MAC adopts the IEEE 802.11e Enhanced Distributed Channel Access (EDCA) with Quality of Service (QoS) support and mainly aims at the following:

1. **Improved Throughput:** Given the characteristics of VANETs, and variable network topology and density, VANETs messages should be transmitted between vehicles at regular intervals, while the communication between vehicle and RSU is also necessary to realize the intelligent transport control. Therefore, an intelligent MAC protocol is required that can dynamically adjust its parameters based on the network status, in order to improve network performance. Wang, Ahmed, Krishnamachari and Psounis (2008), and Jang and Feng (2010) proposed several algorithms in this direction, and further research is required to improve network detection algorithms.
2. **Programming Optimization:** The idea of adapting the intervals in Control Channel (CCH) and Service Channels (SCH) has been proposed by Misic, Badawy, Rashwand and Misic (2010) and Wang et al. (2009). The algorithm proposed by Wang et al. (2009) can improve the utilization of the channel when there is a higher number of service messages and less control messages. Because of the variable density of VANETs, an optimization algorithm for scheduling can adjust the ranges of the CCH and SCH, depending on the network status. Furthermore, the network detection algorithms could also be adapted to detect the network status.
3. **Traffic Control:** Burguillo-Rial, Costa-Montenegro, Gil-Castineira and Rodriguez-Hernandez (2008) report studies which indicate that channel contention increases with traffic load. This, in turn, increases collisions, i.e., the number of collisions increases when the traffic load increases and channel

throughput deteriorates. Therefore, traffic control is the key to solve the problem of conflict in VANETs. The idle times reduce utilization and are very critical in VANETs due to the short time required for a critical security application. Obviously, by comparing the results of Grafling, Mahonen and Riihijarvi (2010), and Eichler (2007) we conclude that high traffic loads lead to longer latency and idle channel periods. Consequently, traffic control is also the key to reduce the delay in VANETs and traffic control algorithms are required.

As mentioned above, the standards only cover interoperability issues, protocol specifications and the selection of basic parameters for system operation. However, several implementation issues affecting performance, as listed above, are open to be dealt with by different system designs following different approaches and optimization algorithms.

Besides the MAC and transport layers, new vehicular routing protocols need to cope with high-speed mobility and potentially large flat networks, while achieving the integration with the current and future Internet. The requirements imposed by vehicular communication for support of highly mobile vehicles, frequent topology changes, and scalability with potentially very large number of nodes can be fulfilled by adopting geographical routing principles originally proposed for wireless sensor networks, which, in addition, can be secured with limited node resource consumption (Karkazis, Leligou & Zahariadis, 2012; Zahariadis, Trakadas, Leligou, Maniatis & Karkazis, 2013). Such an approach has been shown to have good performance not only in simulated but also in realistic environments (Festag, Fussler, Hartenstein, Sarma & Schmitz, 2004). Therefore, the 802.11p Task Group of IEEE has chosen the so-called Geocast routing protocol as the core networking protocol for vehicular communication. This protocol supports opportunistic communication as well as multi-hop communication supporting both a) the forwarding of data towards the geographical position of a single destination node for unicast communication and b) the distribution of data in a geographical region.

Geocast is an ad-hoc routing protocol utilizing geographical positions for data transfer. It is assumed that each vehicle is aware of its geographical position via e.g. GPS and periodically advertises this information to its neighboring vehicles (in the so-called beacon messages). Hence, an IEEE 802.11p enabled device is informed about all other IEEE 802.11p devices located within its direct communication range (one-hop neighbors) and maintains a so-called neighbor table in soft state containing all known neighbors' Identifiers (IDs) and geographical positions.

The Geocast protocol supports point-to-point and point-to-multipoint communication. Leveraging the capabilities to distribute information based on geographical routing, it enables innovative applications mainly by exploiting the protocol capabilities for selective addressing of geographical areas as target of data packets. Thus, a vehicle can specify a well-delimited geographic area to which the messages should be delivered/broadcasted. Intermediate vehicles serve only as message relays and only the vehicles located within the target area terminate messages at the application layer conveying related information to the driver as appropriate. In this way, the vehicles that are actually affected by a dangerous situation or a traffic-related event are notified, whereas vehicles unaffected by the event are not targeted. The geographical area can be rectangular, circular or ellipsoidal. To this end, the following forwarding types of Geocast can be used:

- GeoUnicast, which is a unidirectional data transport service from a single source node (S) to a single destination node (D) (Figure 1a). Node S forwards data packets to the one-hop relay node (F) that is closest to the destination, which in turn forwards them along the path until they reach the destination node D.
- Geographically-Scoped Broadcast (GBC), which is used to transport data from a single node to all nodes within a user-defined range and area (either via unicast as shown in Figure 1b or other forwarding methods like the Contention-Based Forwarding (CBF)).
- Topologically-scoped broadcast (TSB), which provides rebroadcasting of a data packet from a source to all nodes in a limited N-hop neighborhood (Figure 1c), which can reside at a distant geographical area. Single-hop broadcast is a specific case of TSB which is used to send periodic messages (beacons or heartbeats), while setting N large enough so as the whole ad-hoc network can be reached. TSB shall be used very rarely since it may cause network congestion.

NETWORK INFRASTRUCTURE SUPPORTING MOBILE OPPORTUNISTIC COMMUNICATIONS

The vehicular networks can be classified into two main transmission types, namely V2V communication among vehicles and Vehicle-to-Infrastructure (V2I) communication between a vehicle and the infrastructure which is a fixed unit generally located at the roadside (Kadas & Chatzimisios, 2012). In this work, we consider the system architecture depicted in Figure 2. Vehicles with or without OBUs travel on the road. The OBUs consist of a system equipped with GPS and sensing devices (e.g. humidity sensor, temperature sensor, accelerometer, gyroscope, CO_2 sensor) and with a device capable of communicating either with other vehicles or with devices installed in the road (infrastructure) side. A Road Side Unit (called hereafter

Figure 1. The routing schemes supported by IEEE 802.11p routing protocol (a) Geographical Unicast, (b) Geographical Broadcast, (c) Topologically-Scoped Broadcast

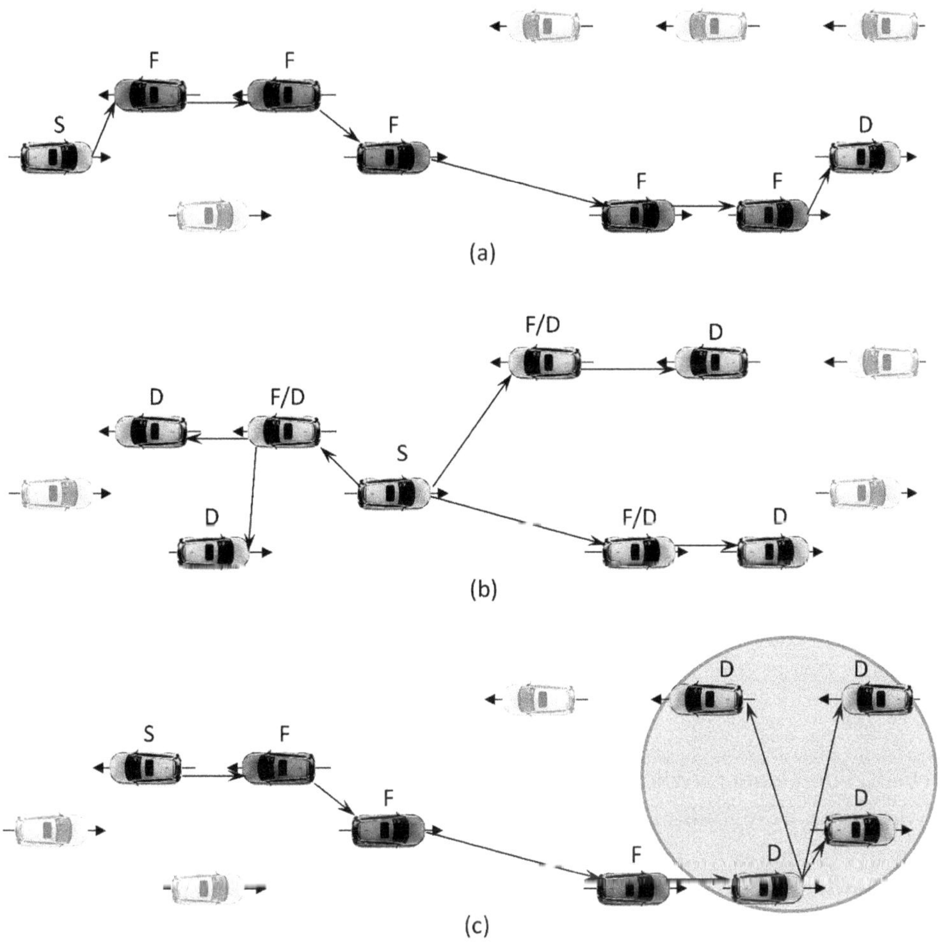

RSU) is capable of communicating (apart from the vehicles in their transmission range) with an application server located at the premises of the traffic management or road operation authorities through any legacy wired or wireless communication technology. We consider that such devices can be attached to traffic light posts or posts holding either lights or cameras in highways where access to power and communication networks is provided.

The communications architecture, involved protocol stacks and potential communication paths are shown in Figure 3. For the communication between the vehicles and between the vehicles and the infrastructure we rely on the IEEE 802.11p protocol which supports wireless communication between nodes moving at high

Figure 2. The system architecture

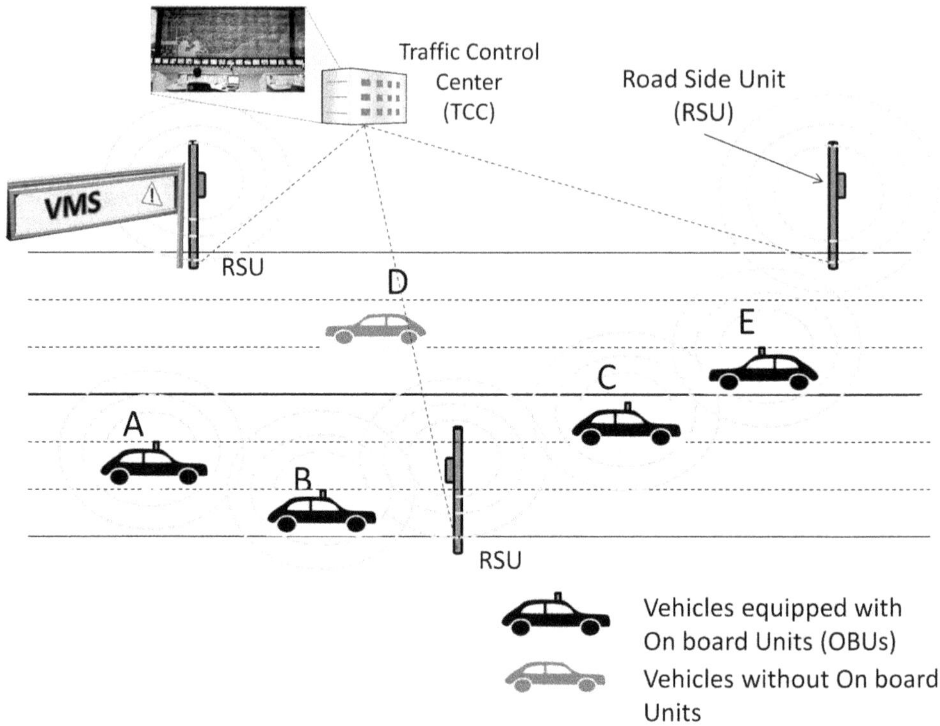

speeds. Short-range wireless communication based on IEEE 802.11 is characterized by low cost, availability and wide deployment. The wireless technology enables a fully distributed vehicular communication network based on self-organization and self-coordination of the network nodes, which is mandatory for car-to-car and car-to-infrastructure communication (called Car-2-X or shortly C2X by NEC (NEC Laboratories, 2013)). The transmission occurs in the 5.9 GHz unlicensed band for both EU and USA with a maximum transmission range of 1000m with 200μs latency and a data rate between 6-27Mb/s.

While car manufacturers install more and more sensors on the vehicles and offer more and more communication interfaces (e.g. Bluetooth, USB), they are reluctant to disseminate the sensor readings (even for a very limited subset of the available sensors). To partially address this issue the US Environmental Protection Agency standardized the On-Board Diagnostics (OBD) interface. OBD systems are designed to monitor the performance of some of an engine's major components including those responsible for controlling emissions. There can be over 300 readings available, depending on the vehicle manufacturer and model. However, vehicles vary in the readings they will support and scanners vary widely in the number of the signals

Figure 3. The communications architecture

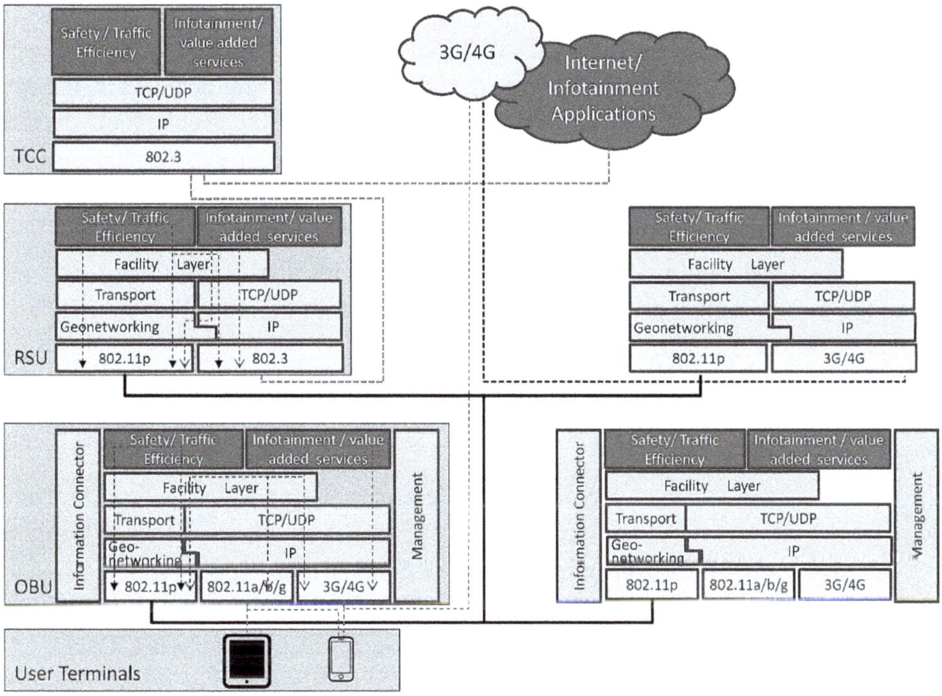

that they can read. Hence proprietary interfaces tend to increase the cost for customization of sensor reading systems per vehicle type and, additionally, sensors critical for a variety of applications (e.g. GPS) are not standard equipment and may not be available. On the other hand, the cost of sensors decreases rapidly as the production volumes increase.

In this view, the development of a single device (OBU) that supports communication based on IEEE 802.11p and that is equipped with a small set of sensors and possibly a display (for communicating with the driver) to implement road safety applications at low cost seems a realistic solution. Statistics show that European citizens are highly interested in safety when buying a car and, thus, it is highly likely for them to be also interested in buying a device that supports safety applications. In the future, leveraging technology consolidation, market adoption and mature (potentially open) interfaces, such a device could be combined /embedded with devices offered by highway operation authorities for electronic toll payment (e.g. e-pass) or automotive control systems. These devices could also support entertainment applications which may offer further incentives for installing such a device in the vehicle even for young drivers who may underestimate safety or traffic management benefits. For example, offering instant messaging with passengers of nearby cars for free could strengthen the incentives. Following this trend, an application for dynamically adapting traffic light timing initially adopted for specific vehicle types, (e.g. for providing priority to ambulances or public transport means) was deployed in Albacete, Spain, albeit using OBUs that communicate over WLAN technologies (INDRA, 2008). In India respectively, in a bid to keep vehicle drivers from violating traffic rules, the Pune Municipal Corporation will be sharing traffic violation data with insurance companies. Through this, insurance players will be able to understand if the accident has happened due to any kind of violation of traffic rules. These initial deployments, as well as several others globally, demonstrate the increasing adoption of smart applications over ITS to enhance safety and transportation efficiency.

In this chapter, we consider the case where OBUs integrate a microprocessor and memory elements with the following limited set of sensors:

- GPS, which provides geographical references, speed and direction information. Such data are required by the IEEE 802.11p standard (i.e. to support node identification and implement geographical routing), as well as to support applications that depend on the detection of direction or speed changes to provide early warning signals and prevent accidents to moving users.
- Humidity and temperature sensors, to enable the detection of severe environmental conditions that may need specific action taking.
- Peripheral devices for informing the driver (e.g. display or audio peripherals).

For experimental purposes, such an OBU device was developed using the NEC Linkbird-MX device connecting to it appropriate external sensors through USB interfaces. Should car manufacturers open part of the car sensor network to third parties, OBUs could be equipped with an additionally interface to communicate with the car sensor network, if this brings reliability or cost benefits.

For the RSU, it is important to support IEEE 802.11p functionality to communicate with the vehicles and another communication technology (wireless or wired) to connect to a server that is capable of processing real-time information and implementing intelligent decision-making schemes. RSUs execute a multitude of functions including forwarding of data (to increase the coverage of the ad-hoc network), transmission and reception of application data from the vehicles (e.g. collecting measurements and disseminating warnings or decisions made at the Traffic Control Center – TCC), communication with the road operation authorities and traffic control center and Internet access support to cars. Furthermore, the RSU could be equipped with sensors (e.g. for weather or CO_2 emissions monitoring) and deliver the readings to decision-making applications.

Exploiting the above architecture, the drivers of all the vehicles, the pedestrians and the road operation authorities can benefit from the sensed data. This architecture comprises a distributed sensing infrastructure (sensing overlay) since the vehicles are capable of detecting events and through their opportunistic communication with the RSU, the sensed data can arrive at the road operator and/or traffic management authorities. This distributed infrastructure senses the whole road network, even neighborhoods or areas where the traffic does not justify investment on monitoring infrastructure. On the other hand, the routing instructions and traffic information can be disseminated towards both pedestrians and drivers through a variety of means which includes (apart from the OBU), Variable Message Signs (VMSs) and mobile phones to reach a wider audience.

APPLICATIONS ENHANCING TRAFFIC MANAGEMENT AND ROAD SAFETY

Leveraging the IEEE 802.11p mechanisms for realizing efficient communication between mobile nodes and between mobile and static nodes (C2X communication), the next step is the design and implementation of intelligent applications than enhance road safety and assist traffic management. In the sequel, we present indicative applications that capitalize on the considered architecture. The applications we implemented as indicative use cases exploit the communication network described above comprising the in-vehicle, ad-hoc, and infrastructure domains. The ad-hoc domain is composed of vehicles equipped with OBUs and stationary RSUs that can be attached to fixed locations. Most prominent locations for RSUs include traffic lights and VMSs. RSUs are responsible for the interconnection between the ad-hoc and the infrastructure domain. Thus, OBUs can directly communicate if direct wireless connectivity exists. In case of no direct connectivity, multi-hop communication is used, where data is forwarded from one OBU to another, potentially via RSUs, until it reaches its destination. By deploying this kind of ITS infrastructure prototype, fixed service-based cooperative information systems have been developed and tested (Leligou, Karkazis, Hatziefremidis, Orphanoudakis & Zahariadis, 2013; Leligou et. al, 2014).

City Traffic Management Based on Distributed Sensing

The traffic lights constitute a key element in traffic management within a city. The coordinated traffic flow control implementing the "green wave" concept was initially proposed to minimize the vehicle stops to red traffic lights within the city (Warberg, Larsen & Jorgensen, 2008). The rationale is that reducing unnecessary stops and increasing the average transportation speed of vehicles reduces both the travel time and the CO_2 emissions. This "wave" is generally enforced in a more or less static manner. Further to static coordination, we describe in this section how sophisticated applications exploiting mobile opportunistic communications can enhance efficiency, without necessarily introducing high complexity or unbearable implementation cost.

Dynamic Traffic Light Cycle Definition

The dynamic adaptation of the traffic light cycles according to traffic conditions has been recognized as a means to achieve the same benefits with the green wave: reduced vehicle stops for shorter travel time and lower CO_2 emissions (Faye, Chaudet & Demeure, 2012; Yousef, Al-Karaki & Shatnawi, 2010). This adaptation requires the detection of the traffic conditions which is currently realized through infrastructure-based systems (e.g. cameras, LIDAR systems, etc., as the approach described by Faye, Chaudet and Demeure (2012)) and the communication with the city traffic management authorities for global city traffic management decision-making. Mainly due to cost reasons, such a traffic monitoring infrastructure has not been installed in all intersections across cities. Respectively, implementations based on wireless communication systems that do not support ad-hoc communications (e.g. as the approach described by Yousef, Al-Karaki and Shatnawi (2010)) limit scalability, efficiency and cost performance as discussed above.

The same effect in a more efficient manner and capable of handling a wealth of information can be achieved by placing RSUs on the traffic light posts where electric power and Internet connectivity is available, as shown in Figure 4. Cycle adaptation can exploit IEEE 802.11p communication facilities, which allow sensing neighbor nodes within a range. In particular, IEEE 802.11p mandates that OBUs periodically transmit beacon messages including their geographical position and heading (i.e. direction). The IEEE 802.11p protocol stack at the RSU side located on the traffic light post receives the beacon messages from all the vehicles within its reception (or a limited defined by the operator) range. The RSU, based on the received beacon messages, should be able to detect the lane, side of the crossroad and direction of each vehicle passing by and maintain counters for each direction. An implementation issue that arises in this case affects the trade-off between resource efficiency and speed of execution and the accuracy of location estimation and classification of vehicle to lanes and respective directions of the crossroad. The computation complexity in this case arises from the fact that under different road network topologies vehicles may follow different turns over time frequently changing direction (e.g. complex multi-lane circular road junctions) or vehicles from nearby streets may fall within range of the RSU. Increased accuracy can be achieved by extrapolating the emitted GPS data with GIS data accessible by the RSU. However, this greatly increases processing and memory requirements at the RSU. An equivalent implementation, which can achieve high accuracy at the cost of a slightly higher latency to converge, is to perform concurrent measurements (in many cases assisted by pre-calculated fixed boundary positions) and compute the vehicle direction and decide on the associated traffic flow by statistically processing a number of the latest reported position and heading values. By performing measurement-based estimation of the load of each traffic flow based on moving averages, light cycle adaptation can be efficiently implemented. Short-term statistics and status information can be forwarded to the TCC for processing to perform longer-term estimations, database maintenance and status visualization.

The message flow is shown in Figure 5. Statistics collection can be a locally-executed application for the dynamic adaptation of the traffic light cycle to the traffic conditions but usually it is expected to reside at the premises of traffic management authorities (state-of-the-art traffic lights already support TCP/IP connectivity for configuration and control) allowing coordination of traffic lights to achieve global traffic control benefits. Application-related beacon messages from OBUs denoted as *MStatus Report* are collected and processed as described above to produce *SStatus Report*. Event report messages facilitate the applications discussed in the following sections.

Figure 4. Exploiting OBU and RSU for dynamic traffic light cycle adaptation

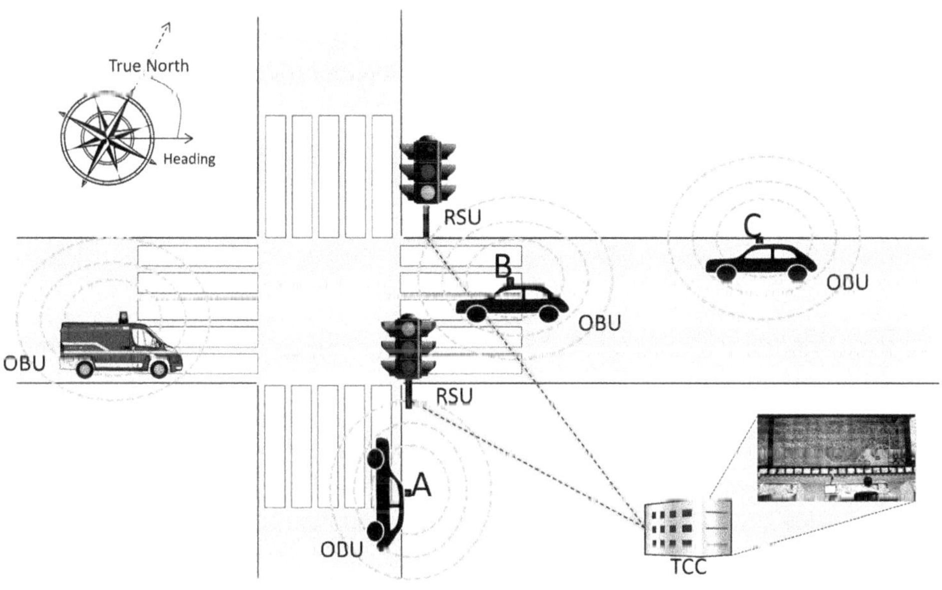

Figure 5. The message flow between the OBU, the RSU on the traffic light posts and the application center

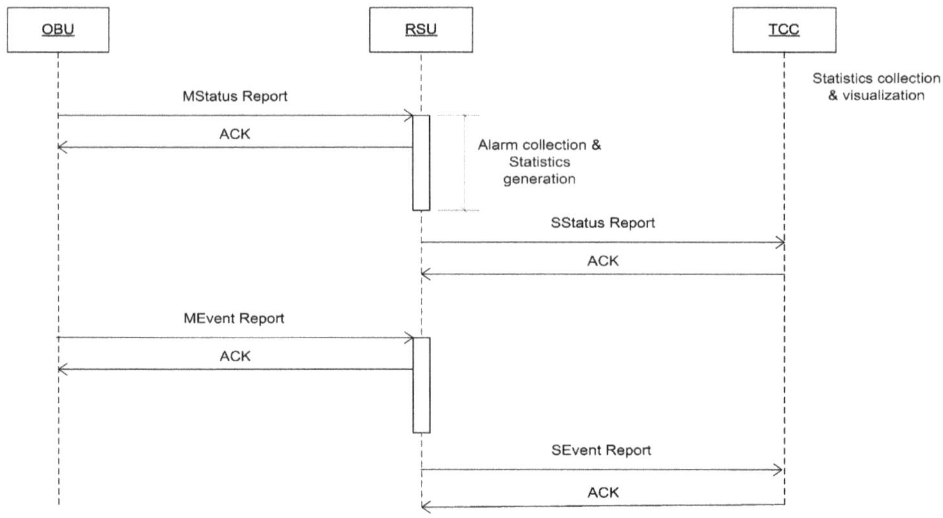

Traffic Light as Sink for Sensed Data

As OBUs are usually equipped with sensors monitoring environmental conditions and the vehicles speed and direction, they are capable of detecting rain (humidity above a certain threshold), ice (a function of humidity and temperature thresholds) or slippery conditions at any location it passes by. With low complexity, it can store this information in order to transmit it to the first available RSU it will be occasionally find within range. The OBU is capable of discriminating other OBUs from RSUs based on the addressing scheme of the IEEE 802.11p standard with discrete address ranges for different types of devices. Based on this communication scheme, road conditions across all locations can be monitored by traffic management authorities. Subsequently, this information can be used to inform drivers and also provide routing advice through VMS or other means.

Improving Safety for Pedestrians

A quite common approach for improving the safety of pedestrians crossing a road is to install a traffic light which becomes red for the vehicles only when a pedestrian indicates its presence by pushing a button on the traffic light post, i.e. the traffic lights provide regularly priority to vehicles. The approach works quite well apart from the case when rain makes it unpleasant both for the pedestrians (to wait until the vehicles stop) and for the drivers (that have to stop on a wet pavement). A lot of times the pedestrians attempt to cross the street running on a wet pavement which may cause severe accidents.

The proposed infrastructure enables the detection of rain and the execution of the reverse logic i.e. regularly providing priority to pedestrians. Rain and special weather conditions in general can be detected either by OBUs or by the RSU (embedding the appropriate sensors). In any case, the RSU controlling the traffic light receives and processes weather condition and traffic load information. In case of low traffic, the default vehicle-priority turns to pedestrian-priority state. In this case, green traffic light for vehicles is triggered by the detection of an approaching vehicle or a timer expiration (to ensure a maximum waiting time for vehicles not equipped with OBUs).

Applications Related to Highway Transportation Systems

The incentives that make citizens willing to pay the associated toll to use a highway (even in presence of alternate routes) include enhanced safety and reduced travelling time. These depend on the construction specifications and quality of the road, as well as on the measures to prevent primary and secondary accidents. A primary accident can be caused by slippery road or careless driving, while a secondary accident is caused by a primary accident which has not been perceived in time. Secondary accidents are very often equally important (and fatal) as primary ones (Zhan, Shen, Hadi & Gan, 2008). In most highways, cameras are installed in regular distances and inspected at the traffic control center by trained staff who react to abnormal conditions by a) informing the interested parties (e.g., first responders) and b) defining appropriate messages that should appear in VMSs. This procedure takes significant time compared to the human reaction time and, as a result, drivers approaching fast to an incident may get into danger.

The reduction of the time between the event detection and information dissemination can be reduced based on the C2X communication model presented in the section describing the network infrastructure, where the OBUs automatically detect abnormal and/or dangerous situations and disseminate information in the appropriate area exploiting either V2V or V2I communication.

In an example scenario shown in Figure 6, the OBU of a vehicle (say car C) detects a sudden change of its speed or slippery road condition. A broadcasted message (event report) then provides the car's geographical position and the detected event (e.g. slippery road/icy conditions). Assuming that car F is close enough to car C (e.g. their distance is less than 1000m) and moving towards the same direction, the OBU of car F receives the event report directly from car C (over IEEE 802.11p) and a suitable warning (either visual on a display or audio) is triggered. Additionally, events are forwarded backwards within a range to inform all approaching vehicles

Figure 6. Basic scenario of event detection and dissemination in a highway

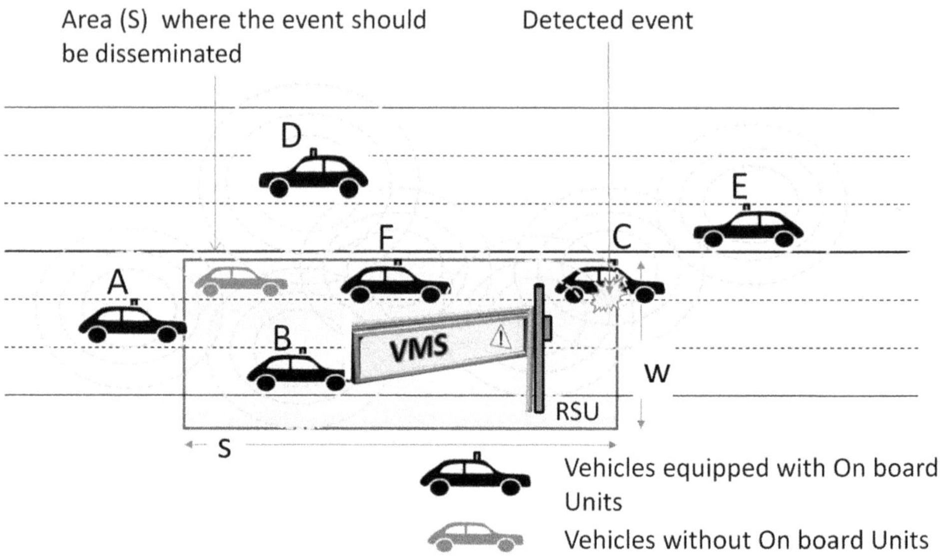

even out of range to vehicle C and can be announced to the RSU which in turn notifies/warns interested vehicles.

Focusing on the implementation aspects, the event detection relies on reading the sensor data and executing simple comparisons with thresholds to decide whether an event that needs further attention/action has been detected, as shown in the pseudo-code describing the application in Figure 7. To avoid fault event detection, the sensors are polled periodically with a period in the range of milliseconds and the event detection is decided and announced only when the comparisons result in the same verdict for a specific number of times.

The definition of the appropriate area for the dissemination of information for abnormal situations (be it relevant to weather or to accidents) is also critical in this case for the effectiveness of the application, because information flooding results in neglecting the received information. Especially inexperienced or elderly drivers are sensitive to information overload thus dictating the need to filter the information and pass to them only the information that is relevant. While it seems that the most complex part is this definition and identification of the vehicles that need to be informed, it is the IEEE 802.11p protocol that undertakes this task once the engineer has decided the interested area from a geometrical point of view.

As an example, let us assume that, as depicted in Figure 6, the appropriate area where the event should be disseminated forms a rectangular with dimensions SxW. Based on the TSB feature of the Geocast protocol, the device that detects the event can issue a data indication message specifying the center of the rectangular (which

Figure 7. Detection logic for a) weather conditions-related event, and b) vehicle behavior-related events

```
// Configuration and initialization routines
// Neighbor table management every T_nodeTableMgt
// sec
// HUM & TEMP register update.
    if [(TEMP<TEMP_C) or (TEMP<TEMP_M AND
    HUM>HUM_M)] then ICE = TRUE
// Alarm reception and forwarding (vehicles within
// range/direction)
// Alarm detection and status update
// Alarm table management every T_alarmTableMgt
```

(a)

```
// Configuration and initialization routines
// Application beacon message generation
// HUM & TEMP register update.
    if [(TEMP<TEMP_C) or (TEMP<TEMP_M AND
    HUM>HUM_M)] then ICE = TRUE
// New Alarm detection and status update
// Alarm Indication inside vehicle
// Neighbor table management every T_nodeTableMgt
// and RSU detection
// Alarm reception and forwarding (vehicles within
// range/direction)
    if [RSU Detect] then
            EVENT_REPORT message generation
            Alarm table reset
    else
            Alarm store
            Alarm forward // within range/direction
// Alarm table management every T_alarmTableMgt
```

(b)

is a function of the geographical position of the source and the dimensions S and W of the rectangular), its dimensions (S, W) and its heading. This information is set in the respective *GeoAreaPos Latitude*, *GeoAreaPos Longitude*, *distance a*, *distance b* and *angle* fields of the IEEE 802.11p message format respectively. Using the Geocast functionality, both cars B and F will be notified while car E will never be bothered and car A will be notified when it enters the interested area, possibly by node B re-transmitting the event report message and if the event persists (e.g. ice). Depending on the detected event and the highway's dimensions, the appropriate area may span both directions of the highway. For example, a weather-related event concerns both directions and thus the rectangle's W dimension has to double.

Similarly to the case where the RSU attached to the traffic light behaves as a sensed data sink, when an OBU of a vehicle moving in the highway detects an event but at the point of detection no connectivity is available, the information related to the detected event is passed to the RSU as soon as the vehicle enters within range, as shown in Figure 8. The RSU communicates this information to the highway operation center, as well as to the interested drivers, i.e. the drivers in the area that can be defined exploiting the topologically-scoped broadcast feature of the Geocast protocol. The communication with the highway operation center enables wider dissemination of warnings since it a) controls the messages appearing on the VMSs and b) can enforce additional means for the dissemination of information e.g. through mobile phones and radio.

Figure 8. Opportunistic communication is exploited for delivering information to the RSUs

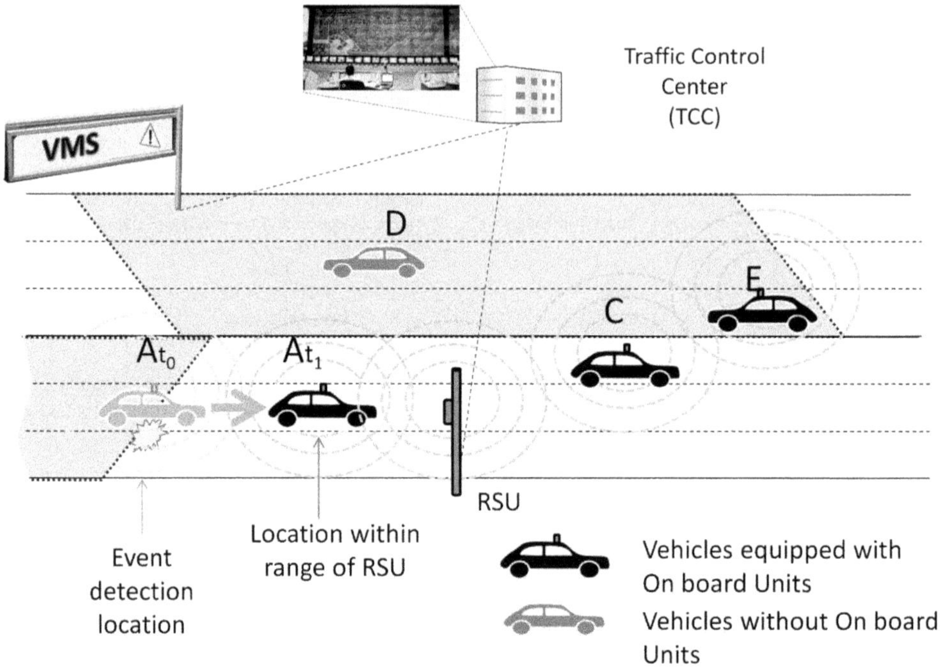

APPLICATIONS DEVELOPMENT

The development of the applications described in this work is based on the NEC's Linkbird-MX platform (Figure 9) which comes in a package of 11x14cm providing multiple serial interfaces (USB and RS232) and is powered by a 64bits MIPS microprocessor running at 266MHz. The Linkbird-MX comes with a Linux Operating System (kernel version 2.6.19) to provide an open development platform and is accompanied by the NEC's C2X-SDK communication system and API, providing geographical routing with advanced features and C-based development tools.

The C2X-SDK provides access to the 802.11p protocol implementation and interfaces to update and retrieve the necessary data structures including position, speed and direction data and mobile node neighbor database management. The applications described above include polling-based (for exchanging regular beacon messages to support the location based 802.11p operation and retrieving sensor data), as well as interrupt-based communications to process asynchronously received messages announcing specific events (Figure 7). Sensed data are retrieved and

Figure 9. Prototype device integrating sensing, communication and application logic

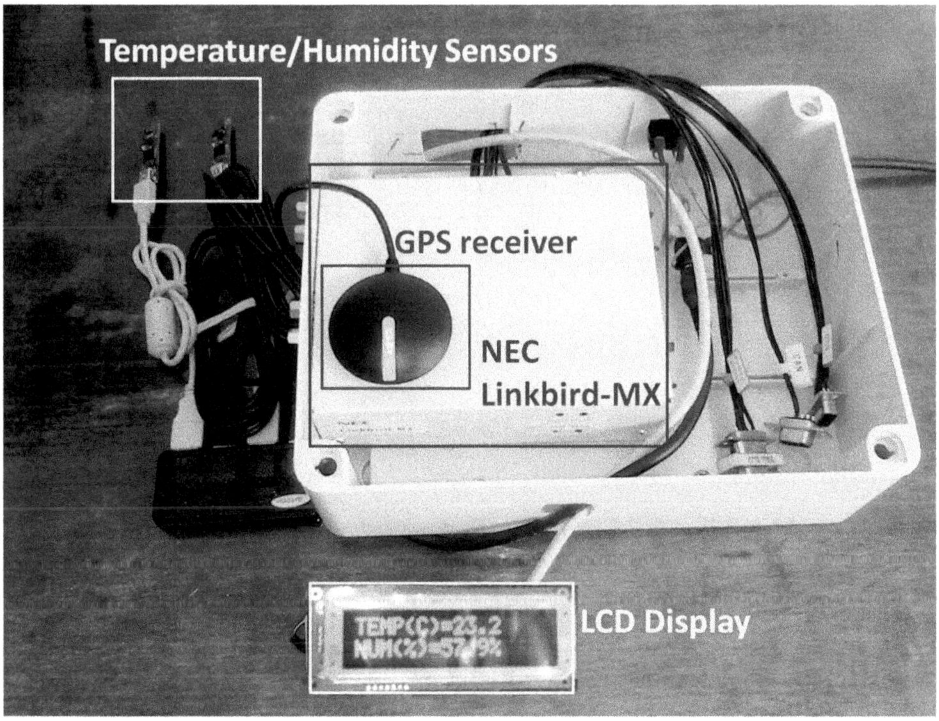

displayed via serial and USB interfaces developing the appropriate drivers to interface to the peripheral vehicle sensors and display unit. The C-based applications running on the OBUs and the RSUs periodically manage the retrieved information and implement the conditional statements to result to smart decisions regarding announcement of early warning messages and configuration of traffic management parameters like traffic light timing.

The information collected at the RSUs is communicated to the TCC using a typical client/server model developed over the TCP/IP protocol stack. At the TCC, this information is visualized through a Graphical User Interface (GUI) which has been implemented in Java (Figures 10 and 11). On the left hand side of each figure, the available crossroads and highways are listed. Choosing among them, the status of the traffic lights is shown in real-time along with information about the weather and traffic conditions and a list of detected events reported by vehicles passing by (Figure 10). On the other hand, Figure 11 shows the graphical representation of events detected by the OBUs in a highway.

Figure 10. Graphical representation of the sensed data at the TCC for improved traffic management

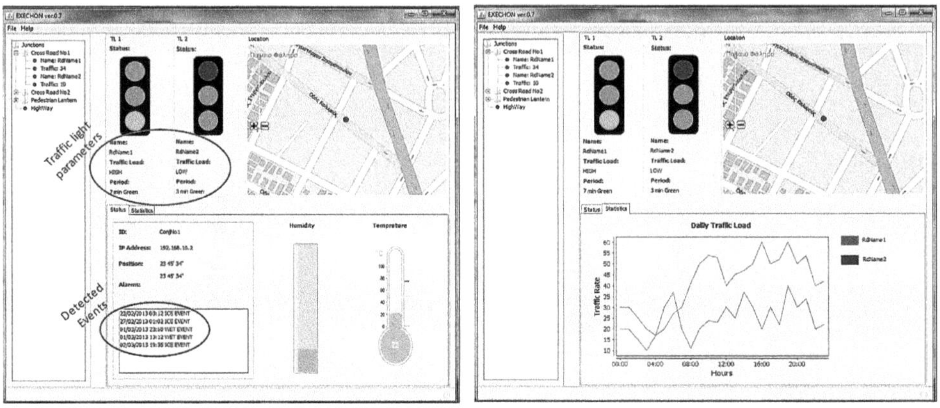

PERFORMANCE ASSESSMENT

Experimental Setup

The capability of the proposed system to support representative applications for road safety and traffic management was assessed through experiments in a laboratory testbed, the topology of which is depicted in Figure 12. The testbed's wireless network, which is based on the IEEE 802.11p technology and exploits GeoNetworking (Geocast) techniques, consists of a number of nodes N that emulate the behavior of devices installed in vehicles (OBUs) and of one node which emulates the behavior of a device installed in a fixed position in the ITS network infrastructure (RSU). The wired network part of the testbed is based on Ethernet connectivity and interconnects the RSU with the fixed node running the software of the TCC, as well as all the nodes of the wireless network (i.e. OBUs and RSU) with the node which controls the execution of the experiments. The realization of the latter connections through a separate wired network, serves to isolate the control traffic, i.e. traffic generated during initialization, monitoring and termination of experiments, from the traffic that is due to the applications under test (Figure 3). Control traffic is transported over secure shell sessions established between the experiment execution controller and the RSU and OBUs. NEC's Linkbird-MX devices, running Linux, were used for the OBUs and the RSU, and laptops, running Linux and Windows 7, were used for the experiment execution controller and the TCC, respectively.

The application used for the assessment of the proposed system assumes that the nodes of the wireless network periodically broadcast a status message (with

Figure 11. Screenshot of TCC application for enhanced road safety listing the detected events on a highway

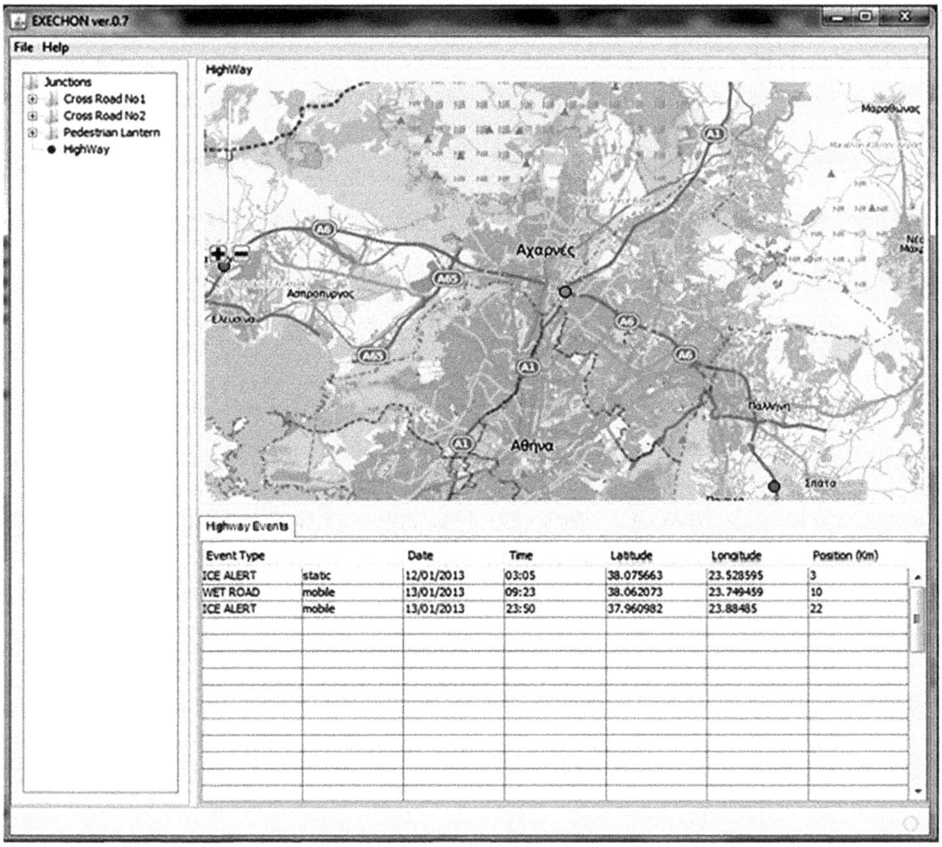

information about their node ID, location, environmental sensor data, speed and heading) and check for local events. The nodes emulating OBUs check for events related to the speed of the vehicle in which they are installed (events are generated when the vehicle has stopped or is moving with very low speed) while the node emulating the RSU checks for events related to the condition (slipperiness) of the road surface. The detection of events related to low speed is based on data provided by a GPS device attached to the OBUs whereas the detection of slippery road surface was based on readings from a humidity sensor which was attached to the RSU (for the evaluation it is assumed that humidity above a predefined threshold corresponds to slippery surface). The fact that the OBUs of the laboratory testbed were placed at fixed positions meant that in every code execution period, in addition to the periodic status messages, they were also transmitting an event related to

Figure 12. Testbed topology

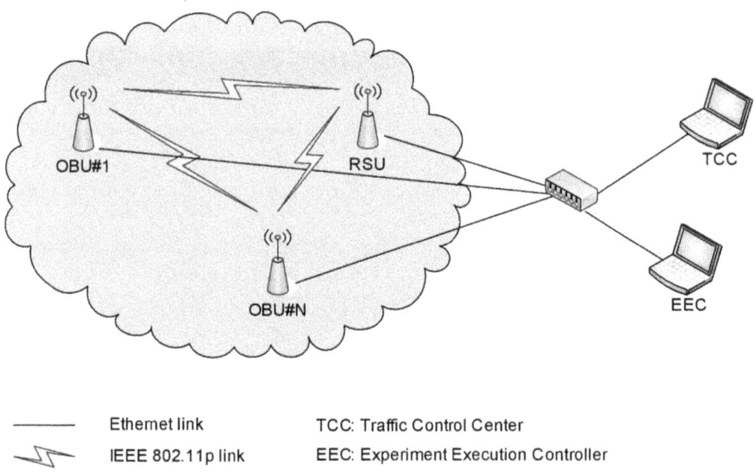

stopped vehicle. A simple behavior was assumed for the RSU, according to which this node detected and subsequently transmitted in every software execution interval an event for slippery road surface.

To emulate the capability of the RSU to serve as event repository, the OBUs store detected events (Figure 8) and transmit them to the RSU when the latter is within their transmission range (the OBUs rely on their neighbor location table, provided by the GeoNetworking protocol, to determine when they can reach the RSU through one-hop transmission). Upon receiving an event from an OBU, the RSU broadcasts it on the wireless medium to notify other OBUs that are not in the transmission range of the node originating the event.

The RSU communicates with the TCC over a wired link and sends, in a periodic fashion, status messages and, in an asynchronous manner, the events that were received from the OBUs. The regularly transmitted status messages include information on a) the environmental conditions (namely, temperature and humidity) in the vicinity of the RSU, b) the situation of the nearby OBUs (number and arrival rate of OBUs per road direction) and c) the status of the traffic lights that are controlled by the RSU (for the experimental study, the traffic lights have been emulated as devices connected to a serial port of the Linkbird-MX node acting as the RSU).

The network and application parameters used in the experiments are listed in Table 3. It is noted that the size of the messages includes all protocol headers.

Table 3. Network and application parameters

Parameter	Value
Number of OBUs, N	3
Distance between nodes in the wireless network	6-10m
Maximum transmission power of IEEE802.11p network card	1dBm
IEEE 802.11p data rate	3Mbps
IEEE 802.11p channel bandwidth	10MHz
Type of GeoNetworking (Geocast) messages	Single-hop topologically scope broadcast
Transmission period of status report messages sent by the RSU to the TCC	10s
Size of status report messages sent by the OBUs and the RSU	111bytes
Size of event report messages sent by the OBUs and the RSU	132bytes
Size of status report messages sent by the RSU to the TCC	210bytes
Size of event report messages sent by the RSU to the TCC	194bytes
Size of GeoNetworking beacon messages	50bytes
Transmission period of GeoNetworking beacon message	5s

Performance Results

The conducted experiments evaluate the data rate of the application described in the previous section and the overall system throughput (throughput of the wireless network and throughput of the wired link interconnecting the RSU and the TCC). For network traffic monitoring, the tcpdump and the ethereal tools were used (they were installed on the Linkbird-MX devices and the experiment execution controller, respectively) while the pktstat tool was used in all the aforementioned nodes for throughput measurements.

The overall throughput of the wireless network (including the traffic that is due to the periodically transmitted beacon messages of the GeoNetworking protocol) for different intervals between periodic execution of the application code is depicted in Figure 13. The results validate the fact that an increase in the rate of code execution (during which each of the OBUs and the RSU transmit two messages) results in increased wireless network throughput. Shorter intervals between code executions correspond to an increased rate of road surface condition and traffic monitoring and can be used when there is a need for more accurate recording of the system state (note that an interval of 1s corresponds to the case where a vehicle moving at

Figure 13. Wireless network throughput

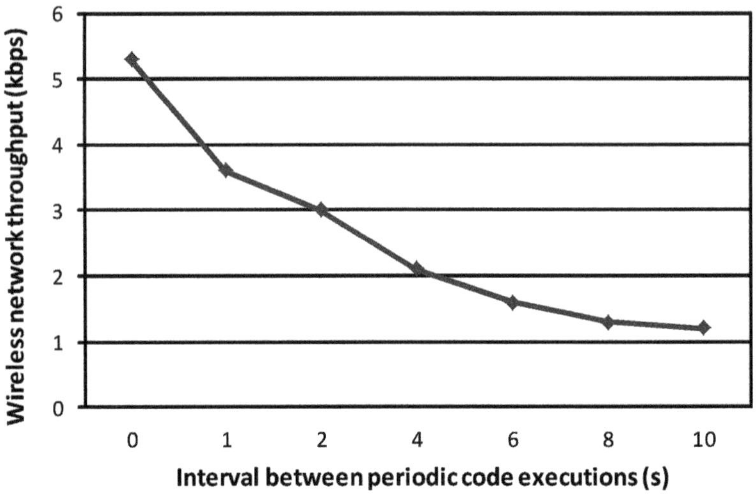

a speed of 100km/h detects and subsequently transmits an event every 28m). The maximum throughput, achieved for zero intervals between periodic executions of the code for event detection and transmission of the status messages, was measured at 5.3kbps. This maximum value is determined by the code execution time, the average value of which was measured at 2.3s (this increased time is due to the need to access peripheral devices like the GPS and the temperature/humidity sensor, as well as to the fact that the Linkbird-MX platform has limited computational resources). Each one of the OBUs contributes an equal amount of data rate while the RSU contributes to the total throughput a larger data rate because of the retransmission (relaying) of the OBU events. Comparisons with the case where the RSU does not implement this extra functionality showed that this form of rebroadcasting increases the network throughput by up to 30%.

The measured and the anticipated throughput of the wired link between the RSU and the TCC for different intervals between the periodic executions of the code at the nodes of the mobile network are illustrated in Figure 14. This throughput includes the data rate that is due to the periodic transmission of status report messages from the RSU to the TCC (which also serve as keep-alive messages) and the rate that is imposed on the link for the forwarding to the TCC of the events detected at the OBUs. The anticipated throughput is calculated based on the size of the messages transmitted to the TCC and on the expected rate with which these transmissions take place. In the case of status report messages their expected rate is determined by the predefined interval that is imposed between successive transmissions (which is set to 10s), while for the event messages the expected rate results from the

Figure 14. Wired network throughput

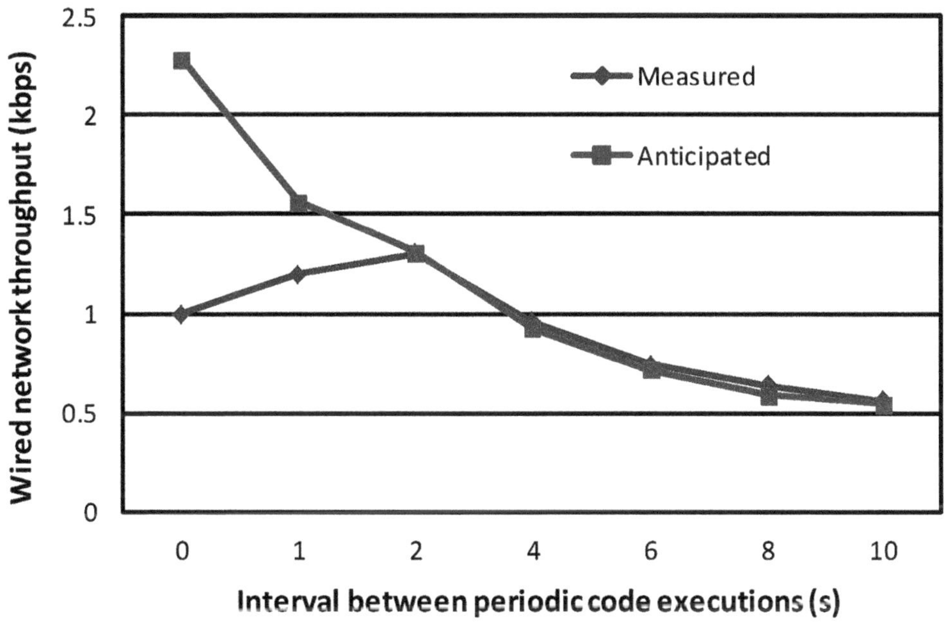

analysis of experimental data regarding the transmission rate of the corresponding messages at the wireless network.

The results of Figure 14 show that as long as the interval between periodic executions of the software (and subsequent transmissions of messages) at the OBUs and the RSU is larger than 2s (which is the case for applications operating in normal conditions), the RSU manages to transmit messages to the TCC at the anticipated rate. For intervals lower than this threshold, the RSU cannot cope with the increased message rate at the wireless network and a fraction of the event messages generated by the OBUs are not forwarded to the TCC (in such settings, which correspond to extreme load conditions, the experiments showed that the utilization of the RSU CPU was close to 100%). These results show that when rebroadcasting is used to increase the network coverage, efficient message filtering is needed to reduce the number of unnecessary retransmissions.

BENEFITS AND FACTORS AFFECTING PENETRATION

The main benefits of the presented architecture are that it contributes in forming a denser and wider sensing (relay) grid all over the road network. While dynamic traffic lights and dynamic VMS are widely deployed these days, making them capable of reacting in real-time to detected events all over the road network is expected to significantly improve safety and road experience. The denser the sensing grid is, the more wise and fresh traffic-related decisions can be made. On the other hand, the width of the grid directly affects the sensed area and consequently the interested drivers, who can then assist in traffic management decision based on global traffic and weather condition knowledge. A dense and wide sensing grid further motivates the purchase of OBUs by citizens and comes at almost no cost for the authorities since the only CAPEX involved is the purchase of RSUs.

The cost of the RSUs but mostly that of the OBUs plays an important role on their penetration. The OBU consists of a chipset realizing the IEEE 802.11p protocol, a processor and memory chipset, few attached or embedded sensors and an audio/visual peripheral for communicating with the driver. Reviewing the OBU operations required for the execution of the applications described above, it follows that the processing load, in normal application conditions with limited information rebroadcasting and efficient duplicate event suppression, is relatively low and, thus, a low-cost processor is sufficient. Apart from the antenna, the rest can be packed on a device with dimensions in the range of few centimeters (Vandenberghe, Moerman & Demeester, 2011). Already, prototypes supporting IEEE 802.11p communication can be found commercially available.

Regarding the peripherals, as GPS and sensing technology becomes cheaper every day, the cost of such a device for mass production is expected to be well below 40$. The cost could be even lower if this device is combined with devices used for e-toll payment, sharing common packaging and other components. Exploiting the path from the TCC to the RSU and the driver, it is possible to cater routing instructions for the route of interest to each driver, strengthening the incentives for purchasing the OBU. This way the driver can enjoy reduced toll fees and routing instructions apart from safety/weather warnings, all in one. While the estimation of the cost of such device is not at the focus of our investigation, implementing affordable systems and/or systems that justify well their cost is mandatory for the growth of realistic applications.

From the traffic control management perspective, the presented architecture caters abundant sensing information and enables more detailed traffic planning and more sophisticated route guidance. The corresponding decisions and advice can be disseminated to the interested drivers and pedestrians through a variety of communication channels: OBUs, VMS positioned in the highways and in bus stops, and more importantly through mobile phones. Setting up applications that provide routing guidance based not only on geographical information and on historical records of the traffic fluctuations but also on real-time traffic and weather conditions has been proven important for both car drivers and bicycle/motorcycle riders as well as for pedestrians. Warning a rider about slippery road and rain can save his/her life exactly as warning a citizen suffering from asthma that CO_2 emissions are above a specific threshold. Assuming that a significant part of the citizens are willing to pay for such a service (setting the route endpoint through the mobile phone and receiving back routing guidance and relevant information), the profit can be re-invested in OBU cost sharing to enlarge the sensing grid.

CONCLUSION

This chapter presented the design, implementation and evaluation of a framework that supports applications targeting road safety and traffic management over mobile opportunistic systems based on the IEEE 802.11p standard. The functional requirements for the support of relevant applications were discussed, critical parameters that can affect system performance were identified and a use case based on the NEC Linkbird-MX hardware platform providing the IEEE 802.11p communication framework was demonstrated. The applications were based on distributed execution of functional modules implemented on the three main entities that interact in the framework of ITS, i.e., the OBUs, RSUs and TCC. For the OBUs and the RSUs, the applications were directly interfaced to the IEEE 802.11p communication stack running on the aforementioned platform and the integrated peripherals supporting sensing and user interfacing. The assessment of the system performance was based on tests carried out on an experimental network. The study evaluated the achievable network throughput for a representative application and identified potential bottlenecks, which may lead to degraded system performance when the overall information generation rate at the vehicular ad-hoc network is high. In such cases, in-network data aggregation and efficient processing are needed to deliver an ITS that can reliably provide real-time information to drivers and transport infrastructure administration authorities for responding appropriately to weather and traffic conditions, enhancing, in this way, transportation efficiency and safety.

ACKNOWLEDGMENT

This work was performed in the context of the TROLLS and EXECHON projects. TROLLS is co-financed by the European Union (European Social Fund – ESF) and Greek national funds through the Operational Program "Education and Lifelong Learning" of the National Strategic Reference Framework (NSRF) – Research Funding Program ARCHIMEDES III "Investing in knowledge society through the European Social Fund". EXECHON was implemented under the NSRF 2007-2013, national action "Cooperation" and was co-financed by the European Union and Greek national funds.

REFERENCES

Bilstrup, K. (2007). *A survey regarding wireless communication standards intended for a high-speed vehicle environment. Technical Report IDE 0712*. Sweden: Halmstad University.

Burguillo-Rial, J. C., Costa-Montenegro, E., Gil-Castineira, F., & Rodriguez-Hernandez, P. (2008). Performance analysis of IEEE 802.11p in urban environments using a multi-agent model. In *Proceedings of the 19th IEEE International Symposium on Personal, Indoor and Mobile Radio Communications (PIMRC)*, (pp. 1–6).

Cayotte, E., & Buchow H. (2009). *Who dies of what in Europe before the age of 65*. Eurostat statistics.

Eichler, S. (2007). Performance evaluation of the IEEE 802.11p wave communication standard. In *Proceedings of the 66th IEEE Vehicular Technology Conference (VTC)*, (pp. 2199–2203).

European Telecommunications Standards Institute (2011). *Final draft ETSI ES 302 663 V1.2.0, Intelligent Transport Systems (ITS); Access layer specification for Intelligent Transport Systems operating in the 5 GHz frequency band.*

Faye, S., Chaudet, C., & Demeure, I. (2012). A Distributed Algorithm for Multiple Intersections Adaptive Traffic Lights Control using a Wireless Sensor Networks. In *Proceedings of the 8th ACM International Conference on emerging Networking EXperiments and Technologies (CoNEXT)*, (pp. 13-18).

Festag, A., Fussler, H., Hartenstein, H., Sarma, A., & Schmitz, R. (2004). Fleet-Net: Bringing Car-to-Car Communication into the Real World. In *Proceedings of the 10th World Congress on ITS.*

Grafling, S., Mahonen, P., & Riihijarvi, J. (2010). Performance evaluation of IEEE 1609 wave and IEEE 802.11p for vehicular communications. In *Proceedings of the 2nd International Conference on Ubiquitous and Future Networks (ICUFN)*, (pp. 344–348). doi:10.1109/ICUFN.2010.5547184

IEA. (2005). *CO_2 Emissions from fuel combustion*. IEA.

IEEE 1609.0. (2013). *IEEE Guide for Wireless Access in Vehicular Environments (WAVE) – Architecture*. IEEE.

IEEE 802.11p. (2010). *IEEE Standard for Information Technology-Telecommunications and Information Exchange Between Systems-Local and Metropolitan Area Networks-Specific Requirements - Part II: Wireless LAN Medium Access Control (MAC) and Physical Layer (PHY) Specifications, Amendment 6: Wireless Access in Vehicular Environment*. IEEE Computer Society.

INDRA. (2008). *Annual Report 2008*. Retrieved from http://www.informeanual2008.indra.es/

Jang, H. C., & Feng, W. C. (2010). Network status detection-based dynamic adaptation of contention window in IEEE 802.11p. In *Proceedings of the 71st IEEE Vehicular Technology Conference (VTC)*, (pp. 1–5).

Kadas, G., & Chatzimisios, P. (2011). Collaborative Efforts for Safety and Security in Vehicular Communication Networks. In *Proceedings of the 15th Panhellenic Conference on Informatics (PCI)*, (pp. 117-121). doi:10.1109/PCI.2011.75

Kadas, G., & Chatzimisios, P. (2012). The Role of Roadside Assistance in Vehicular Communication Networks: Security, Quality of Service and Routing issues. In R. Daher & A. Vinel (Eds.), *Roadside Networks for Vehicular Communications: Architectures, Applications and Test Fields* (pp. 1–37). IGI Global.

Karkazis, P., Leligou, H. C., & Zahariadis, T. (2012). Geographical Routing in Wireless Sensor Networks. In *Proceedings of the 5th International Conference on Telecommunications and Multimedia (TEMU)*, (pp. 19-24). doi:10.1109/TEMU.2012.6294717

Khairnar, V., & Kotecha, K. (2013). Performance of Vehicle-to-Vehicle Communication using IEEE 802.11p in Vehicular Ad-hoc Network Environment. *International Journal of Network Security & Its Applications*, *5*(2), 143–170. doi:10.5121/ijnsa.2013.5212

Leligou, H. C., Chatzimisios, P., Sarakis, L., Orphanoudakis, T., Karkazis, P., & Zahariadis, T. (2014). An IEEE 802.11p compliant system prototype supporting road safety and traffic management applications. *International Journal of Wireless Networks and Broadband Technologies*, *3*(1), 1–17. doi:10.4018/ijwnbt.2014010101

Leligou, H. C., Karkazis, P., Hatziefremidis, A., Orphanoudakis, T., & Zahariadis, T. (2013). Implementation of road safety and traffic management applications over mobile opportunistic systems. In *Proceedings of the 17th WSEAS International Conference on Communications (CSCC)*, (pp. 163-168).

Leontiadis, I., Marfiazy, G., Mack, D., Pauyx, G., Mascolo, C., & Gerla, M. (2011). On the Effectiveness of an Opportunistic Traffic Management System for Vehicular Networks. *IEEE Transactions on Intelligent Transportation Systems, 12*(4), 1537–1548. doi:10.1109/TITS.2011.2161469

Misic, J., Badawy, G., Rashwand, S., & Misic, V. B. (2010). Tradeoff issues for cch/sch duty cycle for IEEE 802.11p single channel devices. In *Proceedings of IEEE Global Telecommunications Conference (GLOBECOM)*, (pp. 1-6).

Murray, C., & Lopez, A. (1996). *The global burden of disease* (Vol. 1). Cambridge, MA: Harvard University Press.

Nantulya, V. M., & Reich, M. R. (2002). The neglected epidemic: Road traffic injuries in developing countries. *British Medical Journal, 324*(7346), 1139–1141. doi:10.1136/bmj.324.7346.1139 PMID:12003888

NEC Laboratories. (2013). *Car2X Communication SDK 2 Userguide v2.3.3.*

Panorama of Transport. (2009). *Eurostat EC.*

Pincus, M. (2011). *Intelligent Transportation Systems Benefits, Costs, Deployment, and Lessons Learned Desk Reference: 2011 Update*. US Department of Transportation, Research and Innovative Technology Administration.

Vandenberghe, W., Moerman, I., & Demeester, P. (2011). Approximation of the IEEE 802.11p Standard Using Commercial Off-The-Shelf IEEE 802.11a Hardware. In *Proceedings of IEEE 11th International Conference on ITS Telecommunications (ITST)*, (pp. 21-26).

Wang, S. Y., Chou, C. L., Liu, K. C., Ho, T. W., Hung, W. J., & Huang, C. F., … Lin, C. C. (2009). Improving the channel utilization of IEEE 802.11p/1609 networks. In *Proceedings of Wireless Communications and Networking Conference (WCNC)*, (pp. 1–6).

Wang, Y., Ahmed, A., Krishnamachari, B., & Psounis, K. (2008). IEEE 802.11p performance evaluation and protocol enhancement. In *Proceedings of IEEE International Conference on Vehicular Electronics and Safety (ICVES)*, (pp. 317–322).

Warberg, A., Larsen, J., & Jorgensen, R. M. (2008). *Green wave traffic optimization – a survey. Technical Report*. Technical University Denmark.

Yousef, K. M., Al-Karaki, J. N., & Shatnawi, A. M. (2010). Intelligent traffic light flow control system using wireless sensors networks. *Journal of Information Science and Engineering, 26*(3), 753–768.

Zahariadis, T., Trakadas, P., Leligou, H. C., Maniatis, S., & Karkazis, P. (2013). A novel trust-aware geographical routing scheme for wireless sensor networks. *Wireless Personal Communications*, *69*(2), 805–826. doi:10.1007/s11277-012-0613-7

Zhan, C., Shen, L., Hadi, M. A., & Gan, A. (2008). *Understanding the Characteristics of Secondary Crashes on Freeways*. Paper presented at the 87th Annual Meeting of the Transportation Research Board, Washington DC.

Chapter 8
A Scheduling Scheme for Throughput Optimization in Mobile Peer-to-Peer Networks

Odysseas Shiakallis
University of Nicosia, Cyprus

Athina Bourdena
University of Nicosia, Cyprus

Constandinos X. Mavromoustakis
University of Nicosia, Cyprus

Evangelos Pallis
Technological Educational Institute of Crete, Greece

George Mastorakis
Technological Educational Institute of Crete, Greece

Evangelos Markakis
Technological Educational Institute of Crete, Greece

Ciprian Dobre
University Politehnica of Bucharest, Romania

ABSTRACT

Mobile Cloud Computing (MCC) paradigm includes all the emerging technological advances, mechanisms and schemes for the efficient resource offloading and the energy-efficient provision of services to mobile users. In addition, the mobile nodes will act as flexible networking points in emerging mobile networking architectures, where several challenges have to be addressed, like the high energy consumption and the data packets transmission failure, under a Mobile Peer-to-Peer (MP2P) approach. Towards addressing such challenges, several factors that contribute to the increased consumption of the energy, have to be investigated, as well as issues

DOI: 10.4018/978-1-4666-9941-0.ch008

Copyright ©2016, IGI Global. Copying or distributing in print or electronic forms without written permission of IGI Global is prohibited.

related with the loss of data during the provision of services. In this framework, a Traffic-based S-MAC scheme is proposed in this chapter, towards increasing the data exchange and minimize the energy consumption, between mobile nodes operating under an Ad-Hoc approach. The performance of the proposed scheduling scheme was thoroughly evaluated, through a number of simulation experiments.

I. INTRODUCTION

The portable devices (e.g. smart phones, tablets etc.) in emerging mobile network architectures have shrunk in size, incorporating advanced functions and mechanisms (Mastorakis, Mavromoustakis, Bourdena, Kormentzas & Pallis, 2013). This allows a node (or a mobile device) to act as a wireless terminal, or as a repeater creating a network for efficient resources sharing (Mavromoustakis, Dimitriou & Mastorakis, 2013). A self-organizing and adaptive collection of such mobile devices connected with wireless links is now referred to as a Mobile Ad Hoc Network (i.e. MANET). An Ad Hoc Network is not in need of any centralized control. The network should be able to detect any new nodes in range and induct them unobstructed (Mousicou, Mavromoustakis, Bourdena, Mastorakis, & Pallis, 2013). Nevertheless, if any node moves out of the range of the network, the remaining nodes should automatically reconfigure themselves, in order to adapt in the new topological scenario. A working group named MANET has been set up by the Internet Engineering Task Force (IETF) for promoting research in this area. Most usually, there are two types of architectures in Ad Hoc Networks: the flat and the hierarchical architecture (Chakrabarti & Mishra, 2001; Toh, 2002). Each node has a transceiver, an antenna and a power source. The properties of these nodes can vary regarding their size, transmission range, battery power and processing ability. It is common that some nodes can be used as servers, by lending themselves, others as clients and even some may be operational to act in conjunction (server and client), depending on the situation. There are specific cases though, where each node may need to act as a router, in order to channel information between nodes (Royer & Toh 1999; Haas & Tabrizi1998).

In this context, a Traffic Sensor Media Access Control (TS-MAC) protocol is examined in this chapter, based on S-MAC, a Sensor Media Access Control that uses three atypical techniques to reduce energy consumption and support self-configuration (Mavromoustakis, Bourdena, Mastorakis, Pallis, & Kormentzas, 2014). In order to bring energy consumption to the lowest level possible in listening to an idle channel, nodes periodically sleep (Mavromoustakis, Mousicou, Papanikolaou, Mastorakis, Bourdena, & Pallis, 2015). Neighboring nodes form *virtual clusters* so that they can auto-synchronize between sleep schedules. S-MAC, which was influenced by PAMAS, also sets nodes to sleep during transmissions of other nodes.

Unlike PAMAS, it only uses in-channel signaling. Finally, S-MAC applies *message passing* to reduce contention latency for sensor-network applications that require store-and-forward processing as data move through the network (Ye, 2002). In addition, the latency, connectivity, energy and memory are the essential elements of today's mobile environments, whose performance may be significantly improved, by caching techniques. The TS-MAC protocol can be used in various topological patterns with varying packet sizes (Andreou, Mavromoustakis, Mastorakis, Bourdena, & Pallis, 2014).

Adaptively, the proposed TS-MAC scheme modifies the duty cycle of a node, adapting the nodes' sleep wake time in order to save power and maximize both the throughput and the network lifetime. Therefore, the proposed TS-MAC scheme manipulates the network state in a distributed form and actively controls each node's sleeping duration, by using specific methods, in order to minimize the energy cost of peer-to-peer communication among mobile terminals. In addition, this protocol provides schemes for both delayed sensitive packets, as well as without deadlines (also known as don't-care packets). Finally, this model could be applied to Ad Hoc networks of various nodes position, number and topology (Mastorakis, Bourdena, Mavromoustakis, Pallis, & Kormentzas, 2013). In this framework, the organization of the chapter is as follows: Section II discusses the related work that has been performed on various MAC protocols, sleeping mechanisms and functions, as well as on implementation techniques. Section III discusses the system model and mechanisms of the S-MAC, which further on introduces the proposed Traffic Sensor Media Access Control (TS-MAC) protocol, followed by Section IV which provides the evaluation and simulation results of the proposed scheme. Finally, Section V concludes with a summary of our contribution and further research.

II. RELATED WORK AND REASEARCH MOTIVATION

This section provides a brief overview of existing energy conserving MAC protocols in asynchronous and synchronized approaches. Medium Access Control Protocols can be classified in Contention-free, Contention-based and Schedule-based Protocols. This chapter will focus on the Contention-based and Schedule-based protocols, which are used in Ad Hoc Networks. Contention-based protocols are the protocols that focus their attention and workability on the contention/traffic of a network. According to the workload of a network, a node can determine how long it will sleep, if the MAC protocol at hand supports that functionality, and in what way it will act in order to conserve energy, improve throughput and minimize latency. In Schedule-based protocols, the nodes are assigned specific time slots, in order to know when to turn ON or OFF the transmission or the reception of data. P.Karn in

(Karn, 1990), introduces MACA, in an effort to solve the problem of the hidden and the exposed terminal. Multiple Access with Collision Avoidance (MACA), makes use of the Ready to Send (RTS) and/or Clear to Send (CTS) messages, in order for nodes to avoid collision of data. Data collision is not ensured to be avoided, as the nodes add in their packet schedule a random amount of time for other nodes to wait before they transmit. If a collision occurs, the nodes wait in a random amount of time, until they retransmit their data. Also, MACA provides poor energy conservation, as nodes do not go to sleep when there is no data for transmission. PAMAS (Singh & Raghavendra, 1998) protocol, which is based on the MACA protocol, introduced the addition of a separate signaling channel while transmitting data packets. In an effort to conserve energy, the PAMAS protocol emphasizes on the node's ability to sleep, while there is no transmission, either sending or receiving data, between itself and a neighboring node. Through the use of a separate signaling channel, nodes can exchange RTS and/or CTS messages without the use of the signaling channel for transmission of data. By applying this, the nodes are able to determine at what specific time and for how long time period they can turn themselves off until the next retrieval or transmission of data. In this case (Yadav, Varma & Malaviya, 2009), the node sensor costs and complexity that arise by the use of two different channels, are enormous as there is excessive energy consumption because of the frequent switching of sleep/wake states of a node.

S-MAC (Joshi, Jaiswal, & Tyagi, 2013), or Sensor-MAC is a simple MAC protocol, inspired by PAMAS protocol, which was designed in an effort to decrease energy consumption of nodes, while trying to send and receive data. The way S-MAC works is, by using a function called Listen-Sleep, integrated with an in-channel signaling methodology. Through this function a node listens for a specific amount of time for any type of RTS, CTS, SYNC or Acknowledgment (ACK) communication method to identify through the network. If the node stumbles upon a RTS or CTS, which is intended for a different node, then it goes back to sleep. When the node is closing by its listen period, it tries to communicate with other nodes for any prospective data packets it may be entitled of. If it finds no data intended for that node, it goes back to sleep. While in Sleep Period, it tries to turn off its radios so that it minimizes the energy consumption and extend its lifetime. The S-MAC uses as well, a Periodic Neighbor Discovery process, which is applied by sending periodic SYNC packets between nodes, in order to determine a nodes' neighboring node, so that they can exchange a synchronization schedule between them, in order to avoid collision. By avoiding collision, the non-corruption of data is ensured. Despite these, the S-MAC trades off energy with throughput and latency, as throughput is decreased because of the single use of the Listen Period of a node and latency is increased because of the extended creation/generation of RTS, CTS and ACK, even while in Sleep mode. T-MAC, or Timeout-MAC (Joshi, Jaiswal, & Tyagi, 2013), is a modified version of the S-MAC protocol, which provides better results in low load conditions, as well as better energy conservation and higher network lifetime. The way the T-MAC functions is that it transmits data packets through bursts of variable length. Between those transmissions, there is a gap called the sleep time/state of a node, and this serves as a function of minimizing idle listening. A frame is created (frame referred to as a complete sleep/wake state), which contains the messages that are about to be transmitted and are currently residing in buffers. The frame is transmitted, when there is an active connection between nodes, and the transmission is over when there are no more frames left to be sent. T-MAC performs poorly in terms of general energy conservation, as it stays idle waiting for data packets that might be transmitted towards it. The amount of energy consumed by staying idle and waiting for data packets is exactly the same with transmitting or receiving data. That means there is a significant energy consumption that minimizes the node lifetime and efficiency. In addition, when neighboring nodes compete for any available medium at the time, data collision and corruption occurs, so packets either get lost or corrupted (Bourdena, Pallis, Kormentzas, & Mastorakis, 2013).

Yang, Tseng, Wu & Chen (2005) propose U-MAC, which stands for "Utilization Based Duty Cycle Tuning" MAC Protocol, based on its predecessor, S-MAC. U-MAC synchronous protocol, emphasizes on lowering the power consumption of a node without increasing latency, by modifying the duty cycles of a node and managing the sleep period after transmission of data packets. Based on U-MAC design, each node does not act according to the same constant duty cycle, but instead, it reconfigures itself according to the last synchronization the node had, and according to the workload it needs to handle. In addition, the node sends its reconfigured schedule to other nodes across the network to its neighboring nodes via broadcasting. All in all, U-MAC can be considered, as a well-thought out protocol that is easily adaptable in different scenarios and can produce desired results. Despite these, although the design is smart and efficient, the overall throughput is not taken under consideration, as well as the data packets transmission between different workloads. Synchronized protocols, such as S-MAC (Wikipedia, 2014), T-MAC (Webopedia, 2010) and U-MAC (Tanenbaum, 2003) can specify when nodes are awake and asleep within a frame. In S-MAC, neighboring nodes form virtual clusters to synchronize the working schedules (Kumar, 2004). B-MAC Protocol (Berkeley-MAC Protocol) was developed in an attempt to decrease energy consumption, providing high avoidance of collisions, as well as efficiently utilizing data transfer channels. As stated by Polastre, Hill and Culler (Polastre, Hill, & Culler, 2004), B-MAC is an asynchronous, reconfigurable MAC protocol that is designed with minimum coding, able to adapt into Wireless Sensor Network environments, providing bidirectional functionalities, towards middleware applications. While B-MAC's smart and simple Low Power Listening provides good results, in terms of energy efficiency, excess latency (Buettner, Yee, Anderson, & Han, 2006) is present by its long preamble throughout each hop a data packet is sent, thus excess energy consumption is wasted through nodes that act as a part of the bridge of transmitting data to its destination. For that reason, X-MAC (Buettner, Yee, Anderson, & Han, 2006) was devised in an effort to improve the asynchronous B-MAC protocol, by introducing a low power communication technique, which would retain and increase energy conservation, as well as all its other advantages discussed above, by shortening the preamble and abrogating the sleep/wake schedules of the transmitting/receiving nodes. X-MAC introduces a shorter preamble in an effort to increase energy efficiency and decrease latency, by embedding the destination node address information in the preamble so that that non-target nodes will go back to sleep, as soon as they transmit the data packets (overhearing problem). Issues faced with X-MAC was that the timing mechanism contained in the protocol could not create functional sleep schedules for periods of less than 20ms, which in fact would not provide any improvement, regarding the optimal energy-consumption scheme, thus forcing nodes to sleep a lot longer than they should. As a result, latency is increased and throughput is decreased because of these.

WiseMAC (El-Hoiydi, Decotignie, Enz, & Le Roux, 2003) or Wireless Sensor MAC, uses the adaptation of preamble sampling by utilizing the neighboring nodes schedule, thus decreasing the overhearing problem. That is applied by listening to the radio for a specific constant *Tw* period. While that, the transmitter node embeds a wake up preamble in front of its data packets, for the destination node to acknowledge and be awaken when the transmission of data occurs. Because of the nodes independent sampling schedule offsets and in an effort to minimize the overhearing problem, which means nodes expecting for a long time period for a preamble that may contain data transmission towards them without any result because of them not being destined to receive data packets, each node learns the offset of their neighbor's sampling schedule, so that it can be in synchronization when it comes to transmitting or receiving data. WiseMAC may be efficient for short length multi-hop data transmission, but when it comes to long distance multi-hop data transmission, it fails to accomplish its cause. B-MAC, X-MAC and WiseMAC asynchronous protocols, make use of the adaptive preamble sampling technique in an effort to increase throughput and minimize latency, duty cycles and idle listening. One of the most widely used Ad Hoc Network Protocols is 802.11. The 802.11 protocol (Kumar, Raghavan, & Deng, 2006), which is a CSMA with Collision Avoidance protocol, offers two modes of MAC protocol. The DCF mode, which stands for Distributed Coordination Function, is used in the Ad Hoc Networks and the PCF mode which stands for Point Coordination Function and is used for infrastructure –based networks. The DCF mode uses the messages that are widely used in other protocols, such as the CTS, RTS and ACK. Also, it makes use of the DATA message. 802.11 does not only make use of a Physical Carrier Sensing procedure, but also it introduces the Virtual Carrier Sensing procedure. 802.11 performs well when it comes to low latency and throughput, but performs poorly in terms of energy conservation because of the idle listing problem. The idle listening problem is when a node stays idle without going to a sleep state in order to save energy, but instead listens to the network for any traffic. It has been proven that the energy consumed for idle listening is the same with the energy consumed for transmitting data. Most of the recent works focus on how to prolong a sensor or network battery lifetime. However, these works do not consider the performance of data transmission when sensors have different workload. Our work differs from all of the above-mentioned approaches, by striking the balance between the battery life and transmission latency according to the ongoing traffic situations. The main difference is that the proposed TS-MAC protocol is based on the demand of data transmission of sensors. The use of a traffic-aware protocol can conserve sensors battery life and reduce the transmission latency caused by sleep (Mavromoustakis, Andreou, Mastorakis, Papadakis, Bourdena, & Stratakis, 2014).

In addition, the recent paradigm of Mobile Peer-to-Peer (MP2P) networks can be exploited in future networking architectures. The mobile devices, such as laptops, smart phones, PDAs etc are becoming very popular, supporting the data exchanging like text files, movies, music and applications of voice over IP and gaming. This is why researchers are very interested on MP2P systems and the related architectures that are able to store and retrieve data in a mobile network. The well-known client/server model has been overcome and the main goal of a MP2P architecture is to distribute the load among all mobile nodes (peers) rather than relying on a central powerful server. MP2P systems are a combination of a mobile Ad Hoc network (MANET) underlay with a distributed hash table overlay. MANETs and MP2P networks share concepts of self-organization and decentralization. They do not require a preexisting communication infrastructure for operation but they both operate without a central, coordinating entity. Considering these similarities, combining MANETs and P2P networks, is reasonable to obtain a fully distributed, decentralized and infrastructure-less communication substrate. They operate with a non pre deployed infrastructure and are different from wired ones because each node is able to move around, having a limited radio range. As peers move, the topology of such a system changes constantly. Due to all these, MP2P networks are suitable not only for file sharing but for collaborative activities, social applications, disaster relief, tactical scenarios and development aid projects.

In addition, the evolution of technology, the demand of the modern user of digital devices and the popularity of the new type devices brought together two quite different fields of computer science. The fast and reliable data exchange in conjunction with the growing use and sales of the new generation of mobile devices brought together the mobile web services and the peer to peer networks. In the mobile web services domain big leaps have occurred lately. The efficiency of the mobile devices (i.e. smartphones) is increasing, while the users request even more multimedia services provision. The web services need more software applications day by day, in order to satisfy the increasing number of the users. The skyrocketing of the web services field recently is due to the latest development of the 3G and the 4G technology in combination with the vast number of the wi-fi installations in almost every market. At the same time, the recent mobile phones present several advantages, such as the interconnection between different networks. Some other new features that we can pinpoint, is the increased storage capacity, the faster processor and the longer-lasting batteries. This exponentially exploding situation of the Internet and the new devices has caused also the need of the accurate and easy exchange of even bigger packets of data. Fortunately, at this exact point, the use of the Peer-to-Peer (P2P) networks technology seems to be matching quite perfectly. In normal wired wide networks (e.g. Internet) the P2P automation seems to provide a stable, reliable and robust architecture for exchanging resources. That is exactly the reason why this kind of file sharing systems have over-passed, in content transferring, technologies such as the email and World Wide Web. P2P is a file sharing system, where devices equally share data directly to each other.

Furthermore, much work has been done in the Peer-to-Peer resource sharing for wireless mesh networks, which in addition is compatible with all the network interfaces. A number of researchers have, also, indicated the need of a classification system in order to categorize the Peer-to-Peer applications, due to their evolution. Some more Peer-to-Peer programs have been created to facilitate the resource sharing over the mobile devices. This kind of programs can be, for instance, the hybrid Peer-to-Peer for mobiles, the symtorrent and the WinMobile Torrent. The most recent mobile devices present a number of new capabilities, such as combining several types of networks, either wide or close distant ones. The Peer-to-Peer networking is taking advantage of it and has over-passed the problem of using networks with different characteristics between them, for instance, the stability of the IP address. At this point, we should pinpoint another restriction that came up while trying to install the Peer-to-Peer application on new devices. On the other hand, there are still some disadvantages that could create unwanted signal reduction while transferring information over the air interface. One of the most popular applications in the Peer-to-Peer networking in static computers is the utorrent. The torrent-based technology is one more tool adopted by the mobile devices. The utorrent program, a little bit differentiated, is ready to provide the exact services in new devices. For the creation of this application, the developers use C++ as a programming language. Though, SymTorrent is developed in a Symbian platform, thus we cannot use it in any other operating system. While implementing it, the restrictions were almost zero, as long as the operating system of Sybian is a multithreaded operating system where the developers can easily reach and work with the advanced services of the platform. This, also, means that there are no difficulties or restrictions for hosting applications which use several sockets, file-access and complex user interface. Applying small changes to torrent-based application, other more similar ones can be successfully installed in alternative operating systems except the Sybian. Another example of a torrent-based implementation is the MobTorrent. It is developed in Java ME since the most of the mobile phones support the Java ME. As a result, it gives us the options of downloading and uploading files and thus we can use the Peer-to-Peer evolution. Maybe the latest of this king of applications is the WinMobile Torrent or else the wmTorrent. The wmTorrent is the first torrent client for mobile devices that is running under windows operating systems. The latest version of this torrent application offers advanced features and supports trackerless torrenting, peer exchange (a protocol that augments bitTorrent), Protocol Encryption and HTTP Seeding, which allows data to be exchanged via the hypertext transfer protocol.

The Peer-to-Peer evolution has also made leaps at the field of web services on mobile phones. Online services are much easier to work efficiently on this kind of phones when the wireless signaling rate has been improved that much. This Peer-to-Peer method is used, in order to advertise the online services and the new creation of the JXTA network is responsible to relate distributed components-services to each other, taking under consideration the existed formal interfaces. JTXA, in fact, is a model of Service Oriented Architecture (SOA), which is implemented to support the Peer-to-Peer networking. In details, a Service Oriented Architecture as the JTXA decides, whether a couple of application should communicate over the web. More specific, we refer to applications that provide services over the internet. The most exciting fact about this particular SOA is the simplification of the Mobile Host's entry to a Peer to Peer domain by using Advertisements. The implementation of the JTXA advertising consists of language-neutral metadata written in XML form. In the end, except the pairing of the web services through the Peer-to-Peer technology, we also gain access to software, by using primarily the SOAP and secondarily the HTTP, BEEP, UDP for the transmission.

Going further to one more technological prodigy, we find the m-Chord. The m-chord, in fact, is a modified chord in order to manage and support the Peer-to-Peer mobility in all IP networks. Before explaining the way the m-chord works, we should take a brief on the mobile IP and how it works. Although Mobile Nodes switch some times their subnets, the Foreign Agent which belong to other networks still keeps a copy of a temporary Care-of-Address. The Home Agent of the mobile Node is informed about the Care-of-Address and it relates it in a table with the permanent IP of the mobile node. So, the Home Agent tunnels to the Foreign Agents all the packets, which are addressed to Mobile Nodes of his home network. The development of the m-Chord idea, is based on the Peer-to-Peer topology. Mobile

Agents are taking advantage of this topology advertising, communicating, exchanging information and in general, using all the features of the peer networking. For instance, m-Chord supports the feature of connecting and disconnecting from the network without affecting the operations, for every Mobile Agent. The researchers of this technology used the Chord networking as a foundation of the new one due to its performance, provable correctness, and the simplicity of the topology. M-Chord uses ring topology of Mobile Agents, agents that could be either Home of Foreign ones. These Agents are capable of hosting more than one Mobile Nodes. Both the agents and the nodes are having a Parent/Child relationship, which is a usual protocol used in hierarchical unstructured Peer-to-Peer networking. M-Chord could be considered as a one-hop wireless transmission technology, where a mobile node has one Mobile Agent and advertises or exchange data in the Chord over the Mobile Agent using just one hop.

In MP2P, it is important to reduce the overhead generated by the lookup algorithms. Each node that is looking for an object compares its own ID with the object ID. The next hop is selected after calculating the matching prefix of those IDs but lookup messages are not always routed on the shortest physical paths. Several architectures have been proposed where each node ID is related with its position. In addition the operation area is splitted in several clusters with their own IDs. The ID of each node in the cluster is defined by this prefix and a random suffix. Clustering level is defined by the depth of the clustering. Furthermore, several researchers propose new routing table methods in order to improve the system performance because existing MP2P methods assume that all peers are on a single layer. These new methods classify peers in two categories. Super peers and sub-peers. Super peers manage the sub-peers that are connected to them and keep information such as id, address, file list of the sub-peers. So each super peer has a routing table with each sub-peer's information. When a sub-peer wants to find a file, it asks its super peer. Super informs the sub-peer where is the file if it is found in its routing table. If file does not exist in super peer's routing table, the super peer informs other super peers with a message in order to locate the file. If any of super peers has the file in its routing table, it informs the requested super peer with a response message. This routing table algorithm prevents multi-broadcasting.

There are also many other studies that propose a double-layered P2P system in which super-nodes are selected based on their mobility pattern. Node energy level is taken in consideration along with mobility factor in order to enhance the system stability and reliability. Super-peer topology construction and maintenance scheme based on network coordinates is also a proposition. On the other hand RobP2P, which is a new proposition that integrates many factors to efficiently select super peers, including the node's current mobility, immediate energy level, network capability, mean uptime, and connectivity degree. There are three main objectives for

RobP2P: 1) develop a robust and efficient super-peer selection protocol; 2) reduce the overhead traffic of network topology maintenance; 3) increase the reliability and stability of the network infrastructure through enabling peers to flexibly change their role. Several regions are created for the P2P network. Each group selects (according to an algorithm) a super peer that represents the group head, while the rest of the peers become ordinary peers. From this point, ordinary peer's messages (hello, advertise and query) are sent to the respective super peer. Super peers collect and index the group information including active peers, advertised resources and offered services in order to manage the group communications and resolve queries addressed to the group. The super peer is also responsible of maintaining the group state including selecting new super peers, in case its context changes or moves away. MP2P networks have often been considered as the "new-comer" of the computer communication family, while the client/server file-share scheme has been the "incumbent." In MP2P nodes span a wide range of mobile device form factors and wireless sensors with different features and variety hardware and software stacks. Some of these peers are high-end mobile devices such as laptops, tablets, smartphones, and vehicles equipped with processing power and network connectivity. Other peers have limited processing power, battery life, storage, and network capabilities, such as featured phones, low-end wireless sensors, cameras, etc. We assume that peers form P2P groups based on location proximity, within a few hops from one another, regardless of their interests and resources they may share. Peers are on the move and their context changes dynamically. This technology will enable a next generation of innovative killer applications, collaboration technologies will promote the creation of worldwide user communities and will permit low cost sophisticated content publishing and distribution tools (net radio, net tv)

MP2P should provide a continuously connection – disconnection and direct communication with other mobile nodes at broadband speeds based on their physical proximity, in a low level overhead traffic. The selection of super peers is the most critical point. Randomly selection of super peers is not a solution so, in order to select super peers we have to take in consideration many parameters, like the distance cost, processing power, content similarity and high capacity. Periodically peers build super peer candidates which are responsible for answering the queries that come from simple peers. If a super peer cannot answer a simple's peer query (file request), it communicates with other super peers in order to satisfy simple's peer query. Finally it is critical for nodes to have the ability to change easily their role from simple peer to super peer and vice versa.

III. PROPOSED TRAFFIC BASED S-MAC PROTOCOL (TS-MAC)

System Model and Mechanisms of S-MAC

Power control in wireless devices aims to determine the transmit periods and the associated power level, such that the energy consumed is steadily reduced, whereas at the same time it aims to guarantee the resource sharing stability. Since nodes in wireless networks typically rely on their battery energy, this work, proposes a mechanism which hosts a traffic-aware scheme for conserving energy in wireless environments using a Clustered-based mechanism. This mechanism is encapsulated in a middleware as in (Mavromoustakis, Dimitriou, Mastorakis & Pallis, 2013) and encompasses a scheme, which evaluates the scheduled activity periods of each node, in order to measure the scheduled time that each node can safely sleep. The proposed framework uses a quantitative model based on the supervision of the Sleep-Proxy (a role-based node) which actively, in a clustered-environment, takes into consideration the cooperative caching and the resource exchange scheme for enabling Energy Conservation in the Cluster.

Nodes in wireless networks typically rely their survivability on their battery energy. Energy consumption should be minimized in order to save power in a decentralized form. The effects on reducing each node's power consumption has been studied thoroughly (Mavromoustakis, Dimitriou, Mastorakis & Pallis, 2013). However an association of the capacity and caching characteristics and measures with respect to efficient message passing for self-tuning energy consumption is explored in this work. The energy conservation mechanism has to be closely collaborative with the upper layer protocols used, to maintain the packet forwarding mechanism, in an error free mode. A best candidate solution must be cost-effective and adaptive, and should be able to keep abreast of any possible changes in the network, in terms of load and failures. The above issues combined with the unpredictable movements of the users, and with each device's technical characteristics (like link asymmetry) should be balanced. This work proposes a way to avoid any technical discrepancies that may exist between different devices, while it balances energy consumption on each device or sudden energy deficiency. In comparison with the CSMA/CA protocol, S-MAC is more energy efficient as it conserves more energy due to the use of duty cycles. Although this is true for low load networks, it is not as efficient with heavy load networks, where sleeping schedules are not applied between nodes due to the constant transmission of data. S-MAC synchronization mechanism is designed in a way that forms virtual clusters in Ad-Hoc networks and utilizes energy conservation, by avoiding overhearing and transmitting packets and long messages in an efficient manner (Mavromoustakis & Karatza, 2004). Although S-MAC synchronization mechanism utilizes a sleep-wake schedule in an effort to conserve energy, it does not provide flexibility when it comes to bursty traffic because of the fixed duty cycles operation and fixed sleep periods of nodes. In a low load network, while each node sleeps through a specific period of time, it has to wake up at the beginning of each frame, in order to receive the SYNC packet, which will determine where and what to send on its next wake period. An important and constant amount of energy is consumed in this procedure because of this fixed duty cycle structure. Nonetheless, if faced with a heavy load network, the listen period may prove to be too short, in order to accommodate traffic, leading to unnecessary delays between packet transmissions amongst Sender and Destination nodes. The sleep time duration can be measured using the following expression (Mavromoustakis & Karatza, 2004):

$$S(t_{new})_D = S(t)_{D(\tau-1)} - S(t)_{D(\tau-1)} \cdot \left[\frac{Packets\ Cached}{Total\ Packets\ destined\ for\ D} \right]$$

While S-MAC uses virtual clusters where all nodes sleep at the same time and transmit data during the listen period in an effort to conserve energy, it actually trades off latency. Although idle listening duration is reduced, it also increases the end-to-end latency amongst nodes and data packets. This makes S-MAC unreliable when it comes to delay-sensitive data. Furthermore, increased collision of data packets has been noticed in heavy load networks, constituting S-MAC protocol as inappropriate for this type of networks, where heavy congestion and high loads are present. In order to enhance the already existing S-MAC protocol for a Wireless network, to save energy and transmit data packets, the scheduling mechanism had to be modified and applied in various topologies, with data packets of variable size. The results of these modifications are shown in part IV, of the experiments. The scheduling mechanism is based on preset duty cycle of each node, using the S-MAC protocol. By changing the duty cycle of each node, the purpose of modifying its sleep and wake state, either by increasing their sleep state or decreasing, is achieved, thus providing results based on different types of topologies that these modifications were tested. The main attempt is to achieve greater and more productive results in low and high density networks, by increasing or decreasing the sleep wake states, in order to provide an enhanced version of the S-MAC protocol.

IV. PERFORMANCE EVALUATION ANALYSIS, EXPERIMENTAL RESULTS, AND DISCUSSION

In this section, four different scenarios are examined, were the duty cycle of all nodes was modified in order to extract results, in comparison with the proposed TS-MAC

Figure 1. Three nodes in random order topology with 512bytes data packets (Scenario 1)

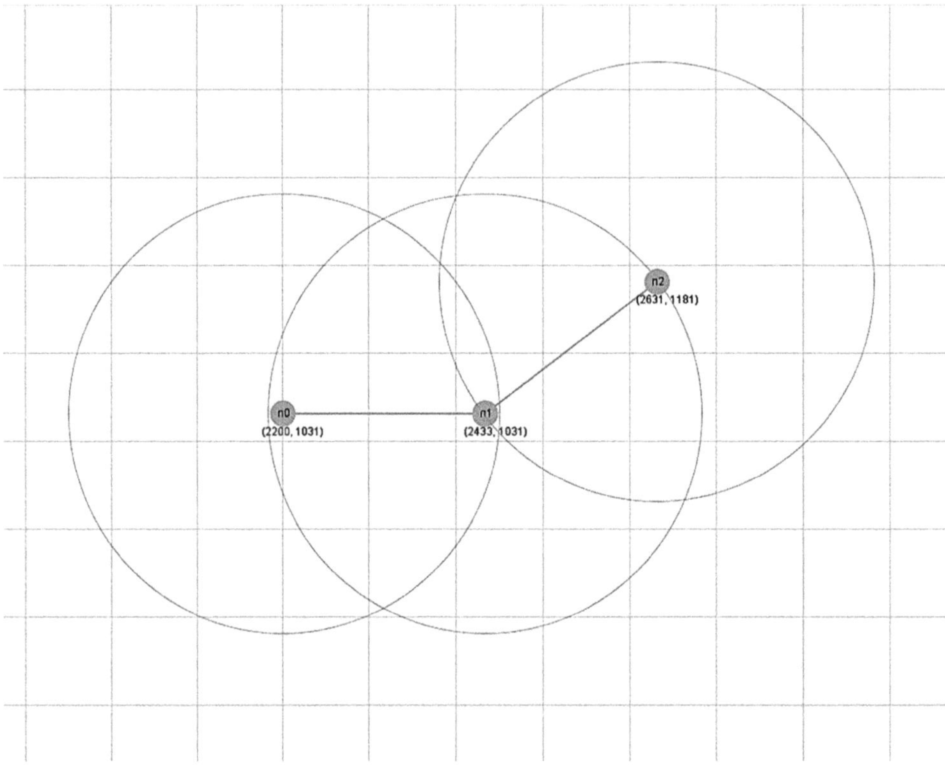

protocol. For this implementation Network Simulator 2 (NS2) was used, a tool that helps simulating different network scenarios and providing the flexibility of selecting protocols, node quantities, topologies and much more. The first scenario consists of 3 nodes in a random order, sending 512bytes data packets as shown in Figure 1.

As seen through all Scenarios, the transmission range of each node is visible, as it covers the neighboring node were the transmission of data packets is feasible. In the second scenario (Figure 2), 5 nodes send 256bytes data packets, again in a random order.

For the third scenario, 7 nodes have been placed in a chain order, sending 512bytes of data packets as shown in Figure 3.

For the last scenario (Figure 4), 16 nodes have been used in a grid order, sending 512bytes of data packets.

For a more detailed description of the scenarios discussed above, Table 1 provides all the necessary details accordingly.

Figure 2. Five nodes in random order topology with 256bytes data packets (Scenario 2)

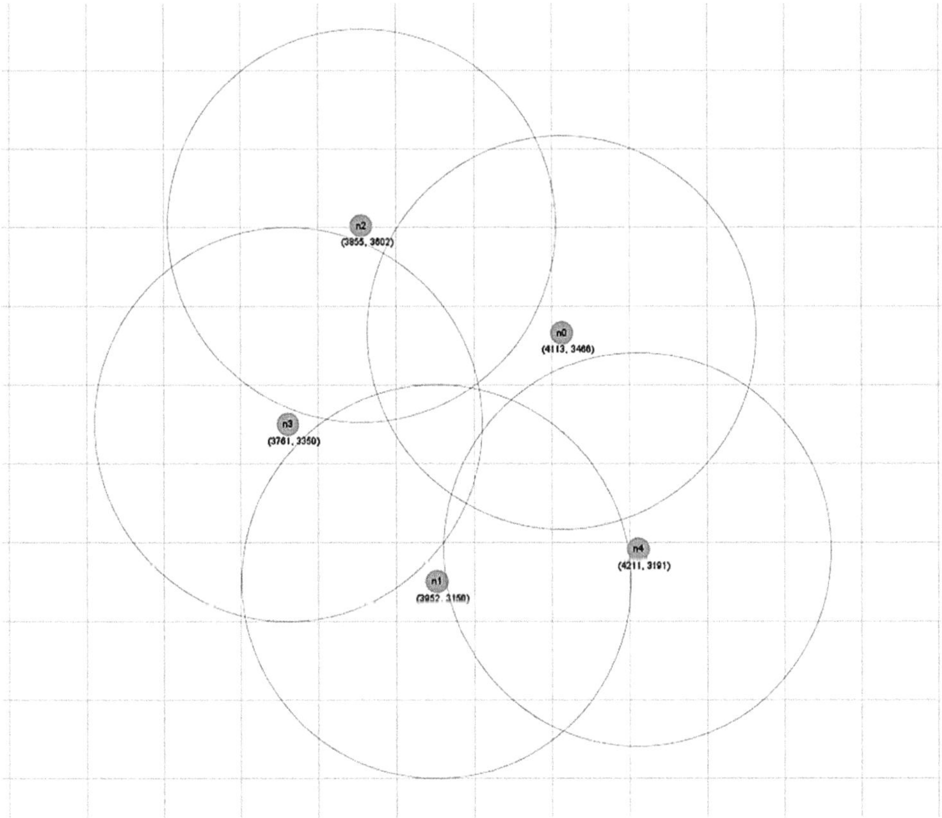

Figure 3. Seven nodes in chain order topology with 512bytes data packets (Scenario 3)

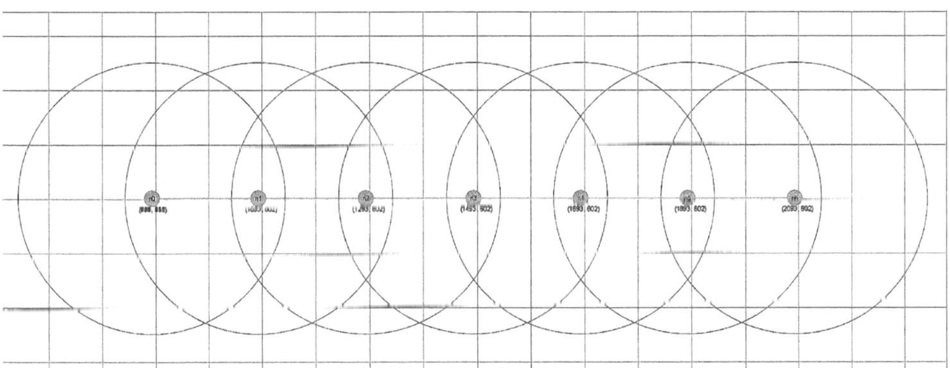

Figure 4. Sixteen nodes in grid order topology with 512bytes data packets (Scenario 4)

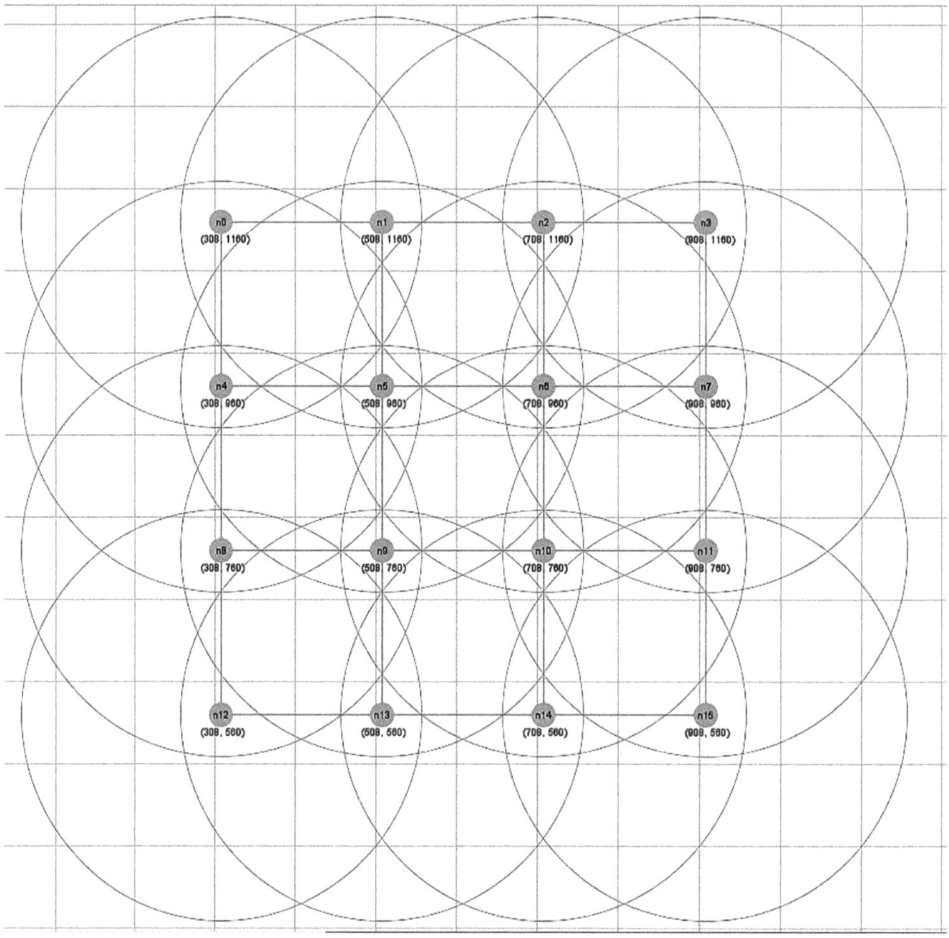

In each scenario, various changes of the Duty Cycle have been implemented ranging from 0% to 100%, incrementing by 10% at a time. For example, a script with 10% duty cycle was compiled, another one with 20%, 30% and so forth. By applying these modifications to the scenarios, one at a time, 40 different samples for the TS-MAC protocol have been created, and another 4 samples using the standard S-MAC protocol for comparison reasons, from which results of the graphs were extracted and analyzed regarding the following:

1. Average Throughput.
2. Average Instant Jitter.
3. Average End-to-End Delay between Sent and Received Packets in seconds.

4. Average End-to-End Delay between Sent and Dropped Packets in seconds.
5. Packet Delivery Ratio for Sent, Received and Dropped packets.
6. Received over Sent Packets Ratio.
7. Energy Consumption (Joules).

More specifically, the Throughput has been obtained below over specific Duty Cycles, through all 4 scenarios. The results have compared between the SMAC protocol and the proposed TS-MAC Protocol. It should be noted that based on the following findings, the differences between the standard S-MAC protocol that has a fixed duty cycle operation, are examined and all the variations of the duty cycles of the TS-MAC protocol are investigated, in order to distinguish which duty cycle of the TS-MAC is most effective, thus using it as the default duty cycle of the proposed TS-MAC protocol. In scenario 1, the Average Throughput (measured in bytes) extracted by the TS-MAC protocol is much higher than the Average Throughput of the S-MAC protocol, as shown in Figure 5. When the Duty cycle of the TS-MAC protocol is modified to 10 duty cycles per sleep state, a significant decrease is observed regarding the Average throughput of Packets. Nonetheless, when the duty cycle is modified to TS-MAC/20DC, there is a significant improvement, much higher than the extracted Throughput of packets by the S-MAC protocol. This improvement is noticed between TS-MAC/20DC, TS-MAC/50DC, as well as TS-MAC/90DC per sleep state.

In scenario 2 (Figure 6) of the Figure below, the same principles apply, but in this case only the TS-MAC/100DC exceeds the Average Throughput of the S-MAC protocol. It can be observed that there is a greater number of nodes (5 in comparison with 3 in Scenario 1), yet in a Random order as before. Nonetheless, the Data

Table 1. Detailed explanation of scenarios

Scenario	1	2	3	4
Nodes (Number)	3	5	7	16
Topology	800 x 800	16321 x 100	14306 x 932	3295 x 959
Order of Nodes	Random	Random	Chain	Grid
Data Packets (Bytes per packet)	512bytes	265bytes	512bytes	512bytes
Simulation Time (seconds)	35000sec	35000sec	35000sec	35000sec
Maximum Packets (quantity)	50000	50000	50000	50000

Figure 5. Average Throughput (Scenario 1)

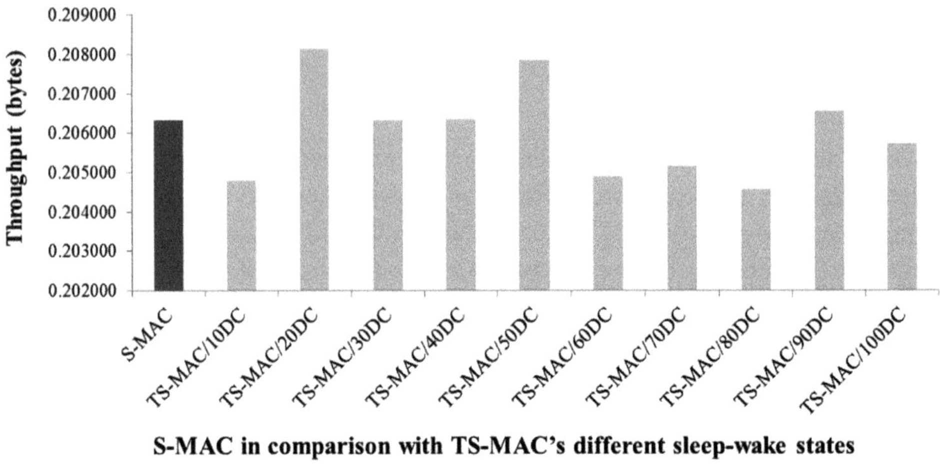

Packets size has been significantly decreased for this Scenario and the possibility of the number of the nodes affecting the Average Throughput, should be taken under consideration, as through this experiment there is a noticeable difference regarding the Average Throughput between the two protocols.

In scenario 3, a different topology has been implemented. In this case, the nodes are 7, placed in a Chain order as seen in the previous page, with 512 bytes of Data Packets and the Simulation lasted for 35000 seconds. In this Scenario, there is no significant change to the Average Throughput regarding the TS-MAC, as the S-MAC exceeds its rival regarding the Average Throughput as seen in Figure 7. Nonetheless,

Figure 6. Average Throughput (Scenario 2)

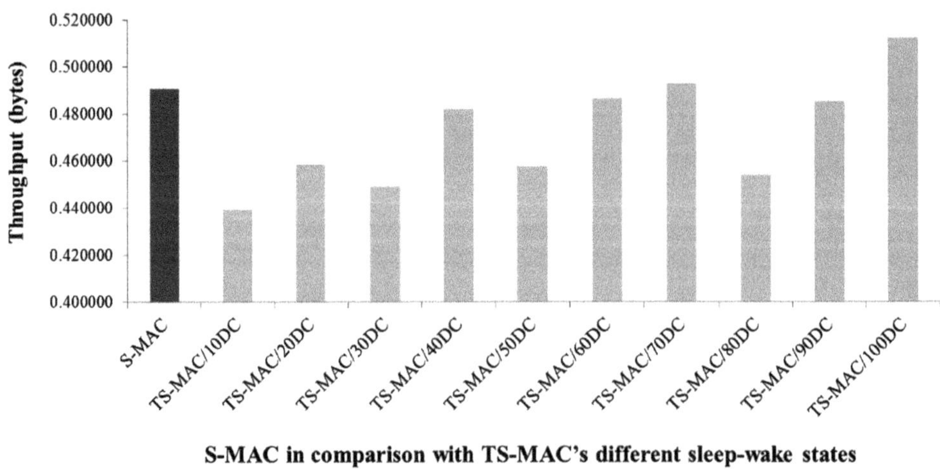

this is not a drawback, as there are also other factors that need to be taken under consideration, in order to evaluate and improve a proposed network transfer protocol, as discussed further on.

In scenario 4, in Figure 8, 16 nodes are introduced in the topology, in a Grid order with the same specifications as used in Scenario 1 and 3, with different dimensions of the topology. By taking a closer look, and studying the extracted results, the TS-MAC exceeds the S-MAC in many aspects. Also the duty cycles that do not exceed the S-MAC protocol are decreased from 10 in Scenario 3, to 6 in Scenario 4, which is considered as a 40% increased throughput. It is of significant value, that it is kept in mind the purpose of these Scenarios, which is to extract the most efficient duty cycle of TS-MAC, in comparison with S-MAC.

Throughout these, the conclusion is made that scenario 4 has the best application of the proposed TS-MAC. In comparison with the other three scenarios, TS-MAC has achieved greater throughput than S-MAC, throughout different sleep-wake states. In scenario 4, the vast majority of these different duty cycles of TS-MAC have produced greater results, than any other scenarios. Nonetheless, scenario 4 will not only be used in the rest of the experiments, but also other scenarios will be included as well, in order to show that positive results are extracted throughout any type of topology or number of nodes. TS-MAC/70DC exploits the default duty cycle scheduling mechanism for TS-MAC, in order to experiment with the compiled scenarios and use the required findings for the further development of the TS-MAC. In addition, the Average Instant Jitter between TS-MAC and S-MAC protocols has been extracted. Different Scenarios will be examined for the rest of the experiments, in order to notice the differences between the applied protocols throughout the 4 scenarios.

Figure 7. Average Throughput (Scenario 3)

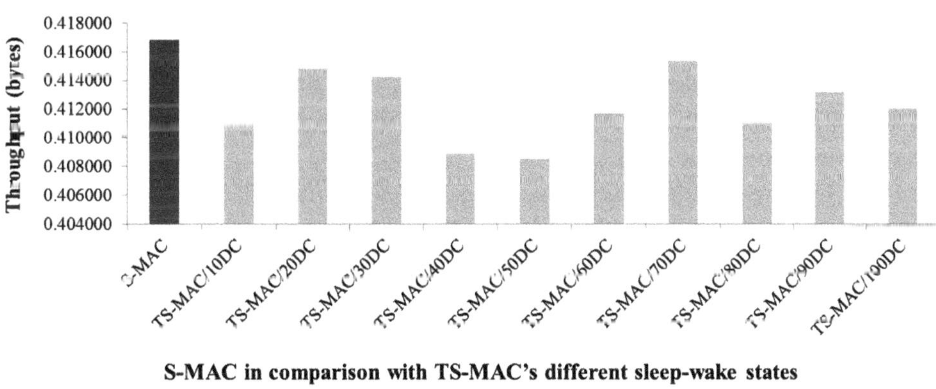

Figure 8. Average Throughput (Scenario 4)

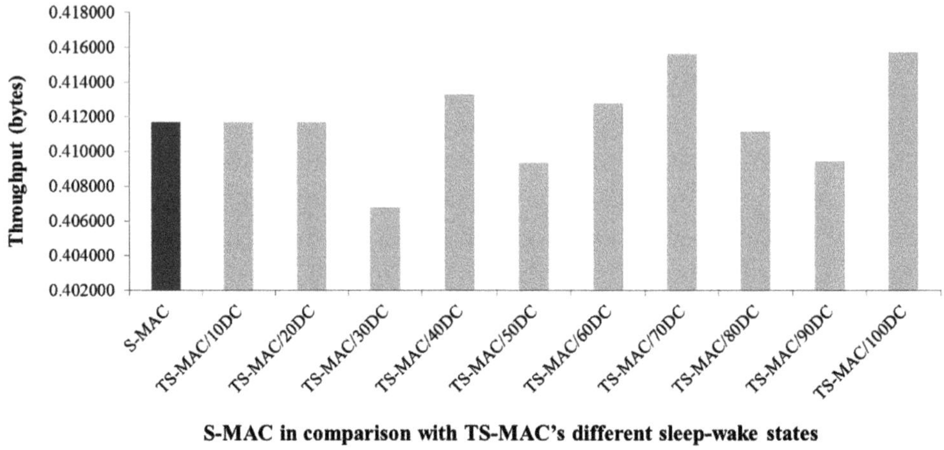

In scenario 4 in Figure 9, it is clearly shown than depending on the time, the Average Instant Jitter (measured in bytes) of the S-MAC, mostly exceeds the one of the TS-MAC. In this case, it is not desirable for the instant jitter of TS-MAC to exceed S-MAC as this plays a determining role in the energy consumption of the nodes. There are also, other duty cycles that produce less jitter than the S-MAC protocol, thus providing less energy consumption for the nodes. In this specific scenario, the 30, 40, 50, 70 (used as the default duty cycle), 90 and 100DC produce the desirable outcome. Furthermore, the end to end delay between the nodes has been examined, as well as the Average Packet Duration (measured in seconds) between the Sent and Received Packets (Figure 10) has been calculated. It is quite interesting that all 4 scenarios produced the same results. There is no specific delay between the Sent and Received packets between nodes, both in TS-MAC and S-MAC protocols.

In addition, the end to end delay between the Sent and the Dropped packets has been examined as well. The outcome was defined between the Default Duty Cycles of S-MAC and the TS-MAC protocol, as there was a difference of 0,11 seconds for the Average Packet Duration between the Sent and Dropped Packets. Nonetheless, this is also applicable for all 4 Scenarios, thus a very important factor is signified, in order to improve the proposed TS-MAC protocol: by taking under consideration the effectiveness of the S-MAC protocol regarding the end to end delay between Sent and Dropped Packets, implementing and modifying accordingly in order to provide greater effectiveness. Further on, the Packet Delivery Ratio (measured in packets) for all Sent, Received and Dropped Packets has been examined. Below

Figure 9. Average Instant Jitter (Scenario 4)

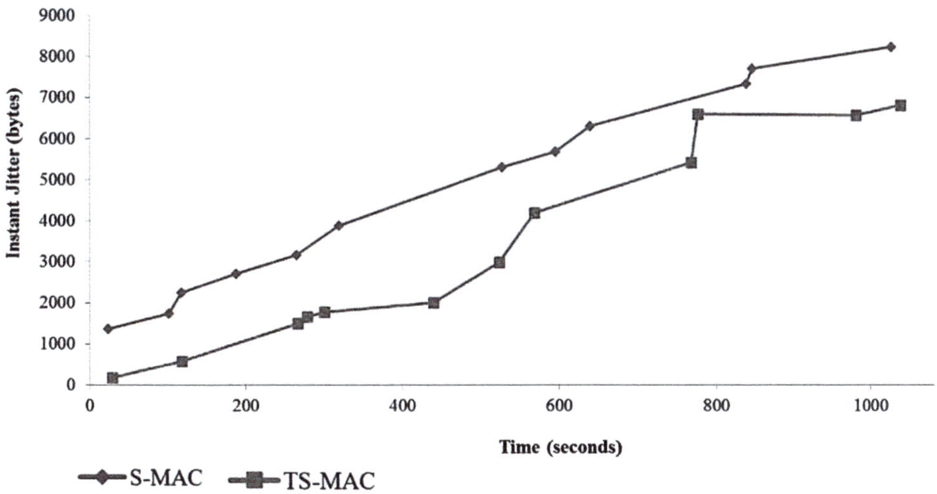

there are some extensive results for scenario 2, portraying the number of packets Sent (Figure 11), Received (Figure 12) and Dropped (Figure 13) between S-MAC and TS-MAC protocol.

In Figure 11, the TS-MAC was captured sending a larger number of packets in comparison with S-MAC, within the same time frame. As the sent packets were increased in order to extract the results for the modified proposed protocol, a vast difference between the numbers of packets sent between the two protocols is visible.

Figure 10. End to End delay_Average Packet Duration_Sent-Dropped (in seconds) (Scenario 2)

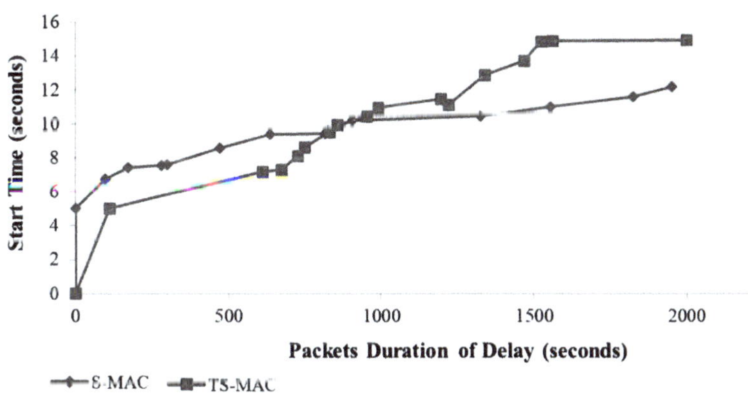

Figure 11. Packet delivery ratio for Sent Packets (in packets) (Scenario 2)

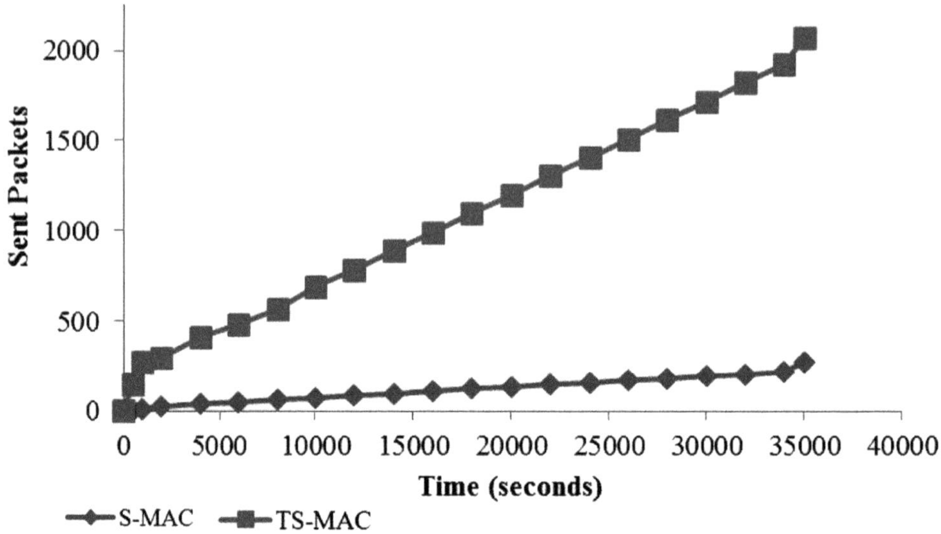

The received packets increased (Figure 12) in an orderly manner as time progressed. Despite the increased number of Sent packages of TS-MAC, it should be mentioned that the Default S-MAC protocol received more packets than its rival, in comparison with the sent and the received packets.

Figure 12. Packet delivery ratio for Received Packets (in packets) (Scenario 2)

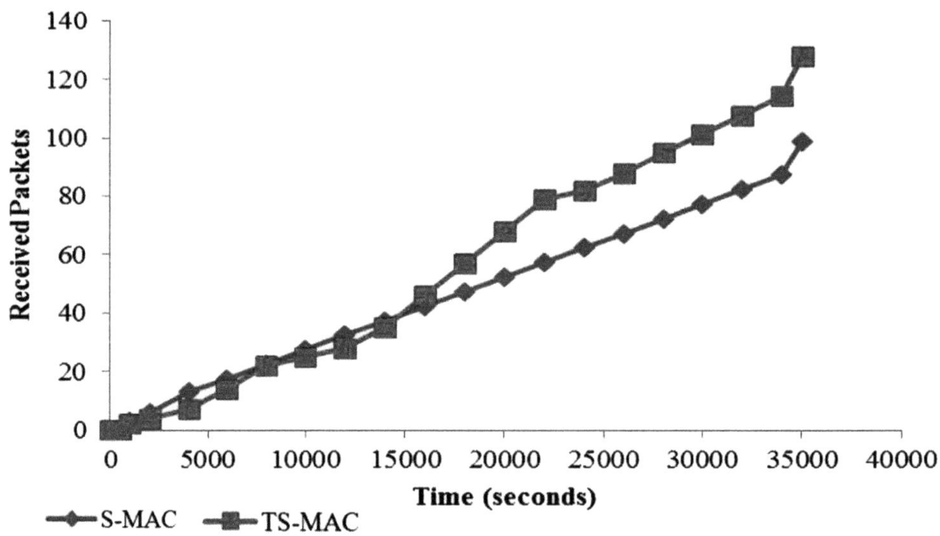

Figure 13. Packet delivery ratio for Dropped Packets (in packets) (Scenario 2)

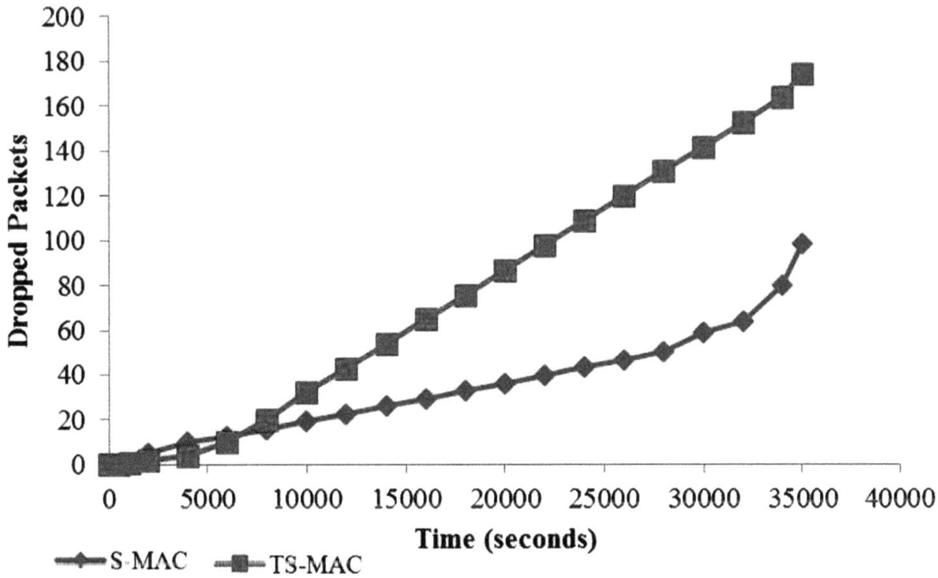

While the time frame progressed, so did the dropped packets of TS-MAC (Figure 13), thus surpassing the received packets. This is a very important pointer as the potential of corrupted packets between nodes makes the protocol not as efficient and productive as it is intended to.

Moreover, in Figure 14, the Received over the Sent Packets Ratio (measured in packets) for the Packet Delivery Ratio (Average) is examined. Once again, it is very important not to overlook the differences between the TS-MAC and the S-MAC protocols. The TS-MAC protocol surpasses the S-MAC protocol regarding the Received/Sent Ratio as it exceeds the values in all attempts to apply the proposed TS-MAC.

The final experiment, for testing and extraction was implemented for the Energy Consumption for both protocols. In Figure 15, scenario 4, the TS-MAC protocol consumed less energy, than the S-MAC, which shows that the nodes lasted a lot longer throughout the experiment. The TS-MAC utilized the nodes energy, 2.5 times more than the S-MAC, which implies that power consumption is handled in a better manner by the proposed protocol.

Based on the graphs and the extracted results, the modification that has been applied to all the scenarios, was alternating throughout the experiments. There were variations, according to the scheme of each scenario. For instance, scenario 2 had the highest throughput than any other, in comparison for the 100DC, but in sce-

Figure 14. Packet delivery ratio: Received over Sent Packets Ratio (Scenario 2)

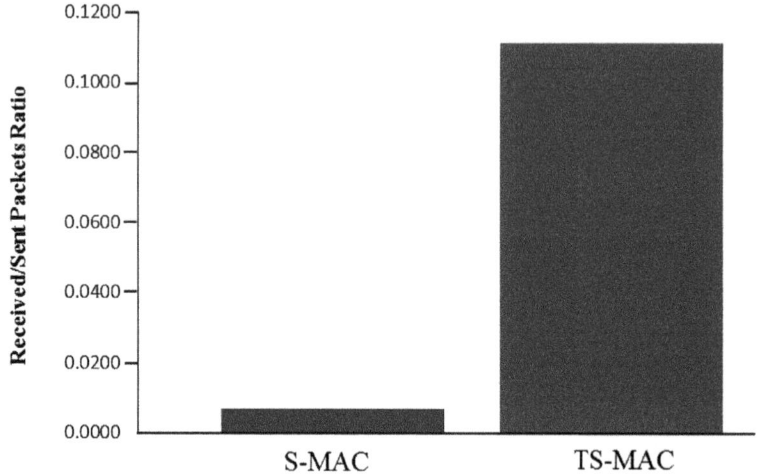

nario 3 and 1, it had an in between value, in comparison with the other examples. Also, the overall throughput for each scenario depicts which of the duty cycles were proven most efficient. Regardless, it should be noted that an important number of advantages is applicable, when it comes to the TS-MAC protocol, as well as some important thought-provoking matters, such as transmission of packets that will help,

Figure 15. Energy consumption in Joules (Scenario 4)

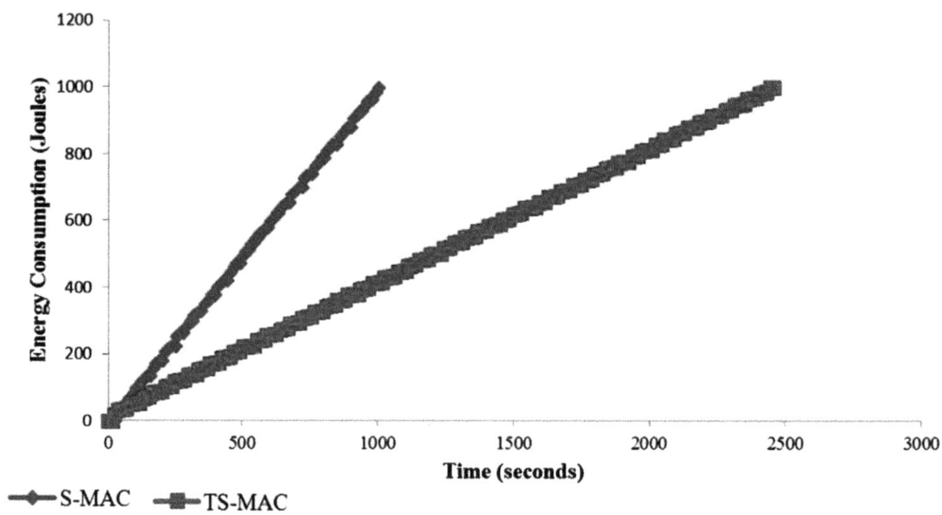

in improving a protocol, which is both efficient and productive for its use. Overall, the TS-MAC has been proven to work best for large data packets transmission, as well as in larger topologies with a vast number of nodes, handling energy consumption in a more efficient way than the rival protocol in comparison (scenario 4).

V. CONCLUSION

This work proposes a reflective scheduling mechanism, by taking into consideration the differentiation of the wake sleep duty cycle of each node. Through the modification of the duty cycles, it has been derived that it can produce both positive and negative outcomes, as some factors can differentiate the results, such as topology, density, number of nodes, data packets size and the duty cycle of each node. If this mechanism is developed in more depth, it will produce good results, by minimizing even more the energy consumption and increasing the throughput, decreasing dropped packets, by modifying the sleep wake state of each node. The development of the protocol must focus on the factors mentioned above, such as the quantity of the sent and the dropped packages, as well as the end to end delay between them. If these important factors, mentioned throughout this research, are taken under serious consideration, the development and implementation of the TS-MAC may prove to be a success, producing desired results and increase the efficiency and the energy of the nodes, thus making an Ad-hoc Network effective towards the current principles of Network Performance and Design. In this direction, future work includes the expansion of the utilization of the traffic to be mapped onto traffic patterns. These patterns will include traffic engineering models (Bourdena et al., 2014), in order to map and figure out the behavior of such dynamically changing scenarios (Dimitriou, Mavromoustakis, Mastorakis & Pallis, 2013). Furthermore, this work is planned to be expanded into a framework (Mavromoustakis, Pallis & Mastorakis, 2014) that uses the combined infinitesimal perturbation analysis so that stochastic algorithms may depict the performance gradient of the system. The above issues comprise of open-end research dimensions with many research concepts for future examination (Mavromoustakis, Mastorakis, Bourdena & Pallis, 2014).

ACKNOWLEDGMENT

The work presented in this chapter is co-funded by the European Union, Eurostars Programme, under the project 8111, DELTA "Network-Aware Delivery Clouds for User Centric Media Events".

REFERENCES

Andreou, A., Mavromoustakis, C. X., Mastorakis, G., Bourdena, A., & Pallis, E. (2014). Adaptive Heuristic-Based P2P Network Connectivity and Configuration for Resource Availability. In Resource Management in Mobile Computing Environments (pp. 221-240). Springer International Publishing.

Bourdena, A., Mavromoustakis, C. X., Kormentzas, G., Pallis, E., Mastorakis, G., & Yassein, M. B. (2014). A resource intensive traffic-aware scheme using energy-aware routing in cognitive radio networks. *Future Generation Computer Systems, 39*, 16–28. doi:10.1016/j.future.2014.02.013

Bourdena, A., Pallis, E., Kormentzas, G., & Mastorakis, G. (2013). Efficient radio resource management algorithms in opportunistic cognitive radio networks. *Transactions on Emerging Telecommunications Technologies.*

Buettner, M., Yee, G. V., Anderson, E., & Han, R. (2006). X-MAC: a short preamble MAC protocol for duty-cycled wireless sensor networks. *4th International conference on Embedded networked sensor systems (SenSys '06).* ACM.

Chakrabarti, S., & Mishra, A. (2001). QoS issues in Ad Hoc wireless networks. *IEEE Communications Magazine, 39*(2), 142–148. doi:10.1109/35.900643

Dimitriou, C. D., Mavromoustakis, C. X., Mastorakis, G., & Pallis, E. (2013, June). On the performance response of delay-bounded energy-aware bandwidth allocation scheme in wireless networks. In *2013 IEEE International Conference on Communications (ICC 2013), Workshops*, (pp. 631-636). doi:10.1109/ICCW.2013.6649310

El-Hoiydi, A., Decotignic, J. D., Enz, C., & Le Roux, E. (2003). Poster abstract: wiseMAC, an ultra-low power MAC protocol for the wiseNET wireless sensor network. *1st International conference on Embedded networked sensor systems (SenSys '03).* ACM.

Haas, Z. J., & Tabrizi, S. (1998). On some challenges and design choices in Ad Hoc communications. In IEEE MILCOM98.

Joshi, S., & Jaiswal, A. K., & Tyagi, P. K. (2013). A Novel Analysis of T Mac and S Mac Protocol for Wireless Sensor Networks Using Castalia. *International Journal of Soft Computing and Engineering, 2*(6).

Karn, P. (1990, September). *MACA-a new channel access method for packet radio.* Paper presented at ARRL/CRRL Amateur Radio 9th Computer Networking Conference.

Kumar, S. (2004). Medium Access Control protocols for Ad Hoc wireless networks. *Survey (London, England)*, 326–329.

Kumar, S., Raghavan, V. S., & Deng, J. (2006). Medium Access Control protocols for Ad Hoc wireless networks. *Survey (London, England)*, *4*(3), 326–358.

Mastorakis, G., Bourdena, A., Mavromoustakis, C. X., Pallis, E., & Kormentzas, G. (2013, July). An energy-efficient routing protocol for ad-hoc cognitive radio networks. In Future Network and Mobile Summit (IEEE FutureNetworkSummit), 2013 (pp. 1-10). IEEE.

Mastorakis, G., Mavromoustakis, C. X., Bourdena, A., Kormentzas, G., & Pallis, E. (2013, September). Maximizing energy conservation in a centralized cognitive radio network architecture. In *2013 IEEE 18th International Workshop on Computer Aided Modeling and Design of Communication Links and Networks (IEEE CAMAD 2013),* (pp. 175-179). doi:10.1109/CAMAD.2013.6708112

Mavromoustakis, C., Andreou, A., Mastorakis, G., Papadakis, S., Bourdena, A., & Stratakis, D. (2014, August). Energy Consumption Optimization through Pre-scheduled Opportunistic Offloading in Wireless Devices. In *EMERGING 2014, The Sixth International Conference on Emerging Network Intelligence* (pp. 22-28).

Mavromoustakis, C., Dimitriou, C. D., & Mastorakis, G. (2013). On the Real-Time Evaluation of Two-Level BTD Scheme for Energy Conservation in the Presence of Delay Sensitive Transmissions and Intermittent Connectivity in Wireless Devices. *International Journal on Advances in Networks and Services, 6*(3-4), 148-162.

Mavromoustakis, C. X., Bourdena, A., Mastorakis, G., Pallis, E., & Kormentzas, G. (2014). An energy-aware scheme for efficient spectrum utilization in a 5G mobile cognitive radio network architecture. *Telecommunication Systems*, 1–13.

Mavromoustakis, C. X., Dimitriou, C., Mastorakis, G., & Pallis, E. (2013). Real-Time Performance Evaluation of F-BTD scheme for optimized QoS Energy Conservation in Wireless Devices. *Proceedings of IEEE GlobeCom, 2013*, 1156–1161.

Mavromoustakis, C. X., & Karatza, H. D. (2004). *Adaptive Energy Conservation Model using Dynamic Caching for Wireless Devices.* IEEE. doi:10.1109/SIMSYM.2004.1299490

Mavromoustakis, C. X., Mastorakis, G., Bourdena, A., & Pallis, E. (2014). *Energy efficient resource sharing using a traffic-oriented routing scheme for cognitive radio networks*. IET Networks.

Mavromoustakis, C. X., Mousicou, P., Papanikolaou, K., Mastorakis, G., Bourdena, A., & Pallis, E. (2015). Dynamic Cloud Resource Migration for Efficient 3D Video Processing in Mobile Computing Environments. In Novel 3D Media Technologies (pp. 119-134). Springer New York.

Mavromoustakis, C. X., Pallis, E., & Mastorakis, G. (2014). *Resource Management in Mobile Computing Environments. Modeling and Optimization in Science and Technologies*. Springer International Publishing.

Mousicou, P., Mavromoustakis, C. X., Bourdena, A., Mastorakis, G., & Pallis, E. (2013, November). Performance evaluation of dynamic cloud resource migration based on temporal and capacity-aware policy for efficient resource sharing. In *Proceedings of the 2nd ACM workshop on High performance mobile opportunistic systems* (pp. 59-66). doi:10.1145/2507908.2507917

Polastre, J., Hill, J., & Culler, D. (2004). Versatile low power media access for wireless sensor networks. *2nd international conference on Embedded networked sensor systems (SenSys '04)*. ACM.

Royer, E. M., & Toh, C. K. (1999). A review of current routing protocols for Ad Hoc mobile wireless networks. *IEEE Personal Commun.*, *6*(2), 46–55. doi:10.1109/98.760423

Singh, S., & Raghavendra, C. S. (1998). PAMAS-power aware multi-access protocol with signalling for Ad Hoc Networks. *SIGCOMM Comput. Commun. Rev.*, *28*(3), 5–26. doi:10.1145/293927.293928

Tanenbaum, A. S. (2003). *Computer Networks* (4th ed.). Pearson Education Inc.

Toh, C.-K. (2002). *Ad Hoc Mobile Wireless Networks: Protocols and Systems*. Prentice-Hall PTR.

Webopedia. (n.d.). *The 7 Layers of the OSI Model*. Retrieved September 24, 2010, from http://www.webopedia.com/quick_ref/OSI_Layers.asp

Wikipedia. (n.d.). *Network Layer*. Retrieved February 18, 2014, from http://en.wikipedia.org/wiki/Network_layer

Yadav, R., Varma S., & Malaviya, N. (2009). A survey of MAC protocols for Wireless Sensor Networks. *UbiCC Journal, 4*(3).

Yang, S., Tseng, H., Wu, E. H.-K., & Chen, G. (2005). Utilization based duty cycle tuning MAC protocol for wireless sensor networks. *Global Telecommunications Conference*, (vol. 6, pp. 32-62).

Ye, W. (2002). *SCADDS: Scalable Coordination Architectures for Deeply Distributed Systems*. Retrieved July 1, 2014, from http://www.isi.edu/scadds/projects/smac/

KEY TERMS AND DEFINITIONS

Ad Hoc Network: A wireless Ad Hoc network (WANET) is a decentralized type of wireless network. The network is Ad Hoc because it does not rely on a pre-existing infrastructure, such as routers in wired networks or access points in managed (infrastructure) wireless networks.

Energy Conservation: It refers to reducing energy consumption through using less of an energy service.

Mobile Cloud Computing: It is the combination of cloud computing, mobile computing and wireless networks to bring rich computational resources to mobile users, network operators, as well as cloud computing providers.

Mobile Computing: It is the human–computer interaction, by which a computer is expected to be transported during normal usage.

Network Management: Network management is the operation, administration, maintenance, and provisioning of networked systems.

Network Performance: It refers to measures of service quality of a telecommunications product as seen by the customer.

Peer-to-Peer: Peer-to-peer (P2P) computing or networking is a distributed application architecture that partitions tasks or work loads between peers.

Chapter 9
A P2P Home-Box Overlay for Efficient Content Distribution

Evangelos Markakis
Technological Educational Institute of Crete, Greece

Daniel Negru
University of Bordeaux I, France

Joachim Bruneau-Queyreix
University of Bordeaux I, France

Evangelos Pallis
Technological Educational Institute of Crete, Greece

George Mastorakis
Technological Educational Institute of Crete, Greece

Constandinos X. Mavromoustakis
University of Nicosia, Cyprus

ABSTRACT

The overlay networks composed of residential gateways (i.e. home-box) leverage their storage and upload capacity to achieve scalable and cost-efficient content distribution. In this chapter, we present the architecture of the home-box overlay for video on demand services, with the network-aware request redirection and content caching strategy that optimizes the resource usage at both network and client side, for reducing the overall distribution cost. The proposed system is compared with existing solutions through comprehensive simulations. The results demonstrate the advantage of network-aware and popularity-based caching strategy, with reduced the overall cost of the VoD services.

DOI: 10.4018/978-1-4666-9941-0.ch009

Copyright ©2016, IGI Global. Copying or distributing in print or electronic forms without written permission of IGI Global is prohibited.

INTRODUCTION

The demand on network resources is growing every day driven by the needs of end user. This increase most times is not followed by the necessary upgrade of core networks capacity due to the economic overhead needed for growing core networks. A lot of technologies and architecture are promising to manipulate in a way the issue of limited resources in the core network such as Content Delivery Networks (CDNs) (Pathan, et al., 2008), Akamai (Nygren et al., 2010), Proxy Servers and others. All of them are using a single intermediate point nearest to the End User in order to achieve a better performance in the last mile connection. These solutions are not taking into account that when a large amount of End-Users located under the same geographic area or network are requesting the content simultaneously, the response of all this technologies will be to provide the nearest to the end user proxy server or in the case of CDNs and Akamai to provide the same CDN server located nearest to the End-Users resulting to content bottleneck (Pathan, et al., 2007). Based on this, a solution that will take into account this redundant issue is needed. A promising solution trying to maximize performance and overcome the inconsistencies of the technologies described above is P2P model. P2P model employs efficient bandwidth monitoring (Markakis, et al., 2012) and routing decision (Gupta, et al., 2004) mechanisms at the application layer applied among nodes at the edges of the network and is tasked of transferring the content to users, making better usage of the underlying "best effort resources". In particular, P2P aware network do not directly reduce the overall network load in comparison with the traditional client-server approach, but they distribute more evenly the network load to the whole network. This leads to single links near to the media source being less congested and smaller distribution time since all peers are also redistributors as well as consumers of the data. More specifically the P2P model minimizes the content bottleneck by using two techniques. The first is a locality aware overlay (Zhang et al, 2004) where peers exchange blocks and metadata with the closest in the underlying physical network and the second, as presented and solved in (Kangasharju, et al., 2007) is caching of data in locations that they are requested with high frequency.

In this respect, we introduce the Home-Box (HB) which is the central element in the service distribution chain, aiming to enhance today's home-gateways, by incorporating to the existing network functionalities, user-generated media processing and distribution. A given End-User is capable to locate, adapt and consume services/content in an efficient way, taking into account context-related issues such as device capability. Additionally, the HB provides means for an efficient content sharing among other End-Users and efficient content exchange mechanisms, based in multicast, unicast DASH HTTP streaming or peer-to-peer (P2P). Most of these features and functionalities rely on content, context and network information gathered at several layers by the cross-Layer monitoring system, whose information will feed the adaptation module in cooperation with the Content-Aware Network (CAN). In order to efficiently distribute services and content within network, a logical interconnection of deployed Home-Boxes establishes an overlay network, also known as the "Virtual Home-Box Layer". It empowers the network system with different modes of media distribution and delivery, including traditional Client/Server mode, multicast mode and peer-to-peer mode. Additionally, the HB has local storage capabilities, which allow for content caching and forwarding and the reception of pushed content from the SP, so that the Home-Box virtual overlay can assist, for instance, the current CDNs (HB-assisted CDN). The HB takes advantage of these distribution mechanisms and content caching functionalities to overcome some of the common issues that arise when live content is being distributed to many consumers, when a new high-popularity VoD is ingested into the system, and in those new scenarios in which the End-User plays also the role of producer (distributing the content via P2P). In this way we propose the use of a mixed CDN/P2P/PROXY scenario that provide an all-in-one solution for the main problems stemming from individual use of the solution described above.

RELATED WORK

CDN is seen as the main approach for delivering video content over the Internet in a managed way. Numbers of researches have focused on their optimization, especially addressing three main technical issues: replicas placement (Qiu, et al., 2001) (Yang, et al., 2003) (Tang, et al., 2004) content clustering (Chen, et al., 2003) and client's requests redirection (Cardellini, et al., 2003). These works consider pure CDNs that rely on powerful server replicas with high connectivity and storage capacity. Compared to them, our proposal introduces new equipment in the SPs' video distribution chain: the HB, capable of content caching and streaming, permitting, through the deployment of the overlay virtual layer, to overcome scalability and deployment cost issues of CDNs.

P2P systems represent another promising solution to overcome the scalability issue faced by VoD services. A certain number of researches have been performed leading to the deployment of many P2P VoD systems (Huang, et al., 2008). However, if P2P systems achieve scalability while keeping service cost low, they also come with some limitations such as high peer churn, lack of control, reliability and network unfriendliness that rapidly result in network congestion, especially at the peering edges. There have been several works on making P2P communications more network-friendly. Solutions such as the P4P architecture (Xie, et al., 2008), the oracle based ISP and P2P collaboration (Aggarwal, et al., 2007) and the ISP-driven informed path selection (Saucez, et al., 2008) have been proposed to perform more network-aware routing in P2P overlays. However, the quality of services and the low maintenance cost still challenging problems in P2P configurations. In our solution, HBs, stable components managed by SPs, are considered as peers. Consequently, they are beneficiating from the scalability of the P2P model while overcoming the lack of control, reliability, peer churn, and heterogeneous caching capabilities that characterize it. Enabling a distributed edge content hosting is considered as the next step in content distribution paradigm (Laoutaris, et al., 2008). Recent works have proposed architectures that rely on boxes deployed at the edges of the network, close to Users' terminals, for live video streaming services (He, et al., 2009) or VoD services (Valancius, et al., 2009). However in the latter, video contents are placed offline, which involves an additional delivery cost. In our proposed solution, the videos are placed during their consumptions by End-Users. In addition, an efficient spread of video content copies among the HBs is also considered.

More recently, some studies related to peer-assisted CDNs have been proposed. This hybrid approach permits, on one hand, to offer scalability thanks to P2P approach and, on the other hand, to compensate the bandwidth imbalance and the churn of P2P systems by relying on the CDNs servers. The great advantage of such solutions have been demonstrated analytically (Xu, et al., 2006), by simulation (Tran, et al., 2005), or by large scale deployment (Yin, et al., 2010). Our proposal differs from these works by relying on the User Environment's storage and connectivity without directly involving clients' nodes participation. The HB located in the User's Environment, is the peer node acting as the always-connected "hub" between the User, Network and Service Environments, respectively being capable of shared ownership between the three actors (End-User, Service Provider, Network Provider) (Mastorakis, et al. July 2013), (Mastorakis, et al. sept. 2013), (Mavromoustakis, et al. August 2014). Therefore, the respective actors can easily manage the HB, feature impossible in case of End-User terminals. Our model is then a fully managed model and consequently overcomes the issues induced by the P2P part in such hybrid systems (e.g. free riding, reliability, peer churn issues, etc.) (Mavromoustakis, et al. 2014), (Mavromoustakis, et al. 2013), (Mavromoustakis, et al. December 2013).

HOME-BOX OVERLAY ARCHITECTURE

The *ALICANTE* system (Borcoci, et al., 2010) aims at allowing SPs to supply End-Users with advanced context-aware multimedia services in a consistent and interoperable way. This will be achieved through a so-called "Virtual Home-Box Layer" (HB layer), which enables uniform access for heterogeneous terminals and supports enhanced Quality of Experience (QoE) (Concolato, et al., 2012, Kilkki,

2008). At the HB layer, the advanced user context management and monitoring functions provides real-time information on user context and network conditions, allowing better control over multimedia delivery and intelligent adaptation. Also at the HB layer, the context management, monitoring and adaptation are key features that enable the support of user mobility using a Connectivity Engine in secure manner. The HB layer is defined as the logical interconnection (in P2P or distributed modes) of deployed HBs (through the Connectivity Engine), which can be seen as the evolution of today's Home Gateways (HGs). Thus, the HB is a physical entity aiming on been deployed inside each user's home.

Located at consumer premises, the today's HG devices aim at providing the broadband connectivity to the Digital Home and the services delivery to, and from, an ever-increasing number of end-devices (mobile phones, TV, laptops, etc.) in the User Environment. However, they are not able to (1) adapt multimedia contents to the capabilities of these devices at home; (2) allow secure access to the adapted to the user context multimedia contents, when he is in mobility. These two functionalities are major requirements for improving the user QoE.

Meanwhile, works are in progress to design future Home Gateways with support of additional services (full interactive IPTV services, online gaming facilities, home management functions, etc). Several initiative groups such as HGI (HGI, 2007), OSGi (Alliance, 2010), published the requirements for designing such future Home Gateway devices. These requirements consist mainly of the support of:

- Centralized access point between the home (with support of multiple LAN/ home networking technologies) & outside world relying on WAN-to-LAN interface (with support of multiple WAN termination, e.g., wireless, xDSL or cable),
- Integrated firewall & security features including VPN connectivity, access control and personal monitoring;
- Efficient delivery of services to the home;
- Remote management in order to enable network configuration, device queries, software upgrades, deployment of new multimedia services to the home, etc.;
- IP telephony, audio/video (IPTV) streaming (ETSI, 2006);
- Quality of service for both unmanaged services (e.g., web, P2P) and managed services (e.g., VoIP, IPTV);
- Performance Monitoring and Diagnostics functions that enable the possibility to diagnose the problem and to take remedial actions;
- Modularity in design;
- Easier use and setup.

In this context, this chapter propos the use of a HB device as a next generation HG providing means for dynamic service management/delivery, monitoring and adaptation support. Building on the above requirements, the HB will extend the existing HG architecture with new context-aware (1) middleware layer that will support the above mentioned functionalities, (2) service layer that will use the middleware layer to enable HB services for the user terminals, and (3) user management layer for session instantiations, according to the User Profile and context. Figure 1 illustrates the architecture of the HB, including the entities needed to support the new features brought by this project. This architecture can be split into the following layers:

- *The Hardware layer* consists of well-known units such as CPU, RAM, Hard drive(s) and controllers and set of network interfaces (Ethernet, WiFi, Bluetooth, etc).
- *The Driver layer* is needed to allow the OS kernel and the other higher-level software (middleware, services) to interact with the hardware layer. Indeed, the purpose of drivers is to simplify programming by acting as an abstraction layer between a hardware device and the applications or operating systems that use it;
- *The Operating System (OS) Kernel layer* is responsible for the regular low level tasks, such as process management and memory management. The HB will rely on the Linux kernel;
- *The OS Service layer* aiming at connecting software components, includes all "low-level" services proposed to the high level applications. These services are, among other, networking related (Web servers, DNS, NAT, DHCP, and similar tools) and other multimedia technologies related (e.g., DLNA, UPnP).

Note that some approaches in the literature merge the driver layer and the OS service layer to the OS layer, referred as OS layer.

- *The Middleware layer* will implement the core functionalities of the HB. Thus, it will include entities for (1) resource management, (2) overall connectivity, (3) QoE-aware monitoring, (4) advanced context-aware adaptation, (5) user mobility and (6) security support;
- *The HB Service layer* includes "high-level" Media Services proposed to the users' terminal devices: Service/Content Discovery, Delivery, Publishing, streaming Gateway and enhanced interactive service features. The overall aim is to support the provisioning of Network and context-aware media services to the users;
- *The User Management layer* is the top level in the HB architecture that instantiates user sessions, based on the User Profile and context management information.

Figure 1. The high level architecture of the ALICANTE Home-Box

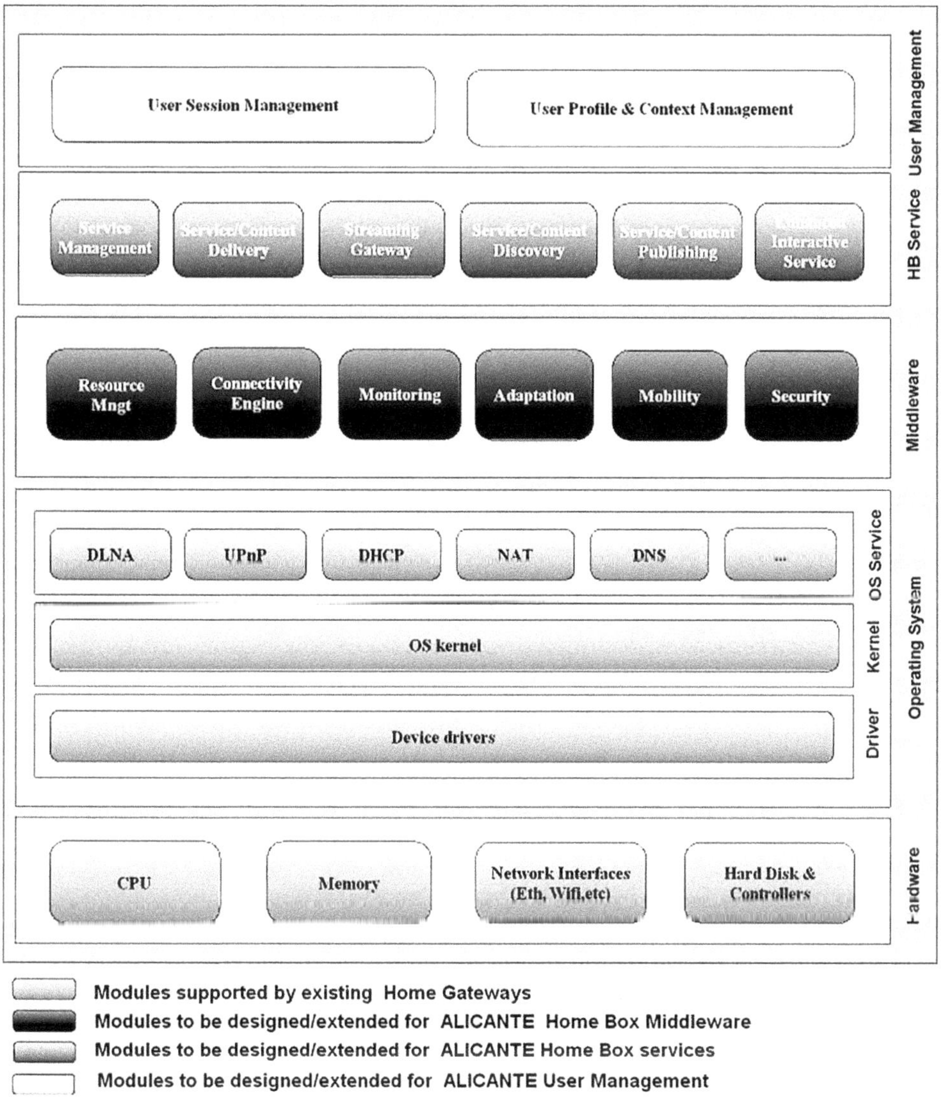

P2P Approach

Peer-to-Peer is considered to be a scalable and cost-effective solution for content distribution and is becoming widely deployed over the current Internet. Some commercial deployments already exists providing most of them VoD services and IPTV to their customers. Even companies such as Skype, which have its P2P approach for distributed VoIP calling online, are supposedly developing a peer-to-peer IPTV service with the intention of distributing video online in the same way. The major reason of this is cost-effectiveness. When compared with CDN-assisted distributed content, or other managed network-based solution, over-the-top P2P is a very attractive solution, as some studies indicate (Chen, et al., 2007). Taking in consideration that the Home-Box entity has built-in storage, stable presence, monitoring capability and management interface, it makes the HB a good candidate for participating in P2P networks. The HB' P2P Engine module is used by the SP to assist the content distribution. Instead of retrieving content from the SP/CP servers solely, the Home-Box can download content from other Home-Boxes, which are caching and forwarding / seeding the content. This approach alleviates the load at the SP/CP servers and simultaneously the costs to maintain and deploy and infrastructure to achieve this goal. The P2P Engine, being part of Connectivity Engine module, provides the necessary functionalities in order to support all aspect of the HB connectivity over the peer-to-peer network. Putting it simple, the P2P Engine is the module that is responsible for content retrieval and seeding on P2P mode distribution mode. P2P-Next team have proposed a new and very promising P2P protocol, called Swift, that has being specified as an Internet-Draft (June 20, 2012) - Peer-to-Peer Streaming Peer Protocol[1].

The Swift protocol is designed to be capable of integration into browsers or operating systems and is able to serve almost all the current Internet traffic. Swift is a multiparty transport protocol, which mission is to disseminate content among a swarm of peers. It might be understood as BitTorrent at the transport layer. Basically, given a Root Hash (the result of a hashing function over the content) it provides the data for it. Ultimately Swift aims at the abstraction of the Internet as a single big data cloud. Most features of the protocol are defined by its function as a content-centric multiparty transport protocol. Entirely drops TCP's abstraction of sequential reliable data stream delivery: for Swift this is redundant. Moreover, it provides a set of features such as: 1) Atomic datagrams, not data stream; 2) Scale-independent unit system; 3) Datagram-level integrity checks; 4) NAT traversal by design; 5) Sub setting of the protocol; 6) Push and pull; 7) No transmission metadata. The Swift protocol, apart for its cousins (Bittorrent, etc.) do not require any associated metadata to work, such as traditional .torrent file. All that is needed is the Root Hash for the file. NextShare Swift library already provides a set of tools that may be very handy during the integration of this platform into the HB. Namely, it provides a Swift module that is able to act as a content seeder, a P2P tracker or both; a Swift-to-HTTP proxy, that may shield the entire P2P network details by providing a simple and fully compliant HTTP interface. The latest is still under experimental development, however.

P2P Swift Integration

As referred above, the Swift library provides means for an easy HTTP integration with other HTTP-enabled modules. This is particularly useful taking in account that the requests will be make using the MPEG-DASH (Sodogar, 2011) standard, which is HTTP-based. From the functional point of view, the modules inside the HB are organized as depicted in Figure 2 in order to achieve the P2P Swift protocol integration. At the control plane, the P2P Engine provides interfaces with Delivery Manager for accepting requests to initialize the P2P Swift-to-HTTP modules, interfaces with the Monitoring Manager, in order to rank the list of available peers, interfaces with the Streaming Gateway and/or Caching module, so the HTTP requests for MPEG-DASH may be properly addressed, with the Resource Manager in order to physically access the local resources to be streamed, and finally an interface with the Publishing Manager so it can request for Root Hashes and to start seeding content back to the P2P network. At the data plane, the P2P Engine interfaces with the P2P network via Swift protocol and with the HTTP client/server, or with the HTTP proxy provided by the Caching Module.

At run time, the instantiation of the different P2P modules, involved in the HB layer establishment using P2P distribution mode, are exemplified in Figure 3.

Figure 2. The main internal blocks of the P2P Engine

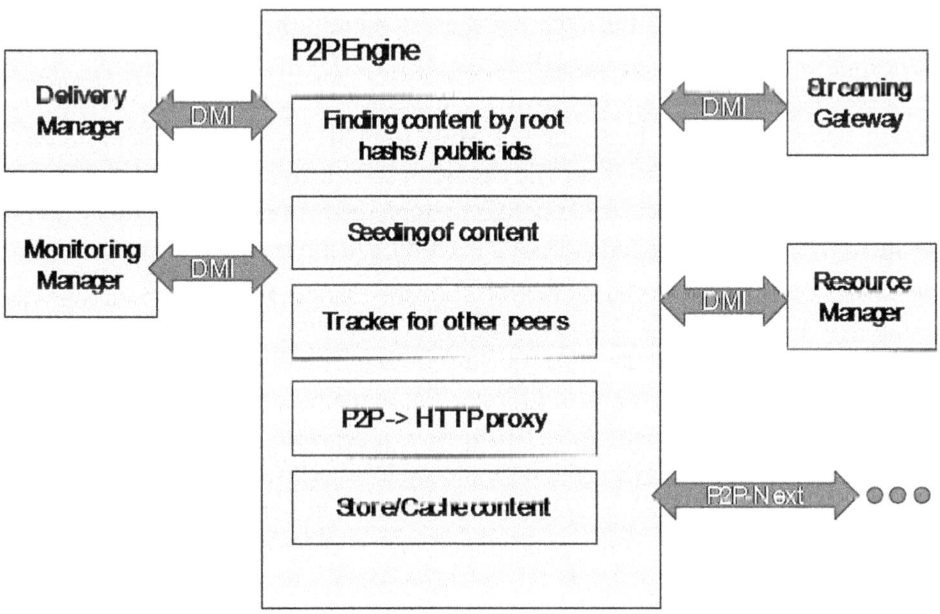

Figure 3. The P2P network instantiation inside the HB

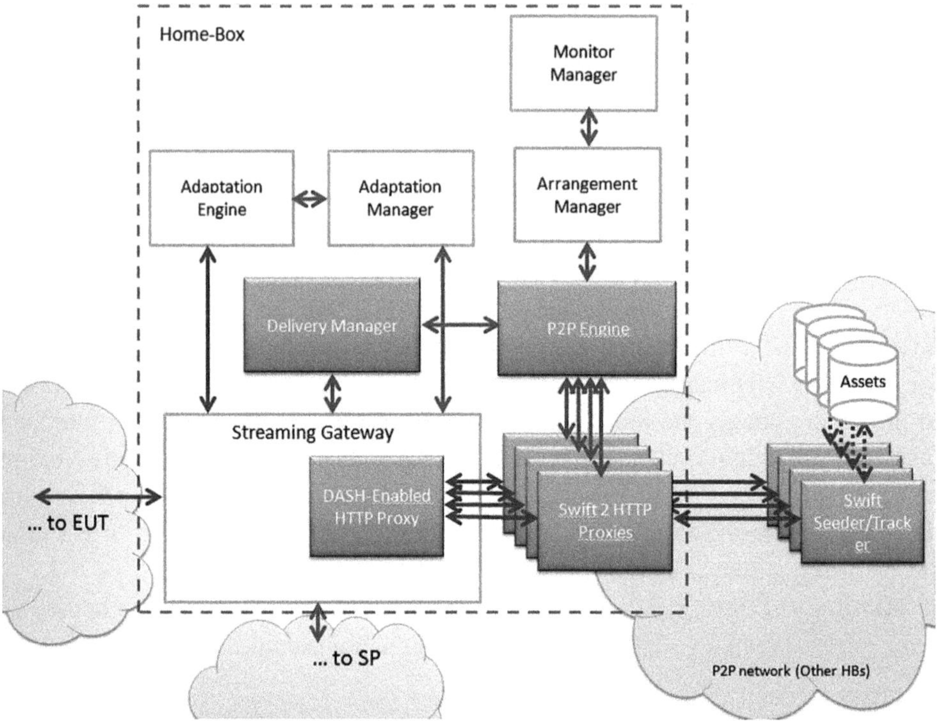

In the figure above the dashed lines delimits the HB boundaries when acting as a P2P consumer node. At the right one can encounter a representation of several seeding nodes, which are in fact HB in the *ALICANTE* integrated solution. At the seeder HB (on the right of the dashed line), there must be, at least, on seeder instance per file being exchanged. In the case of MPEG-DASH segments, the current solution is to provide on instance per file segment, so a different Root Hash will exist per segment. This principle would permit to implement a similar mechanism, as the one used by DASH, to adapt the content according to the network and consumption characteristics. In fact, when a given DASH segment is requested by the consumer HB, it will in fact map to a different file at the seeder and, therefore, to a new Root Hash. If for some reason, the client decides do improve or decrease the quality of the video, then a new request to a different file segment (encoded with other quality parameters) is made and, again, mapped to a separate P2P file, which have a different Root Hash. The HB provides a second level of adaptation, that in practice transcodes the content according to the UP and EU preferences, before leaving the HB towards the EUT. From the point of view of the MPEG-DASH client module at the HB, the content is available in segments, which are requested from the local

Swift-to-HTTP proxy. The HB provides a second level of adaptation, that in practice transcodes the content according to the UP and EU preferences, before leaving the HB towards the EUT. From the point of view of the MPEG-DASH client module at the HB, the content is available in segments, which are requested from the local Swift-to-HTTP proxy. At the requesting (consuming) HB, multiple Swift-to-HTTP proxy instances cohabite so the inner details of the P2P network may be hidden from the remaining modules inside the HB. The Swift-to-HTTP provides a unified HTTP interface than can easily integrated with the Caching module and the DASH-enabled HTTP proxy that already exists inside the Streaming Gateway.

Peers Selection Algorithm

In a traditional P2P network, such as BitTorrent, a *.torrent* file contains the URL(s) of the P2P tracker and hash information for the pieces/chunks, which will be used to verify the integrity of the content. Peers will use the information in the .torrent file to check the content pieces they have downloaded. However, in P2P most of peers don't possess any knowledge about the seeding peers. Peers who have downloaded complete pieces of the content will assist the distribution process by uploading the pieces they have to other peers in the system, and that will significantly reduce the load of the seeders. When a new peer first joins the P2P network, it contacts the tracker. The tracker decides then which subset of peers he known is also downloading/seeding the content and returns the list of peers to the requesting peer. After receiving the list of peers, the requesting peer will contact its neighbours directly without involving the tracker in order to exchange data information. The list of peers in the network is constantly refreshed their neighbour peers in a periodic fashion.

In P2P networks, determining which peer to select in order to optimize the data exchange throughput is not always trivial. The peer selection algorithm typically returns a ranked set of peers. The rank value is equal to the similarity value provided by the similarity function. From this set of ranked peers one can, for example, select the best n peers, or all peers whose rank value is above a certain threshold, etc. Some studies propose the usage of genetic-algorithm based strategies to achieve this objective, while others (including the current implementation of BitTorrent) propose a random strategy instead, with implication both in terms of system throughput and average downloading experienced by users. In particular, studies where the genetic-algorithm based strategies are use claim to improve the efficiency of distribution for peers with slow connections by better utilizing the resources contributed by high-connectivity peers (Koo, et al, 2004).

ALICANTE proposes to enhance the random peer selection strategy, as it is implemented by NextShare Swift library, with the network information and HB load (P2P peer nodes). For that the following procedure has been proposed to be implemented at the P2P Engine module, inside the HB:

1. The Swift-to-HTTP proxy module initializes by connecting to the initial Swift tracker/seeder peer;
2. The Swift-to-HTTP proxy module, that runs at the requesting peer, receives updated information regarding his neighbouring peers via PEX Swift messages;
3. The Swift-to-HTTP proxy module passes on the list of new peers to the P2P Engine module;
4. The P2P Engine obtains the HB distances for each peer on that list and ranks it with the help of the Monitoring Manager;
    a. The HB distance is represented as a vector of several network and system parameters, such as one-way delay, jitter, loss, CPU load, available memory, etc.
    b. The HB Monitoring Manager computes a single value representation of this HB distance using a weighted mean formula:

A non-empty set of data $\{x_1, x_2, \ldots, x_n\}$, representing each one of the HB distance vector dimensions

With non-negative weights $\{w_1, w_2, \ldots, w_n\}$

The mean HB distance value is obtained as

$$\bar{x} = \frac{\sum_{i=1}^{n} w_i x_i}{\sum_{i=1}^{n} w_i}$$

or

$$\bar{x} = \frac{w_1 x_1 + w_2 x_2 + \ldots + w_n x_n}{w_1 + w_2 + \ldots + w_n}$$

The weights $\{w_1, w_2, \ldots, w_n\}$ will be evaluated empirically during the ALICANTE system deployment and field trials.

5. The P2P Engine passes the ranked list of peer back to the Swift-to-HTTP proxy module;

6. During the next file chunk request the Swift-to-HTTP proxy module uses to topmost peer as the server for that chunk.
7. If the topmost peer on the list is not responsive or does not contain that particular chunk, the Swift-to-HTTP proxy module uses the next peer on the list, and so on, and so on;

HB Assisted Content Delivery

Figure 4 depicts the general architecture of the HB-assisted content delivery: the HB overlay includes all participating HBs with caching capabilities. The HB overlay is built on top of the VCAN network, to achieve the network-awareness. The core function is the collaborative caching strategy, which leverages the caching capability of all HBs. Similar to P2P and CDN approaches, the storage capacity of edge nodes are aggregated in order to bring content closer to the end-users and optimize the use of network resources. Other services and applications such as content publication and discovery, adaptive content delivery… are running on the HB overlay and take advantage of the HB caching. The HB caching module will act as a proxy server inside the HB. It intercepts the request from the streaming gateway, redirects it either to the local cache if the content has been requested before, or to the suitable connectivity engine to download the content using DASH or P2P engine. A collaborative caching algorithm is running on the HB caching, in order to efficiently

Figure 4. Overview of the HB-assisted content delivery

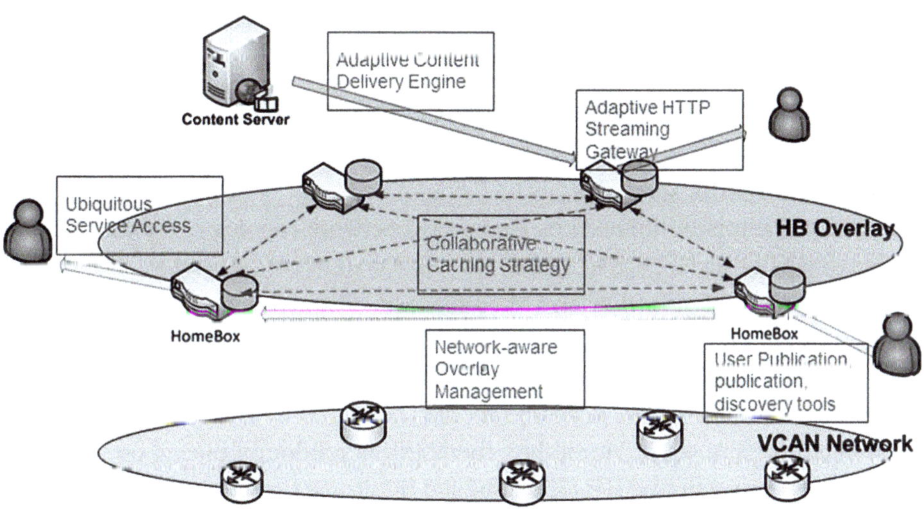

aggregate the HB storage capacity and maximize the cache hit. The rationale of the collaborative popularity-based caching is explained below:

- **Video Popularity Prediction:** The VoD video catalogue has Zipf-like popularity distribution. Moreover, the popularity of a given video varies during its lifecycle. Given a catalogue of size ***M***, the Service Provider is able to estimate the popularity of each video ***i***, denoted by ***P(i)***.
- **Home-Box Planning:** Based on the video popularity, we calculate the minimum number of video copies **k(i)** that should be cached by the Home-Box. The set of all Home-Box can be clustered into **k(i)** clusters, using approximation algorithms for k-median problem. The average diameters of all cluster is denoted by **d(i)**.
- **Content Access:** The request is redirected to the nearest (in terms of network distance and load) Home-Box hosting the content, based on the user context information. If the content cannot be found in the home-box, the request will be redirected to CDN.
- **Content Replacement**: The Home-Box adopts the "*Least Popular*" replacement strategy. The content with least popularity in the cache will be replaced by the new video **i**, if the new video is not cached by any home-box within the distance **d(i).**

The proposed Home-Box-assisted content distribution can efficiently reduce load on the server, delay for end-users and overall distribution cost.

PERFORMANCE EVALUATION

Evaluation Methodology

To evaluate the proposed HB-assisted Content Delivery system, a realistic simulation environment is built based on real-world network topology, statistical studies on existing VoD services, and widely adopted system configuration. The network topology used in our simulations (Magoni, et al., 2005) is obtained through active traceroute probing. This router level Internet map contains 47k nodes and 119k links. This graph is representative enough to produce realistic results, because it has the same power law degree distribution, also observed by other Internet topological studies. A large set of Home-box (n=5000) is randomly placed on this graph. The distance between two HBs is calculated in terms of router-level hop count. To simulate the VoD service, we define the number of available videos in the system, known as the catalog size (c=500). These videos have different popularities. In the simulations, we use the popularity distribution published by Netflix (Bennett, et al., 2007). User requests for the video are randomly generated according to the Netflix' heavy-tailed popularity distribution (Figure 5). Each HB implements the LRU (least Recently Used) caching strategy. In the simulations, we vary two major parameters of the collaborative caching system, i.e. node degree and cache size. The node degree (d = [5, 60]) is the number of active connections a HB should maintain with its neighboring HBs, in order to exchange cache index and forward user requests. The cache size is the storage capacity of each HB in terms of number of videos (s = [5, 50]). The metrics we observe are: 1) the hit ratio: the percentage of user requests fulfilled by the HB cache; 2) the hop count: the average distance between the client and his requested cache object. 3) overall cost for the VoD service.

Evaluation on Caching Efficiency

The caching efficiency is measured in terms of the hit ratio: when a HB receives a request from a client, it verifies firstly whether the request can be fulfilled with locally cached objects (local hit). If not, the request is forwarded to its neighbors (peering hit). If the requested video is neither cached by peering HBs (cache miss),

Figure 5. Netflix video popularity

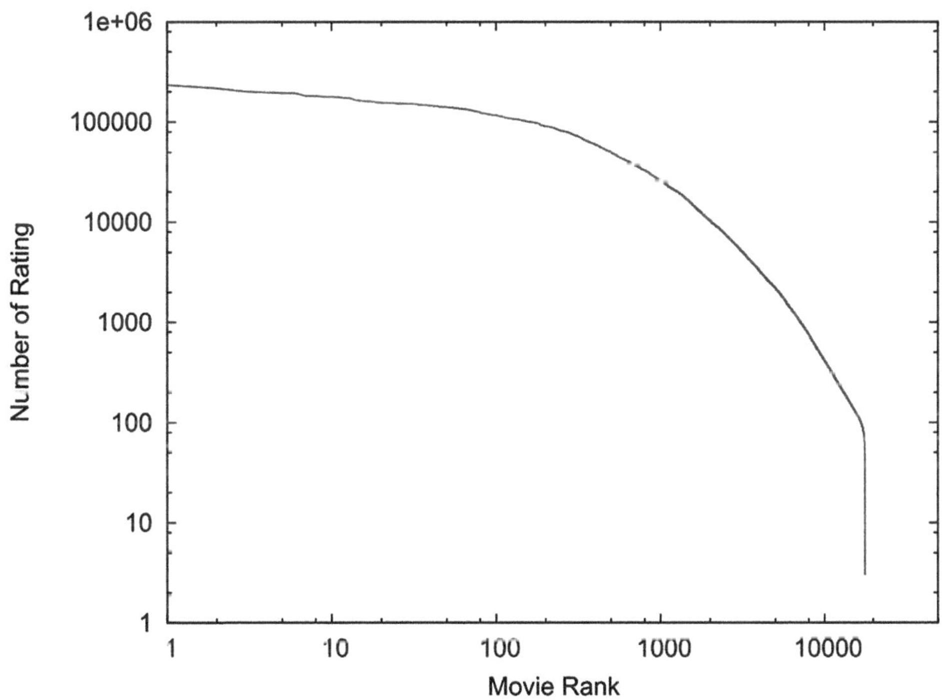

the request will be forwarded finally to the origin server, where the entire VoD catalog is always available.

Results in Figure 7 show that increasing the node degree is an efficient way to increase the hit ratio of the collaborative caches. Without the collaborative caching, the local hit ratio is only 4,5%. The hit ratio increases quickly with the number of neighbors, for example, with d=40, the total hit ratio (local+peering) is above 80%. However, due to the duplicated objects in different HBs, the hit ratio does not increases proportionally with d. On the other hand, a bigger cache size (s) will lead to both better local hit and better peering hit. But, the increase on the hit ratio is less significant compared to the previous case with node degree. We conclude that the node degree (d) has more impact on the cache efficiency that the individual cache size (s).

Evaluation on Network-Awareness

Similar to other distributed systems as P2P, the proposed HB content distribution system should be "network friendly", i.e. inducing minimum unnecessary network traffic especially inter-domain traffic. To achieve this, network-aware peer selection is implemented in our simulations, the neighbors of each HB are always the nearest ones (in terms of hop count). Here we observe the "network friendliness" of our system, by measuring the network traffic generated by different requests. The y1-axis of Figure 8 is the average router-level hop count for each video request, which

Figure 6. Hit ratio vs. number of neighbors

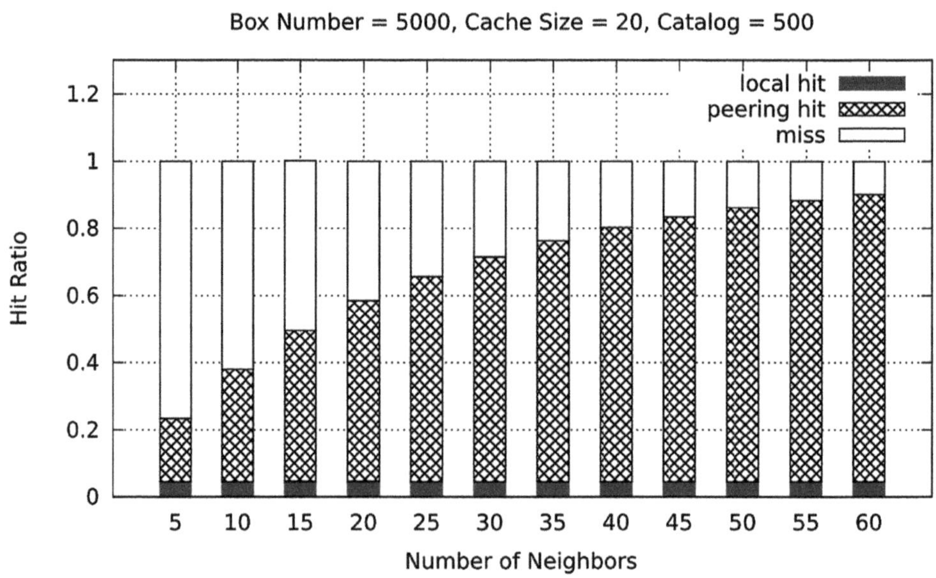

Figure 7. Cache efficiency

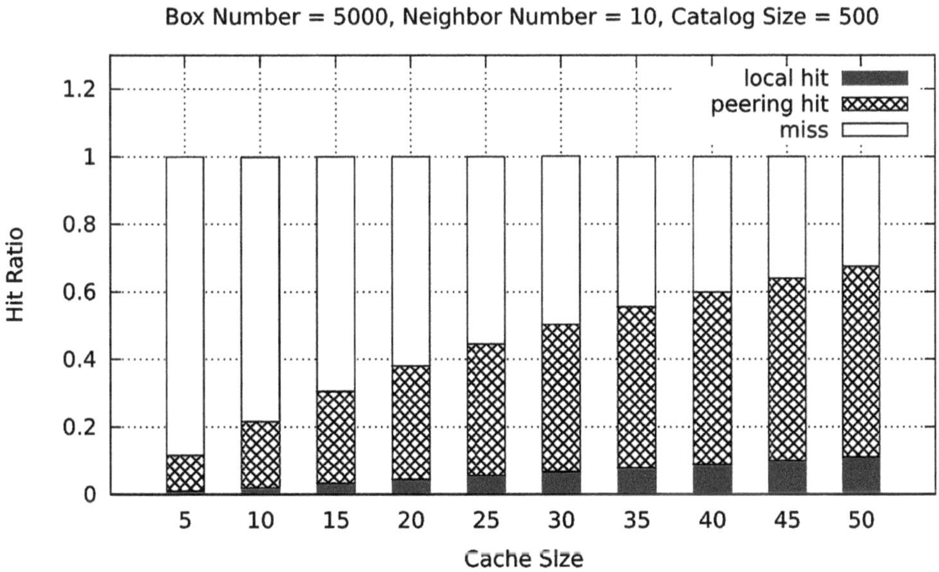

gives a good estimation of the network traffic among different HBs. The y-2 axis is the hit ratio. Both hop count and hit ratio increase along x-axis, meaning that better hit ratio will lead to more network traffic. In the first case, when a HB has

Figure 8. Hop count vs. number of neighbors

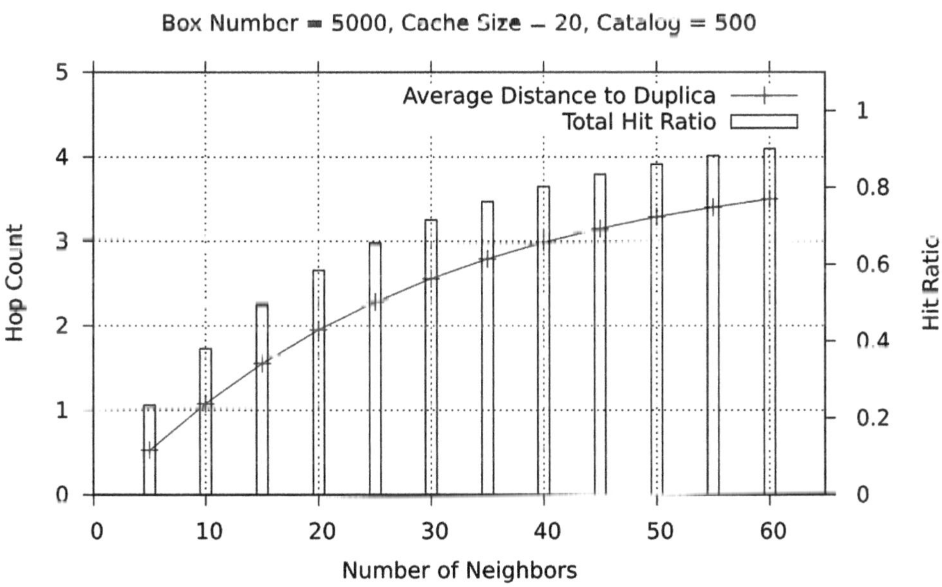

more neighbors, i.e. increased reachable area, the request will be fulfilled more easily (increase hit ratio), but redirected to a further HB (increased hop count). In the second case, when a HB has more storage capacity, more requests can be fulfilled by the same set of neighbors. Therefore, the hop count increase less quickly than in the first case.

CONCLUSION

Our proposal relies on the User Environment's storage and connectivity without directly involving clients' nodes participation. The HB located in the User's Environment, is the peer node acting as the always-connected "hub" between the User, Network and Service Environments, respectively being capable of shared ownership between the three actors (End-User, Service Provider, Network Provider). Therefore, the respective actors can easily manage the HB, feature impossible in case of End-User terminals. Our model is then a fully managed model and consequently overcomes the issues induced by the P2P part in such hybrid systems (e.g. free riding, reliability, peer churn issues, etc.).

REFERENCES

Aggarwal, V., Feldmann, A., & Scheideler, C. (2007). Can ISPs and P2P users cooperate for improved performance? *Computer Communication Review*, *37*(3), 29–40. doi:10.1145/1273445.1273449

Alliance, O. (2010). *OSGi alliance*. Available at http://www.osgi.org/

Bennett, J., & Lanning, S. (2007, August). The netflix prize. In *Proceedings of KDD cup and workshop* (Vol. 2007, p. 35).

Borcoci, E., Negru, D., & Timmerer, C. (2010, June). A novel architecture for multimedia distribution based on content-aware networking. In *Communication Theory, Reliability, and Quality of Service (CTRQ), 2010 Third International Conference on* (pp. 162-168). IEEE. doi:10.1109/CTRQ.2010.35

Cardellini, V., Colajanni, M., & Yu, P. S. (2003). Request redirection algorithms for distributed web systems. Parallel and Distributed Systems. *IEEE Transactions on*, *14*(4), 355–368.

Chen, Y., Qiu, L., Chen, W., Nguyen, L., & Katz, R. H. (2003). Efficient and adaptive Web replication using content clustering. *Selected Areas in Communications. IEEE Journal on*, *21*(6), 979–994.

Chen, Y. F., Huang, Y., Jana, R., Jiang, H., Rabinovich, M., Wei, B., & Xiao, Z. (2007, June). When is P2P technology beneficial for IPTV services. In *Proceedings of the 17th International Workshop on Network and Operating System Support for Digital Audio and Video* (pp. 04-05).

Concolato, C., Dufourd, J. C., Le Feuvre, J., Park, K., & Song, J. (2012). Communicating and migratable interactive multimedia documents. *Multimedia Tools and Applications*, *61*(3), 599–622. doi:10.1007/s11042-011-0805-2

ETSI. (2006). *282 001: Telecommunications and Internet converged Services and Protocols for Advanced Networking (TISPAN)*. NGN Signalling Control Protocol.

Gupta, A., Liskov, B., & Rodrigues, R. (2004, March). Efficient Routing for Peer-to-Peer Overlays. In *NSDI*.

He, J., Chaintreau, A., & Diot, C. (2009). A performance evaluation of scalable live video streaming with nano data centers. *Computer Networks*, *53*(2), 153–167. doi:10.1016/j.comnet.2008.10.014

Home Gateway Initiative. (2007). *Home Gateway Technical Requirement*. Author.

Huang, Y., Fu, T. Z., Chiu, D. M., Lui, J., & Huang, C. (2008, August). Challenges, design and analysis of a large-scale p2p-vod system. *Computer Communication Review*, *38*(4), 375–388. doi:10.1145/1402946.1403001

Kangasharju, J., Ross, K. W., & Turner, D. A. (2007, May). Optimizing file availability in peer-to-peer content distribution. In *INFOCOM 2007. 26th IEEE International Conference on Computer Communications*. IEEE. doi:10.1109/INFCOM.2007.229

Kilkki, K. (2008). Quality of Experience in Communications Ecosystem. *J. UCS*, *14*(5), 615–624.

Koo, S. G., Lee, C. G., & Kannan, K. (2004, October). A genetic-algorithm-based neighbor-selection strategy for hybrid peer-to-peer networks. In *Computer Communications and Networks, 2004. ICCCN 2004. Proceedings. 13th International Conference on* (pp. 469-474). IEEE. doi:10.1109/ICCCN.2004.1401710

Laoutaris, N., Rodriguez, P., & Massoulie, L. (2008). ECHOS: Edge capacity hosting overlays of nano data centers. *Computer Communication Review*, *38*(1), 51–54. doi:10.1145/1341431.1341442

Liben-Nowell, D., Balakrishnan, H., & Karger, D. (2002, July). Analysis of the evolution of peer-to-peer systems. In *Proceedings of the twenty-first annual symposium on Principles of distributed computing* (pp. 233-242). ACM. doi:10.1145/571825.571863

Magoni, D., & Hoerdt, M. (2005). Internet core topology mapping and analysis. *Computer Communications*, *28*(5), 494–506. doi:10.1016/j.comcom.2004.09.002

Markakis, E. K., Pallis, E., Skianis, H., & Zaxaropoulos, V. (2012). Dynamic TCP Window Size Adaptation for Improving Network Performance of P2P-Aware Interactive Broadcasting Environments. *Journal of Communication and Computer*, *9*, 1395–1403.

Mastorakis, G., Bourdena, A., Mavromoustakis, C. X., Pallis, E., & Kormentzas, G. (2013, July). An energy-efficient routing protocol for ad-hoc cognitive radio networks. In Future Network and Mobile Summit (IEEE FutureNetworkSummit), 2013 (pp. 1-10).

Mastorakis, G., Mavromoustakis, C. X., Bourdena, A., Kormentzas, G., & Pallis, E. (2013, September). Maximizing energy conservation in a centralized cognitive radio network architecture. In *2013 IEEE 18th International Workshop on Computer Aided Modeling and Design of Communication Links and Networks (IEEE CAMAD 2013)*, (pp. 175-179). doi:10.1109/CAMAD.2013.6708112

Mavromoustakis, C., Andreou, A., Mastorakis, G., Papadakis, S., Bourdena, A., & Stratakis, D. (2014, August). Energy Consumption Optimization through Pre-scheduled Opportunistic Offloading in Wireless Devices. In *EMERGING 2014, The Sixth International Conference on Emerging Network Intelligence* (pp. 22-28).

Mavromoustakis, C., Dimitriou, C. D., & Mastorakis, G. (2013). On the Real-Time Evaluation of Two-Level BTD Scheme for Energy Conservation in the Presence of Delay Sensitive Transmissions and Intermittent Connectivity in Wireless Devices. *International Journal on Advances in Networks and Services, 6*(3-4), 148-162.

Mavromoustakis, C. X., Bourdena, A., Mastorakis, G., Pallis, E., & Kormentzas, G. (2014). An energy-aware scheme for efficient spectrum utilization in a 5G mobile cognitive radio network architecture. *Telecommunication Systems*, 1–13.

Mavromoustakis, C. X., Dimitriou, C., Mastorakis, G., & Pallis, E. (2013, December). Real-Time Performance Evaluation of F-BTD scheme for optimized QoS Energy Conservation in Wireless Devices. *Proceedings of IEEE GlobeCom, 2013*, 1156–1161.

Nygren, E., Sitaraman, R. K., & Sun, J. (2010). The akamai network: A platform for high-performance internet applications. *Operating Systems Review, 44*(3), 2–19. doi:10.1145/1842733.1842736

Pathan, A. M. K., & Buyya, R. (2007). *A taxonomy and survey of content delivery networks*. Grid Computing and Distributed Systems Laboratory, University of Melbourne, Technical Report.

Pathan, M., Buyya, R., & Vakali, A. (2008). Content delivery networks: State of the art, insights, and imperatives. In *Content Delivery Networks* (pp. 3–32). Springer Berlin Heidelberg. doi:10.1007/978-3-540-77887-5_1

Qiu, L., Padmanabhan, V. N., & Voelker, G. M. (2001). On the placement of web server replicas. In INFOCOM 2001. Twentieth Annual Joint Conference of the IEEE Computer and Communications Societies. *Proceedings of the IEEE, 3*, 1587–1596.

Saucez, D., Donnet, B., & Bonaventure, O. (2008). *Idips: Isp-driven informed path selection*. Academic Press.

Sodagar, I. (2011). The MPEG-DASH Standard for Multimedia Streaming Over the Internet. *IEEE MultiMedia, 18*(4), 62–67. doi:10.1109/MMUL.2011.71

Tang, X., & Xu, J. (2004, March). On replica placement for QoS-aware content distribution. In *INFOCOM 2004. Twenty-third AnnualJoint Conference of the IEEE Computer and Communications Societies* (Vol. 2, pp. 806-815). IEEE.

Tran, M., & Tavanapong, W. (2005, November). Peers-assisted dynamic content distribution networks. In *Local Computer Networks, 2005. 30th Anniversary. The IEEE Conference on* (pp. 123-131). IEEE. doi:10.1109/LCN.2005.102

Valancius, V., Laoutaris, N., Massoulié, L., Diot, C., & Rodriguez, P. (2009, December). Greening the internet with nano data centers. In *Proceedings of the 5th international conference on Emerging networking experiments and technologies* (pp. 37-48). ACM. doi:10.1145/1658939.1658944

Xie, H., Yang, Y. R., Krishnamurthy, A., Liu, Y. G., & Silberschatz, A. (2008, August). P4P: Provider portal for applications. *Computer Communication Review, 38*(4), 351–362. doi:10.1145/1402946.1402999

Xu, D., Kulkarni, S. S., Rosenberg, C., & Chai, H. K. (2006). Analysis of a CDN–P2P hybrid architecture for cost-effective streaming media distribution. *Multimedia Systems, 11*(4), 383–399. doi:10.1007/s00530-006-0015-3

Yang, M., & Fei, Z. (2003, May). A model for replica placement in content distribution networks for multimedia applications. In *Communications, 2003. ICC'03. IEEE International Conference on* (Vol. 1, pp. 557-561). IEEE.

Yin, H., Liu, X., Zhan, T., Sekar, V., Qiu, F., Lin, C., & Li, B. et al. (2010). Livesky: Enhancing cdn with p2p. *ACM Transactions on Multimedia Computing, Communications, and Applications, 6*(3), 16. doi:10.1145/1823746.1823750

Zhang, X. Y., Zhang, Q., Zhang, Z., Song, G., & Zhu, W. (2004). A construction of locality-aware overlay network: mOverlay and its performance.Selected Areas in Communications. *IEEE Journal on, 22*(1), 18–28. doi:10.1109/JQE.2003.820842

ENDNOTE

1 www.ietf.org/id/draft-ietf-ppsp-peer-protocol-02.txt

Chapter 10
Causal and Total Order in Opportunistic Networks

Mihail Costea
University Politehnica of Bucharest, Romania

Radu-Ioan Ciobanu
University Politehnica of Bucharest, Romania

Radu-Corneliu Marin
University Politehnica of Bucharest, Romania

Ciprian Dobre
University Politehnica of Bucharest, Romania

Constandinos X. Mavromoustakis
University of Nicosia, Cyprus

George Mastorakis
Technological Educational Institute of Crete, Greece

ABSTRACT

Opportunistic network applications are usually assumed to work only with unordered immutable messages, like photos, videos or music files, while applications that depend on ordered or mutable messages, like chat or shared contents editing applications, are ignored. In this chapter, we examine how causal and total ordering can be achieved in an opportunistic network. By leveraging on existing dissemination algorithms, we investigate if causal order can be efficiently achieved in terms of hit rate and latency compared to not using any order. Afterwards, we propose a Commutative Replicated Data Type algorithm based on Logoot that uses the nature of opportunistic networks to its advantage. Finally, we present the results of the experiments for the new algorithm by using an opportunistic network emulator, mobility traces and chat traces.

DOI: 10.4018/978-1-4666-9941-0.ch010

Copyright ©2016, IGI Global. Copying or distributing in print or electronic forms without written permission of IGI Global is prohibited.

I. INTRODUCTION

In the recent years, Opportunistic Networks (ONs) have become an important research field as a viable solution for mobile networks. ONs use the store-carry-and-forward (SCF) paradigm to facilitate communication between mobile devices as smartphones or laptops, even if these devices might never be directly accessible (Pelusi, Passarela, & Conti, 2006). Nodes can join and leave the network at any time, the content creators and the nodes interested in the data might never have a direct contact or they not might even be aware of each other existence. A node is aware only of those that come in direct contact with it. Due to these reasons, there is no central entity that knows about the network topology and how each node is connected to each other.

Most of the work in ONs has been directed towards achieving message forwarding or dissemination, where messages are usually considered immutable. Only a few researchers have tackled the problem of mutable content that requires a given order, even though these represent important user applications. Consider the case of discussion forums, where user post news or questions in discussion threads, and other users can respond to them. While different discussion threads are independent from each other and do not require any order, messages in the same thread should respect a total order, so that every user sees the same discussion.

Similarly to discussion forums, we have social networks, where users post something about themselves, or an interesting article, melody or citation they have found on the internet or somewhere else, which also requires a total order for future responses. Other types of applications that require a total order are collaborative editing applications, like Google Docs (Google, 2015b) or wiki pages. Every participant in a collaborative editor, or even a simple viewer of a wiki page, should have a consistent view of the contents. It is not permitted to have strings placed in different orders at different users (e.g. "This is a wiki page" at $user_1$ vs. "This a page wiki is" at $user_2$ – the misalignment here being result of users seeing text change updates in wrong order), as future updates will not correctly propagate to other nodes.

Though we have explained the importance of total ordering for applications with mutable contents, one might ask why these applications should even be supported in an ON environment, when they already exist in traditional environments. Though there are multiple reasons, we refer the reader to the recent example of FireChat, the chat application being used during the Hong Kong's protests to broadcast on mesh networks the revolution live between participants (Open Garden, 2015)(Fitzpatrick, 2014) (Daileda, 2014). The application requires only an ON environment to work, without any need to run over a traditional network infrastructure. Still, on such an infrastructure, if becomes more important to show an order or a relation between distributed messages being exchanged.

A first reason to support such kind of applications is *mobility*, a base element of ONs. In recent years, users have shifted from desktops to laptops and smartphones, and more recently even to wearables as smartwatches. An ON infrastructure should support, in fact, as many types of applications as possible. Another reason is that regions in emerging countries (i.e. India or countries in Africa) are not properly connected to the global network, but, as smartphones have become affordable through programs as Android One (Google, 2015b), ONs represent a viable solution for connecting these devices, solution that will lead to a request of all types of applications similar to the ones mentioned above. Lastly, but not lastly, an important reason is privacy. ONs are able to work without connecting to the global network, providing an isolated environment. If participant nodes in ONs are not allowed to read the contents of the messages they carry unless they are the destination, government agencies or hackers will not be able to easily eavesdrop, as messages are exchanged only between the nodes themselves, without the help of an infrastructure that can route messages wherever it wants.

+ the problem of ordering ONs messages so they respect a given order before being delivered to the application layer. Our contributions consist in three parts. Firstly, we investigate if causal order can be efficiently achieved in ONs in comparison to the hit rate and delivery latency of existing dissemination algorithms. Our experiments run in a ON simulator on three mobility traces obtained in different environments prove that the hit rate and delivery latency to the application layer perform well even if messages must be withheld to respect causality. Secondly, we propose a Commutative Replicated Data Types / Conflict-Free Replicated Data Type (CRDT) (Shapiro & Preguiça, 2007)(Shapiro, Preguiça, Baquero, & Zawirski, 2011) based on the Logoot algorithm (Weiss, Urso, & Molli, 2008) to achieve total ordering. Using the nature of the ONs, our proposed algorithm reduces the metadata that must be carried with each message. Thirdly, we run experiments in the same simulator as for causal order to test how the proposed algorithm performs. For testing the algorithm we have used ten chat traces.

The rest of the chapter is organized as follows: Section II presents related work, and Section III details experimental results related to causal order. In Section IV we present our proposed algorithm for total ordering in ONs. Section V contains the experiments related to our proposed algorithm, and in Section VI we present the conclusions and future work.

II. RELATED WORK

With the increase of mobile devices in the latest years, ONs have become an important research topic. However, researchers have focus on message routing and dissemination. Conti et al. (2010) provides an analysis of message forwarding, security, data dissemination and mobility models in ONs. One of the first algorithms for data routing is Epidemic (Vahdat & Becker, 2000). The algorithm simply floods the network with a message until it reaches the destination(s) - this achieves maximum hit rate in case of infinite data memory, though at the cost of a high overhead and network congestion. A more sophisticated algorithm is ML-SOR (Socievole, Yoneki, De Rango, & Crowcroft, 2013), which uses social network information and encountered nodes information such as node centrality, tie strength and link prediction on social network layers, to download messages when in contact with another node. In the ContentPlace algorithm (Boldrini, Conti, & Passarella, 2008), community information such as users social relationships is used to disseminate data. Similarly, the ONSIDE algorithm (Ciobanu, Marin, Dobre, Cristea, & Mavromoustakis, 2014) uses social information to disseminate data by using nodes that are not necessarily interested in a given message, but they have high chance of encountering nodes that are.

When it comes to causal order, Birman & Joseph (1987) propose the ISIS system, one of the first implementation of messages causal ordering. Every message must carry with it all (or information about, as in recent implementations) the messages that precede it causally. Before the message is delivered to the application layer, all the causal predecessors carried with it will be delivered in the appropriate order, and then the message itself is delivered. From time to time a garbage collection must be executed to remove the messages that no longer require to be carried with the new messages. Schiper et al. (1989) and Birman et al. (1991) use vector clocks (Fidge, 1988)(Mattern, 1989) to ensure causal order. Instead of carrying all the causal predecessors, a message only carries ordered pairs of the type *(destination site, vector time)*. By using this information, even if a message arrives at its destination earlier than its causal predecessors, it can be withheld until all messages arrive. Afterwards, the message can be delivered. An important algorithm is represented by causal barriers (Prakash, Raynal, & Singhal, 1997), which needs to carry only direct dependency information between messages to achieve causal order, reducing communication overhead. No prior knowledge of the network topology or of the communication pattern is required, making it an ideal causal ordering algorithm for ONs.

When it comes to total ordering, the Operational Transformation (OT) (Ellis & Gibbs, 1989) algorithm and/or Commutative Replicated Data Types (CRDT) (Shapiro & Preguiça, 2007) algorithm might provide viable solutions, as both of them usually require only causal order. Any algorithm that requires either a central node or synchronization between nodes cannot be considered, due to the CAP theorem (Gilbert & Lynch, 2002), where consistency, availability and partition tolerance cannot be simultaneously provided in a distributed system. As we target an ON environment, it is impossible to have a central node because at any moment it can get out of direct contact with any other node, or it might leave the network and never return, while using a synchronization protocol between nodes as the three-phase commit protocol (Skeen & Stonebraker, 1983) or Paxos (Lamport, 1998) is extremely improbable because it is a norm to have messages that never reach their destination in ONs.

As mentioned in the previous paragraph, OT algorithms might represent a viable solution. Though, due to some limitations we shall now present, we have chosen to use a CRDT-based algorithm. OT promises to achieve convergence, causality and intention preservation in distributed systems without using locking and serialization. They are intended for share editors, where users can edit the data at any time and in any part, and use operation transformations to achieve consistency, without waiting for messages from other nodes before applying any user change, aspect that makes it ideal for ONs. The earliest OT algorithm is dOPT (Ellis & Gibbs, 1989), which supports transformations for insertion or removal of a single character. Operations are performed immediately at the originating site, offering good responsiveness. But there are some special cases where concurrency control for operation transformation fails, and two nodes can end up with different views for the shared contents. Some algorithms, such as NICE (Shen & Sun, 2002) and SOCT3/4 (Vidot, Cart, Ferrié, & Suleiman, 2000), require a notification server or sequencer to maintain a global total order of execution, making them difficult to implement in ONs; while algorithms such as TIBOT (Li, Li & Sun, 2004) and GOT (Sun & Ellis, 1998) fail to provide the correct result in some corner cases. The SDT algorithm (Li & Li, 2008) promises to ensure consistency in peer-to-peer real time group editors by formalizing alternative conditions that can be proven, while previous work did not formally specify a model for formal proofs, resulting in corner cases where the algorithms fail.

Though some OT algorithms seem promising, most of them have a series of common problems, besides the failures present in some cases. One example is the fact that most of them present transformations for strings on only one character (even the SDT algorithm, though the authors assumed that it can be easily extended to strings). Using one message per one character in ONs is out of discussion. We would like to maximize the actual data transported by each node in regards to metadata as nodes in ONs must carry a message for as long as possible to ensure that it reaches its destination(s), without wasting space that can be used for other messages. In addition, decentralized OT algorithms require a garbage collection mechanism to reduce the operation log size, which is improbable to obtain in ONs. Another problem is that most of these algorithms are complex and hard to extend to other applications besides share editors. We would like to use algorithms that can be used in more application types with minimal changes. Lastly, the complexity of the OT algorithms requires high computation for executing the operation transformations, while ONs nodes are usually mobile devices with limited resources in terms of processing power, memory and battery.

The other viable solution we considered for obtaining total ordering in ONs consists in CRDT based algorithms. A CRDT is a data structure designed to achieve strong eventual consistency (SEC) (Shapiro et al., 2011) in distributed systems. SEC is a subset of eventually-consistent systems where nodes that have received and delivered the same set of messages will immediately reach an equivalent state. As conflicts are not present in SEC, there is no need for a special process to handle them, and nodes end up with the same view of the contents, making them ideal for ONs. The first CRDT algorithm is WOOT (Oster, Urso, Molli, & Imine, 2006). WOOT relies on the property that in a document every character is located between other two characters. By knowing the neighbors, WOOT is able to perform character insertion and removal. In order to ensure consistency for concurrent edits, a deterministic algorithm based on the same rules at all nodes is used to order them. But there is a problem related to this algorithm: it requires *tombstones* to correctly ensure consistency. Tombstones are the elements removed from the document, and when they are deleted they are only made invisible in the user view, without actually being released from memory. A garbage collection is required to remove tombstones, which is impractical in ONs as executing synchronization algorithms as Paxos is improbable, just as mentioned in a previous paragraph.

An important CRDT algorithm is Logoot (Weiss et al., 2008), where each character gets assigned to a unique identifier. Identifiers are generated from a *densely ordered* set, where for a given set S and any two elements x and y in S, there is another element z in$S$ such that $x < z < y$, and, by ensuring causal order, consistency can be maintained at all participant nodes. Logoot is able to create unique identifiers between any two characters by using a clock value that is incremented with every insert operation and a list of pairs of integers made of a random value x and a site id s such that the previous identifier x value is smaller than the new x, which itself is smaller than the next identifier x value. In case x is equal for both previous and next identifiers, then the site ids are used to order the characters. Though it is possible that the site ids to not respect the "<" relation, and in this case a new pair of integers made of another random value x and the site id is appended to the pairs of the previous identifier in order to create a new identifier that can be positioned between its neighbors. Due to this reason, the Logoot identifiers can grow infinitely. If we add the fact that an identifier is required for every different character, we will end up with high overhead for every message, which is undesirable in ONs as mentioned at the end of the paragraph about OT. But an important fact related to Logoot is that it does **not** require tombstones. Characters can be removed at any time and they do not need to be kept in memory in order to maintain consistency, eliminating the need for a garbage collection. The fact that a garbage collection is not necessary makes it an ideal candidate for ONs in case the previous limitation related to message overhead can be overcome.

As an alternative, Logoot, Letia et al. (2009) present the Treedoc algorithm, where document identifiers are represented by a binary tree. The tree nodes contain the document characters, while the path from the root of the tree to the node with the actual character represents the character identifier. The left child of a node represents the previous character, while the right child represents the next character. By using this approach, Treedoc obtains a more compact space for the characters identifiers, though in some cases, like insertions at the end of the document, the tree becomes unbalanced. Also, some tombstones are required because a node cannot be removed from memory unless he does not have any children, though it is safe to remove nodes with no children. Due to tombstones and the fact that the tree can become unbalanced, leading to high overhead for messages made of a single character, a garbage collection mechanism is unfortunately required. Just as Logoot, Treedoc also requires causal order in order to ensure consistency.

The problem with most CRDT algorithms is that their overhead is significant if working with small payloads for messages (i.e. characters). Yu (2012) presents a string-wise CRDT where a string can be represented by a single identifier. Strings can be split and new characters can be inserted or existing characters can be deleted, but it still requires tombstones in order to correctly work, leading unfortunately to a garbage collection requirement.

A really promising CRDT algorithm where multiple characters can be supported by a single identifier is LogootSplit (Andre, Martin, Oster, & Ignat, 2013). The algorithm is based on Logoot, eliminating the need for garbage collection. It supports sequence of characters using a single identifier by adding a new field to the Logoot identifier. Let us consider the same example presented by the authors in their article, where a new string "HEY" has been created. In Logoot, the three characters would have identifiers similar to these *<daa,H>* *<dab,E>* *<dac,Y>* (the authors use a notation different from the original Logoot algorithm, but the *daa* string can be considered the equivalent of the logoot identifier, while "H" is the first character in "HEY"). In the new algorithm, the "HEY" string would have the identifier *<da[.c],HEY>*. In case of a further append of the characters "WO", then these characters will have the identifier *<da[d.e],WO>*, and once its corresponding message is received by other nodes, the "WO" string is merged with "HEY" resulting in a single string with the unique id*<da[.e],HEYWO>*. Instead of having 5 different identifiers that would have a size of 3 * 5 = 15 characters, the new string will have an identifier of only 4 characters. Inserting new characters in a string with a new identifier is as simple as splitting the existing string and creating a new identifier using the same rule as the Logoot algorithm. Removing characters is done in the same manner as Logoot, though reusing identifiers is not permitted.

All of the previous presented algorithms for total order were created for environments different from ONs. Most of them aim to offer real-time group editing with multiple users in distributed systems or peer-to-peer networks. Kärkkäinen & Ott (2014) present an alternative solution for a shared editor in ONs. Their approach is to use revision control mechanism (merging) or to adopt or discard whole versions of the documents. Though this might look promising, it poses a series of limitations which we consider to be too restrictive for ONs. One of their assumptions is that every node in the ON carries a copy of the document contents in order to be able to modify it. This ensures that at any contact between two nodes, changes can be propagated from a node to another and the contents of the document can be updated in order to display a consistent view to the user. The problem is that not every node in an ON might be interested in the contents of that document. Requiring nodes that have no interest in the document to hold a copy of it represents an overhead that cannot be ignored in ONs with limited resources, including memory. Nodes should carry only the contents they are interested in and a limited amount of messages that are not intended to them and which are routed using a dissemination algorithm as in the first part of this section. These messages not intended for the user can be removed at any time in case there is no more free memory, being replaced with new data or new messages, which lead to scenarios where the removed messages might never reach their destination. Another problem is that some sort of merging is needed in the case of merge conflicts, leading to cases where useful data might be discarded or user intervention is required in order to avoid the data loss. Both of these scenarios are undesirable.

Opposed to previous work, our proposed algorithm takes advantage of the nature of ONs, where real-time communication is not required and message loss and direct contact between nodes is not assured in order to exchange messages. The base of the algorithm is Logoot, and it works similarly to LogootSplit in the regard that we also support a single identifier for a sequence of characters. Instead of using the LogootSplit notation, we use the original Logoot notation, where we add a new field to identifiers in order to support strings by using a single identifier. Also, in contrast to LogootSplit, we support reusing identifiers just as in the original Logoot. The clock value is used to differentiate between the original identifier and any reuse of the same identifier after removal. Merging works is different in our case, in the sense that messages are mutable until they are downloaded by another node after an opportunistic contact with another node. We aim to reduce not only the document metadata size as in LogootSplit, but also the metadata that is sent with each message. Instead of placing a message in a buffer from where the routing layer distribute it to other nodes (a.k.a. broadcasting in LogootSplit), we share the messages between the routing layer and the actual application. While users insert or remove characters from a string, the message is updated in real-time until it is downloaded by another node, when it becomes immutable. This way, in case we have characters that are inserted and just after a few moments they are removed (i.e. grammatical corrections which are common in text editors) we would never send an insert and then a remove message for those operations, reducing not only the network overhead, but also the number of messages that are sent to the other nodes, an approach that saves important resources for limited mobile devices present in ONs. Another difference is that our algorithm also works for cases in which a message might never be received by downloading the missing messages from the application itself on direct contact. As we already mentioned, in ONs it is possible to never receive a message and that the message might be removed from the routing layer memory, in which case we can ensure consistency only on direct contact between nodes that run the same application. We think that any algorithm that ensures consistency should be able to cope with this case scenario. The details of the algorithm are explained in Section IV.

III. CAUSAL ORDER EXPERIMENTS

In this section we present our first contribution related to causal ordering in ONs. We prove that, by using a vector clocks algorithm with some small changes, we obtain acceptable results compared to the hit rate and delivery latency of bare dissemination algorithms. The vector clocks algorithm is run on top of these dissemination algorithms with the difference that messages are withheld at a node until all the messages that causally precede it have been received and delivered to the application layer.

A. Experimental Setup

The experiments were run using MobEmu (Ciobanu, Dobre, & Cristea, 2012), an opportunistic network emulator that uses real mobility traces to simulate the behavior of different ONs routing and dissemination algorithms. The mobility traces used in the tests are Infocom (Hui & Crowcroft, 2007), Sigcomm (Pietiläinen, Oliver, LeBrun, Varghese, & Diot, 2009), and UPB (Marin, Dobre, & Xhafa, 2012). The mobility traces contain both the node contacts and information about their interests. Details can be found in Table 1. Sigcomm and UPB also contain information about social connections between nodes, while for Infocom, the social aspect needed by some algorithms had to be disabled.

The model for generating messages is similar to the experimental setup presented in Ciobanu et al. (2014). The data is modeled as messages that are generated through channels to which nodes subscribe to. Each channel is represented by a topic of interest. When a node subscribes to a channel, it is interest in any message

Table 1. Information about the mobility traces

Trace	Nodes	Duration	Type	Topics
Sigcomm	76	4 days	Conference	720
UPB	24	64 days	Academic	5
Infocom	99	4 days	Conference	35

that is generated on that channel. Nodes can generate data only in the channels they are subscribed to, and data intended for a channel cannot be received on another channel. Every node has at least one interest and it generates 30 messages per day, with the exception of last day, in the two-hour interval when most of the contacts happen (around midday in all three traces) by randomly selecting one of its interests as a channel for a message. We have to mention that for UPB and Infocom, some participant nodes did not have any interests. (Sigcomm has interests for all participants). Those nodes do not generate any messages, but they can carry messages for other nodes.

The actual number of messages that can be stored at a node for routing or dissemination is limited in order to simulate limited resources in ONs. In case the data memory is full, on receiving a new message that is or not intended for the receiver, but that might be required by another node in the future, the first received message in the data memory is removed in order to make space. As data memory values, we have used 20, 100, 400, 2500 and 5000 messages / node memory.

B. Causal Order Algorithm

Our aim is to ensure causal order for messages generated on the same channel. The tested causal order algorithm is based on the CBCAST algorithm (Birman, Schiper, & Stephenson, 1991). The differences consist in the fact that a node does not require a priori information about the destination nodes in order to send the corresponding vector clocks. In case of receiving a message from an unknown node that participates in the same channel, then its entry in the local vector clock is equivalent to a value of 0 as no previous messages have been received. At this moment, the receiving node creates a new entry for the unknown node and the algorithm continues to works as in the original version.

The memory for withholding messages that do not yet respect the causal order is limited. This memory, which we shall refer as *causality memory*, is different from the data memory, though it is common to all channels to which a node is subscribed to. Instead of removing the first message, we remove a message on which no other message in our memory depends on from the channel with the most withheld messages that do not respect the causal order. If the message is also present in the data memory, then it is also from there. Our assumption is that messages that depend causally on fewer messages should be withheld for a longer time than the ones that depend on more messages, as fewer messages must be received in order to deliver them. Because we work with an ON environment, it is possible that the preceding messages might never arrive, though this problem is addressed in our proposed algorithm presented in Section IV. The tested data memory values for withholding messages that do not respect causal order are similar to the values presented above: 20, 100, 400, 2500 and 5000 messages / causality memory.

In future work, we might test an algorithm based on causal barriers (Prakash et al., 1997) in order to see the amount of decrease in metadata size, though only if implementing an algorithm that works similar to our proposed algorithm for total ordering. That is, the metadata for a message situated at the routing layer might be used for multiple messages situated at the application layer. Though, in the experiments presented in this section we are only interested in the hit rate and the delivery latency for the application layer, not in the amount of metadata that can be saved.

C. Results

We have run the experiments for two algorithms: Epidemic (Vahdat & Becker, 2000) with limited memory and ONSIDE (Ciobanu et al., 2014), firstly without ensuring any order (i.e. messages are delivered to application layer as they are received), and then with our causal order algorithm on top of them. We focus on hit rate and delivery latency. Their definitions are similar to the ones proposed in (Ciobanu et al., 2014), where hit rate represents the ratio between the number of messages successfully delivered at interested nodes, just that for the application layer instead of the routing layer, while delivery latency is the average amount of time passed between message generation and application layer delivery time. We are not interested in other metrics as hop count, which is the average number of nodes that carried a message until the destination, and delivery cost, which is the ratio between the total number of exchanged messages over the total number of generated messages, as their values are similar for both no order and the causal order algorithm, with differences under 0.5%. The reason for the difference consists in the fact that, once the causality memory gets full, we remove a message from it based on the rules presented at Section III-B, which can lead to the removal of a message from the data memory. This removal of a message from the data memory leads to the small variation in the results, but the variation is insignificant.

Figure 1 contains the experiments for the Sigcomm trace. The total number of generated messages is 5100. The Epidemic algorithm refers to the possible maximum hit rate that can be achieved by any algorithm as it was run without memory limitations. In case of running the algorithms with *no order*, messages are delivered to the application layer as soon as they are received, while running the algorithms with *vector clocks* on top, messages are withheld until causality is respected. We can observe that, for both Limited Epidemic and ONSIDE algorithms, the hit rate / latency results are acceptable for 20 and 100 messages / data memory, good for 400 messages / data memory and really good for 2500 and 5000 messages / data memory. The causality memory has the same values as the data memory at the routing layer (20, 100, 400, 2500, 5000). From now on, when we mention we have *N messages / memory*, we refer to having *N* messages for the data memory, and another *N* messages for the causality memory.

When it comes to the application layer hit rate, for a memory of 20 messages, the difference between Limited Epidemic with vector clocks is about 1.85 times less than Limited Epidemic without any order, 1.5 times for 100 messages, 1.25 times for 400 messages, 1.005 for 2500, and the same for 5000 messages. The reason for almost the same values for 2500 and 5000 messages consists in the fact that the memory is big enough to hold most of the generated messages, ensuring that most causal preceding messages will be received. For ONSIDE, we have a difference of 1.8 times for a memory of 20 messages, 1.3 times for 100 messages, 1.03 times for 400 messages, and the same for 2500 and 5000 messages. The results for ONSIDE are a lot better than Epidemic, especially for 100 and 400 messages, and the reason consists in the fact that ONSIDE obtains, in average, better hit rates than Limited Epidemic, leading to the reception of more messages that causally precede other messages.

Figure 1. Sigcomm trace results

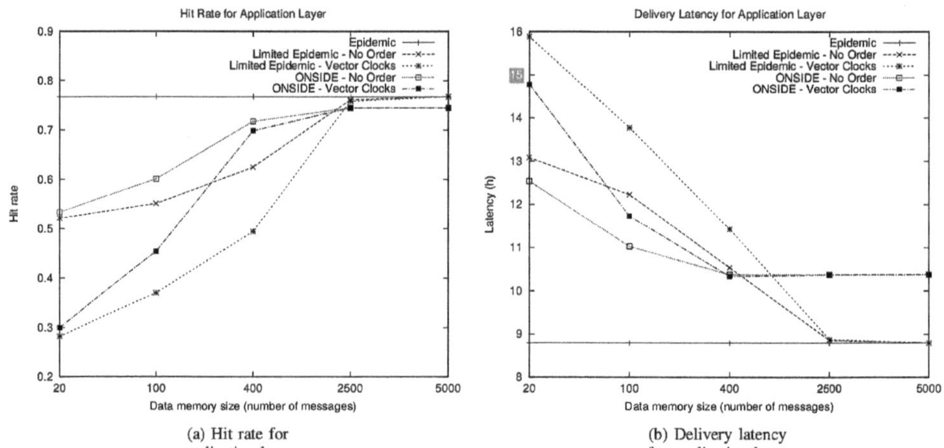

(a) Hit rate for application layer

(b) Delivery latency for application layer

When it comes to the application layer delivery latency, the differences are smaller. For Limited Epidemic, the difference is around 1.21 times for 20 messages, 1.12 times for 100 messages, 1.08 times for 400 messages, 1.002 times for 2500 messages, and the same for 5000 messages. For ONSIDE, the differences are 1.18 times for 20 messages, 1.06 times for 100 messages, and around the same for 400, 2500 and 5000 messages. The reason for better results for ONSIDE is the same as in the case of the hit rate.

Figure 2 contains the experiments for the UPB trace. Because the UPB trace took 64 days compared to 4 days in the other two cases, the total number of generated messages is much higher, reaching 37200. The differences between running the algorithms with *no order*, and running them with *vector clocks*, are higher because the UPB network is sparser (fewer contacts). For Limited Epidemic with 20 messages per memory, the hit rate for vector clocks is 3.1 times less than no order, 3.2 times for 100 messages, 3.05 times for 400 message, 2 times for 2500 messages, and 1.7 times for 5000 messages. For ONSIDE, it is 3.1 times for 20 messages, 3.15 times for 100 messages, 2.85 times for 400 messages, 1.85 times for 2500 messages, and 1.5 times for 5000 messages. We can observe that, if we increase the memory, than we will obtain smaller differences between running the algorithms with no order vs. vector clocks. As for a comparison between ONSIDE and Limited Epidemic, see Ciobanu et al. (2014). In this chapter we are interested to compare causal order algorithms vs. no order algorithm for message delivery to application layer.

For the application layer delivery latency, we obtain smaller differences just as in the case of Sigcomm. For Limited Epidemic, we have a difference of 1.23 times for 20 messages, 1.29 times for 100 messages, 1.19 times for 400 messages, 1.04

Figure 2. UPB trace results

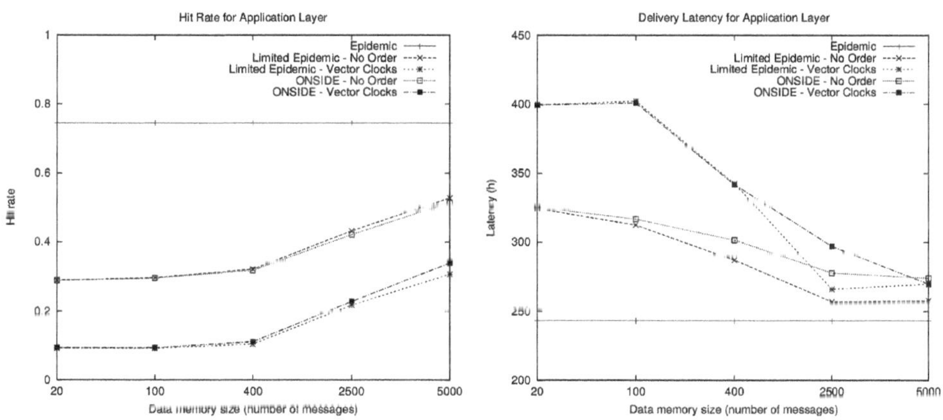

(a) Hit rate for application layer

(b) Delivery latency for application layer

times for 2500 messages, and 1.05 times for 5000 messages. For ONSIDE, we have 1.23 times for 20 messages, 1.27 times for 100 messages, 1.13 times for 400 messages, 1.07 times for 2500 messages, and less than 1 for 5000 messages. We get less than 1 (i.e. smaller delivery latency for vector clocks vs. no order) because delivery latency is calculated only against the total number of delivered messages, which is less for vector clocks.

Lastly, Figure 3 contains the experiments for the Infocom trace. The total number of generated messages is 5670. For Limited Epidemic with a 20 messages memory, the vector clocks algorithm has a 1.95 times lower hit rate than the algorithm without any order. For 100 messages, the difference is 1.4 times, for 400 messages it is 1.05, and for 2500 and 5000 messages the results are the same. For ONSIDE, the difference is 1.83 times for 20 messages, 1.29 times for 100 messages, 1.01 times for 400 messages, and the same for 2500 and 5000 messages.

As for delivery latency, the difference is about 1.24 times for 20, 100 and 400 messages, 1.08 times for 2500 messages, and around the same for 5000 messages when using Limited Epidemic. When using ONSIDE, the difference is 1.16 times for 20 messages, 1.18 times for 100 messages, 1.1 times for 400 messages, and around the same for 2500 and 5000 messages.

As a conclusion, in most cases, the hit rate and delivery latency for the application layer achieve good results, even though messages are withheld until causality is respected. Sigcomm and Infocom have acceptable results for both 20 and 100 messages per memory, while for 400 messages and above the results are good or really good. For UPB, which is a sparser ON, the results are better when having 2500 and 5000 message per memory.

Figure 3. Infocom trace results

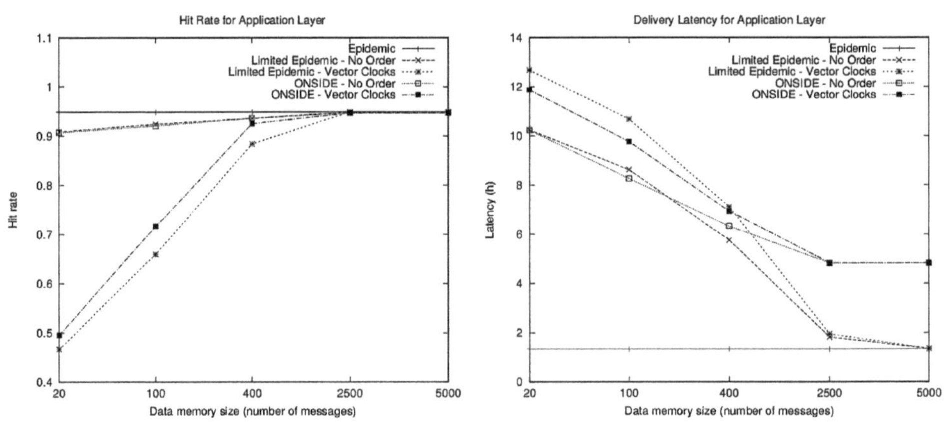

(a) Hit rate for application layer

(b) Delivery latency for application layer

IV. OPPORTUNISTIC LOGOOT FOR TOTAL ORDERING

In this section we propose a new algorithm called *OpportunisticLogoot* to ensure total order in ONs. The algorithm is based on Logoot (Weiss et al., 2008) and it works similarly to LogootSplit (Andre et al., 2013), though it has been adapted to work in ONs where direct and real-time communication is not a given fact.

A. OpportunisticLogoot Operations

Out algorithms supports two operations for modifying the contents of an application:

1. *insert(pos, base_element)*, which inserts a *base_element* at position *pos;*
2. *remove(pos)*, which removes a *base_element* stored at position *pos.*

A *base_element* represents the smallest possible element in the contents of an application. For example, in a text editor, the *base_element* is a character, while in a chat it is the whole message until the user presses *enter*. Applications will work only with this API. It is not up to the application layer to handle merges between multiple *base_elements*, with this task being ensured by the order algorithm (i.e. our proposed algorithm). The reason for this approach is to provide a simple API which could be used for more types of applications than only text editors. The application just have to provide *base_elements* or split the contents in *base_elements* (e.g. for a text editor, the string "abc" would be split in "a", "b" and "c").

B. Content Model

In order to support a total order between all *base_elements*, we use identifiers similar to the ones presented in Logoot. As mentioned in related work, Logoot uses unique identifiers generated in a *densely ordered* set in order to ensure global order between all participant nodes. At any moment it is possible to create a new identifier between any two existing identifiers, and deleted elements are removed immediately from memory, without being transformed into tombstones. Just as Logoot, we require only causality, and it can be obtain with any causal order algorithm based on vector clocks (Schiper, Eggli, & Sandoz, 1989)(Birman et al., 1991) or causal barriers (Prakash et al., 1997). As for the identifier notation, we use the same convention as in the original Logoot article, which is a bit different from the LogootSplit article.

Our content model is identical to the Logoot document model, just that, instead of *lines*, we use a sequence of *base_elements*, which we shall simply refer from now on as *sequence*. The construction of sequences will be explained in a subsequent paragraph. A sequence is defined by a pair <*id, sequence_content*>, where *id* represents the unique identifier and *sequence_content* is a concatenation of base_elements representing the sequence itself. There are two default sequences: *<id_left_limit, null_sequence>* and *<id_right_limit, null_sequence>*. These sequences cannot be removed and they are used to make sure that we always create identifiers only between them.

The *id* field is represented by a list of pairs $<x_i, s_i, l_i, r_i>$ and a logical clock clk_s, where x_i is an integer, s_i is the unique site id, l_i is the id of the first base_element in the sequence (or the range left limit), and r_i is the id of the last base_element in the sequence (or the range right limit). Both l_i and r_i are shorts, $r_i \geq l_i$ and a sequence has $r_i - l_i + 1$ *base_elements*. The clk_s is the logical clock of the site *s* which created the id and it is used to make sure that identifiers are unique in case of reusing the same list of pairs after removing an element, a behavior similar to the original Logoot. The difference from the original Logoot is that we added the fields l_i and r_i to support sequences of *base_elements* (i.e. *lines* in the Logoot model), which is an addition similar to LogootSplit. In summary, an *id* is defined as:

Definition 1: $id = \langle x_0, s_0, l_0, r_0 \rangle, \langle x_1, s_1, l_1, r_1 \rangle, \ldots \langle x_{n-1}, s_{n-1}, l_{n-1}, r_{n-1} \rangle, \langle clk_s \rangle,$ where n is the number of pairs $\langle x_i, s_i, l_i, r_i \rangle$.

In order to obtain a total order between identifiers, we define the comparison relation between them. First we define the relation $<_{pair}$ that compares two pairs, and then we define the relation $<_{id}$ that compares two identifiers.

Definition 2: *Let* $pair_1 = \langle x_1, s_1, l_1, r_1 \rangle$ and $pair_2 = \langle x_2, s_2, l_2, r_2 \rangle$. Then,

$$pair_1 <_{pair} pair_2 \Leftrightarrow x_1 < x_2 \vee (x_1 = x_2 \wedge s_1 < s_2) \vee (x_1 = x_2 \wedge s_1 = s_2 \wedge l_1 < l_2)$$

Definition 3: *Let* $id_1 = p_0, p_1, \ldots, p_{n-1}, \mathrm{clk}_1$ *and* $id_2 = q_0, q_1, \ldots, q_{m-1}, \mathrm{clk}_2$ *be two identifiers, where* p_i *or* $q_i = \langle x_i, s_i, l_i, r_i \rangle$. *Then,*

$$id_1 <_{id} id_2 \Leftrightarrow \exists k \leq m, (\forall l > k, p_l = q_l) \wedge (k = n + 1 \vee p_k <_{pair} q_k)$$

It is important to mention that, for any two consecutive identifiers id_1 and id_2, if all pairs until the last pair are equal (i.e. same site identifier generation), then $r_1 < l_2$, where r_1 is the range right limit for id_1 and l_2 is the range left limit for id_2 for the last pairs. It is not possible for the same site to have superposition for two different sequences as this will lead to have *base_elements* on the same positions.

C. Insertion

In order to execute an insertion $insert(pos, base\_element)$, we need to find either the sequence where to insert the new *base_element*, or the previous and next sequences in case *base_element* must be inserted between two sequences or merged to one of them, search that depends on the value of *pos*. In order to find the correct place for *base_element*, we first explain how we calculate the number of *base_elements* in the application contents.

Definition 4:

$$num\_of\_base\_elements = \sum_{i=0}^{n-1} last\_pair\_range_{id_i},$$

where n *= the total number of sequences and* $last\_pair\_range_{id_i} = r_{m-1} - l_{m-1} + 1$, *and* m *= the total number of pairs for* id_i.

Finding the correct sequence(s) where to insert the new *base_element* is done in a linear time in terms of the number of sequences, i.e. $O(n)$, which is different from the number of *base_elements*. Depending on how long each sequence is and the value *pos*, we can end up in two scenarios:

1. Insertion between two sequences;
2. Insertion in the middle of a sequence.

1. Insertion between Two Sequences

Let us assume we have two sequences seq_1 and seq_2 between which we have to insert a new *base_element elem* at a site with the id *s*. We shall also assume that a sequence can hold a maximum number of *SEQ_LIMIT base_elements*. *SEQ_LIMIT* will be used to limit the size of messages that will be sent through the routing layer in the ON.

When inserting a new element *elem* between two sequences seq_1 and seq_2, we can distinguish four cases:

1. Both seq_1 and seq_2 identifiers where generated at sites with ids different from *s*;
2. seq_1 identifier was generated at the current site with id *s* and seq_2 identifier was generated at a different site;

3. seq_1 identifier was generated at a different site, while seq_1 identifier was generated at the current site with id *s;*
4. Both seq_1 and seq_2 identifiers where generated at the current site with id *s*.

In the first case, we will create a new sequence *new_seq* with a single *base_element elem* that will have an identifier between seq_1 and seq_2 identifiers. The rule to generate this identifier is similar to Logoot:

Definition 5: *A new identifier new_id inserted between* $id_1 = p_0, p_1, ..., p_{n-1}, clk_1$ *and* $id_2 = q_0, q_1, ..., q_{m-1}, clk_2$, with p_i *or* $q_i = \langle x_i, s_i, l_i, r_i \rangle$, *must have a shortest number of pairs* $p_0, p_1, ..., p_i < x, s, l, r >$, $i \leq n$ *such that* $id_1 <_{id} new\_id <_{id} id_2$.

The *'x'* value can be generated using multiple rules, but we have used only two. The first rule is to generate it in increments. That is, $x = x_{i+1} + 1$, where x_{i+1} is the first integer of the previous sequence id_1 as mentioned above, and $i + 1 < n$, or $x =$ 1 if $i + 1 = n$. This rule is useful for applications that usually append elements at the content end, just as chat applications. The second rule is useful for share text editors, and it creates a random *x* between $[x_{i+1} + 1; \max\_value)$ if $i + 1 < n$, or $[1; \max\_value)$ if $i + 1 = n$.

In the second case we will try to append *elem* to seq_1 and increment the last pair range right limit *r* of seq_1 identifier. This is possible only if the number of elements in seq_1 is < *SEQ_LIMIT* or if *r* < *max_value*. Otherwise, instead of merging, we will create a new sequence with only *elem* inside, just as in the first case.

The third case is similar to the second case, just that instead of appending we will try to place *elem* at the beginning of seq_2 and decrement the last pair range left limit *l*, as long as *l* > 0 and the number of elements in seq_2 is < *SEQ_LIMIT* .

In the last case we will try to either place *elem* at the end of seq_1 or at the beginning of seq_2. It is important to notice that, in case all pairs of seq_1 and seq_2 are equal, with the except of last pair that is different only in terms of range left and right limits, then we are allowed to merge the new element with seq_1 or seq_2 only if $r_1 + 1 < l_2$, where r_1 is the range right limit of seq_1 last pair and l_2 is the left right limit of seq_2. We are not allowed to have *base_elements* that might end on the same position.

2. Insertion in the Middle of a Sequence

Let us assume we have a sequence *seq* and an element *elem* which we try to insert in the middle of *seq* (i.e. between the *l* and *r* value of the last pair of the *seq* identifier). We can distinguish two cases:

1. The number of elements in *seq* is < *SEQ_LIMIT*, which means that we can directly insert *elem* at its right position inside *seq*without splitting it;
2. The number of elements in *seq* is equal to *SEQ_LIMIT*, which means that we have to split *seq* in two sequence; the sequence can be split using any rule, though we have chosen to use random split.

In both cases we have to make sure that our new identifier does not superposition with the previous or next sequences. Otherwise, we split the sequence in two, and, where *elem* must be inserted, we create a new sequence with a single element *elem* using the same mechanism as in the previous Section IV-C1.

It is important to mention that merging is allowed only for *mutable messages,* otherwise we will threat insertion as in the first case presented in Section IV-C1. The concept of mutable messages will be explained in a further paragraph. Also, with every insertion that does not end in a merging operation, we will increment the clk_s value of the site *s*.

D. Removal

Removal is a lot simpler. In case a sequence has only one element, we just delete it. In case the sequence has more than one element, we have two cases:

1. Removal at the beginning or the end of the sequence, which means that we just accordingly update the range left or right limit of the last pair;
2. Removal in the middle of the sequence, which means that we split the sequence in two, create two identifiers that are identical to the first one and update the range right limit of the left sequence and the range left limit of the right sequence to not include the removed element.

In case the site id *s* is different from the site id that generated the sequence, or the message is immutable, then we will create a new mutable removal message that will be placed on the routing layer to be disseminated to the other nodes in the ONs. Consecutive removals are merged together until the message becomes immutable. In case the message is mutable, we just remove the element from the corresponding sequence, without generating a new removal message. This is one of the *key aspects* of our algorithm, as *base_elements* that are just inserted, and after a few moments they are removed (i.e. grammatical corrections), will never generate a message that waste resources in the ON.

The complexity to execute the remove operation is identical to the insert operation. That is $O(n)$ in terms of the number of sequences.

E. Mutable Messages

Mutable messages represent one of the two contributions we have added to the Logoot algorithm in order to cope with the environment of ONs.

Definition 5: *A **mutable** message is a message shared between the routing and the application layers, in which the application can insert or remove elements, without generating a new message for the routing layer. A message becomes **immutable** once it is downloaded for the first time by another node based on the dissemination algorithm used by the ON.*

An immutable message cannot be converted back into a mutable message. The scope of the mutable messages is to reduce the metadata of the application contents that travels with each message, and to avoid generating messages for elements that will be removed in the close future before having an opportunistic contact after which the message is downloaded. Attempts of downloading a message that do not succeed (e.g. the contact duration is too short to download the message, or the dissemination algorithm decides not to download our message), will not change the state of the message. Only a successful download can make a message immutable.

The conditions to insert or remove elements from a mutable message are the same as the conditions to merge or remove a *base_element* from an existing sequence, as explained in the previous paragraphs.

It is important to mention that our algorithm supports reusing the same identifiers as the original Logoot, while LogootSplit does not support it. In order to avoid reusing the same identifiers, LogootSplit keeps a track of the minimum and maximum offsets generated for each identifier. The benefit of this approach is that identifiers can be merged at the destination, though this does not help with reducing the metadata size for messages that are already at the routing layer. The cost is that we have a number of identifiers that cannot be reused, even if we have space to add *base_elements* to them, which will lead in some cases to larger metadata size for the messages to disseminate.

We have chosen to be able to reuse identifiers in order to have an identifier space as compact as possible. By reusing identifiers, when we want to create a new identifier between two existing identifiers that are different only in terms of the last pair left and right limits, then we will create it right there without adding a new pair. In LogootSplit, we would have to add a new pair as we are not allowed to reuse that space. This will lead to extra metadata that must travel with the message. Our target is not to reduce the metadata for the end document, but to reduce the metadata for the messages that travel at the routing layer and add overhead to the ON. Though, in future work, it would be useful to implement both of them and compare them.

The reason our algorithm does not support merging at the destination site will be explained in the next Section IV-F.

F. Downloading Missing Messages

The second contribution of our algorithm consists in downloading missing messages on direct contact between two nodes that run the same application with the same contents. In contrast with traditional networks where lost messages are sent again, in an ON we can end up with messages that are never received, even though two nodes that run the same application might have a direct contact in the future. This scenario is possible because the data memory for the routing layer is limited. As mentioned in Section III-A, once the data memory is full, and the routing algorithm decides to download a new message, the first received message is removed in order to make space. Due to this reason, some messages might end up being lost forever. As this scenario is the norm rather than the exception in ONs, we have to design algorithms that can cope with messages lost forever.

Because messages might be lost forever, we can end up with different views for different users, even if the total order is still respected. Unfortunately, this means we are in an inconsistent state. We shall now define the condition under which our algorithm, with the downloading missing message feature, ensures consistency:

Definition 5: *In an opportunistic network, consistency is ensured only after two or more nodes that run the same application are in direct contact for a sufficient time to synchronize their contents views.*

In other words, this is a case of eventual consistency. In order to ensure this consistency, instead of only downloading messages at the routing layer, we shall go up to the application layer and download the actual missing parts. This is possible only on direct contact between two nodes participating in the same application because, even if the routing layer has removed the message from its data memory, the application will keep the contents as long as the user wants to have it.

Downloading the missing parts of the contents is done only after the routing layer has downloaded its messages, in the limits of remaining resources (i.e. contact duration or battery). The routing layer has priority because it downloads messages for any application, including previous messages intended to us that have not been removed from its data memory. Plus, we aim to ensure consistency only if the target nodes spend a long enough time in direct contact.

The download algorithm works as follows:

1. Firstly, we download all or part of the identifiers from the other node (part of them if the contents are too large);
2. Secondly, we compare the identifiers one by one to see which message are missing;
3. Thirdly, we download the missing messages;
4. Lastly, we repeat from step 1 until all messages have been downloaded.

Though the algorithm seems simple, there is a problem we must take into consideration: how do we identify the messages that were removed from one node contents, and which are still present in the other node contents. We do not want to end up with adding again data that a user removed. At this moment, we shall use the logical clock value *clk*. A node remembers the highest clock value clk_{node} of a sequence identifier for each node he has received a message from. Because messages are delivered only in causal order by the routing layer, we can be sure that, if the clk_{node} value stored for a given node is higher than the clock value of a sequence present at the other node, and we do not have that sequence present in our contents, then that sequence has been removed. The reverse also applies: if the clk_{node} value is smaller than the clock value of the sequence identifier, then that sequence must be downloaded.

Both insertion and removal can be done at the level of *base_elements*, as the clock value does not change for their identifier once their corresponding message has been placed on the routing layer and it is transformed into an immutable message. Because the clock value *clk* cannot be changed in order to correctly ensure consistency on direct contact, we cannot perform identifier merging at the destination site. By merging the identifiers for different sequences we have to remove one of the clock values, which will make our algorithm incorrect.

V. TOTAL ORDER EXPERIMENTS

In this section we present our last contribution related to total ordering in ONs by using the algorithm in Section IV. We prove that our algorithm not only reduces the metadata size that travels with each message, but also improves other metrics as hit rate, delivery latency, hop count and delivery cost, when considering a message as the basic unit that is exchanged between nodes.

A. Experimental Setup

The experimental setup is similar to Section III-A setup, and we used again the MobEmu emulator with the same mobility traces. The data memory size and the causality memory size have the same values: 20, 100, 400, 2500 and 5000 messages / memory, though the message generation is different. For some nodes interests, we have used a series of chat traces to generate messages instead of the mechanism initially used, in order to compare the size of the metadata against the size of the messages payload. Unfortunately, even if the chat traces we have used contain real discussions, they were generated in a server-client architecture where all messages are already totally ordered. In order to cope with this problem, we have created a model that generates concurrent messages, as explained in the next paragraph. Another problem is that chat applications support only the append operation, and there is no removal. There are still enough application that work using this model (e.g. posting on social networks), but we are working on integrating Wikipedia pages with MobEmu in order to support both insertion and deletion operations at any position, which will be present in a subsequent paper.

The number of chat traces we have used is 10 and the number of participants varies between 4 and 5 people per chat, while the size a chat message can have from 0 characters (blank message when a user only presses enter) to 759 characters. In order to have parallel chat discussions during every run of the emulator, each chat has been assigned to one interest (a.k.a. channel). Chats are randomly assigned to channels, by starting from the channel with the most subscribers in order to make sure that all chat traces will be used. The condition to assign a chat is to have at least 4, respectively 5 subscribers per used channel, and the chat participants are chosen randomly from the subscribers in case their number is greater than the number of chat participants. The nodes that are subscribed to a channel, but are not part of a chat discussion, only receive messages, without generating any, and they have the download missing messages component of our proposed algorithm disabled. One node can participate in multiple chats by using different channels.

To make sure that all of the chat messages are used and evenly spread, they are generated at fixed times: a message is generated at the time

$$i * (traceEndTime - traceStartTime) / \max NumOfMessagesPerParticipant,$$

where i is the message index for a chat participant. In other words, all chat messages are independent from each other, with the exception of the messages generated by the same user and the messages generated after receiving a message from another user on the same channel. These later messages depend causally on each other. Our approach to generate chat messages has limitations, but unfortunately we cannot model real chat scenarios in ONs as we are unaware of any such traces. The traces should not only be distributed, but they should also support delays, as real time discussions are not a given fact in ONs.

Not all messages are generated using the above rules, as not all channels have been assigned to chat discussions. The remaining channels are used to simulate network traffic, by using the same generation mechanism as in Section III-A.

B. Results

We have run the experiments for both Epidemic with limited memory and ONSIDE, just as in Section III, but, due to space constraints, we present the results only for ONSIDE. The results for Epidemic tend to be similar to ONSIDE, with a few changes in case of delivery latency, delivery cost and hop count. The differences mainly appear from the fact that Limited Epidemic has worse values for these metrics when running with 20, 100 and 400 memory, just as highlighted in (Ciobanu et al., 2014), but this section focuses on comparing our proposed algorithm against the original Logoot, the vector clocks based algorithm and the routing algorithm with no order, and not between routing algorithms.

Our proposed algorithm OpportunisticLogoot has been run five times with 1,2,3,5 and ∞ as the maximum number of messages allowed to be merged (i.e. the *SEQ_LIMIT* parameter from Section IV-C). MobEmu considers a message as the basic unit exchanged between nodes, and not the size of the message. The reason we have chosen 1,2,3 and 5 as limits is because a chat message has 759 characters in the worst case, which is 3795 characters in case of a merge limit of 5 messages, the equivalent of less than 4 KB. Exchanging 4 KB instead of 0.8 KB per message payload, without adding the message overhead incurred by other layers and the transport protocol, does not represent a significant difference, though we recognize that this is still a limitation in comparison to real world scenarios. ∞ was chosen as a theoretical limit.

Another parameter we varied is the number of messages that can be exchanged during a contact (which we shall refer as the network capacity), between 1 and ∞ messages per second. Figures 4 and 5 contain the results obtained for ∞ as the network capacity, while the routing algorithm below the order algorithms is ONSIDE, and the mobility trace is Sigcomm. The total number of generated chat messages is 3343, while the number of extra messages is 2730, resulting in a total of 6073 generated messages.

Figure 4. Sigcomm trace results, ONSIDE routing algorithm, ∞ network capacity

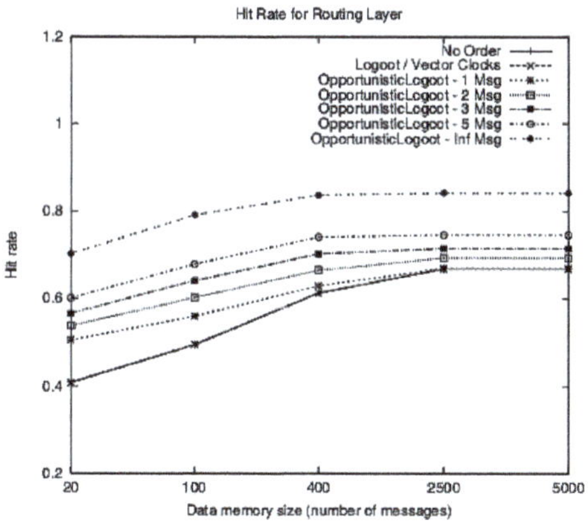

(a) Hit rate for routing layer

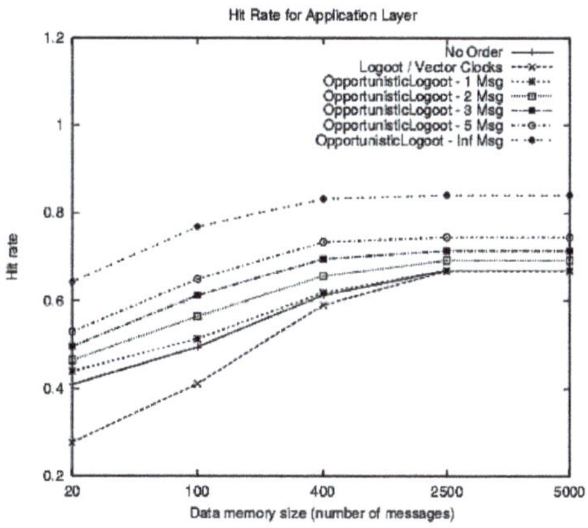

(b) Hit rate for application layer

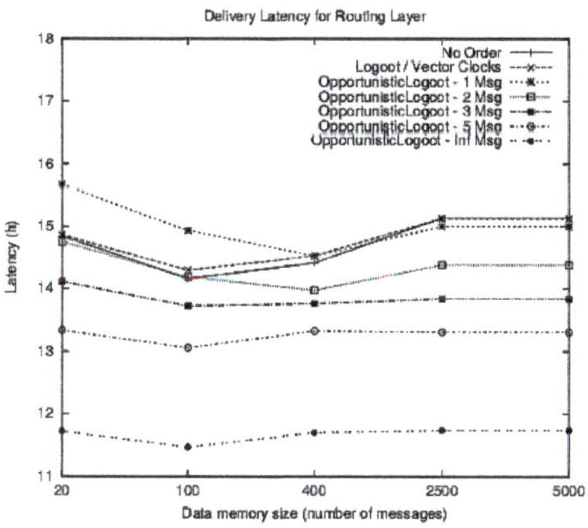

(c) Delivery latency for routing layer

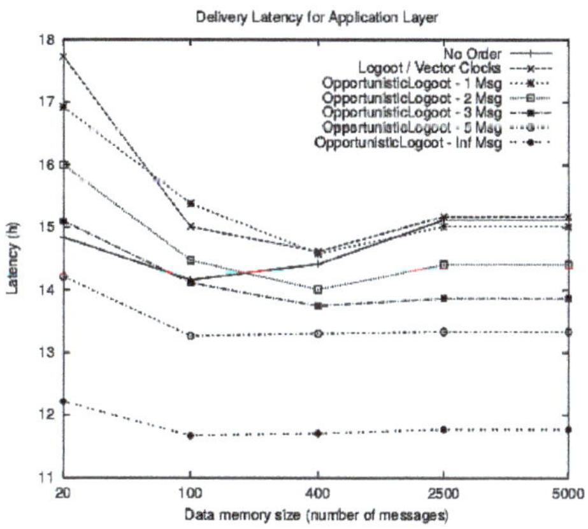

(d) Delivery latency for application layer

We can observe in almost all cases that the hit rate and the delivery latency are better when increasing the number of messages that can be merged, while delivery cost and hop count are better for 400 messages. For the hit rate, our proposed algorithm obtains better hit rates when compared to Logoot and vector clocks, and even when compared to the routing algorithm with no order. There are two reasons for this effect. Firstly, when increasing the number of messages that can be merged, we decrease the total number of messages on the routing layer, just as observed in subfigure 5h, resulting in less messages to transport while the data memory remains the same. Secondly, we support the download of missing messages on direct contact, while the routing algorithm does not have enough application layer information to know which messages to download, resulting in lower hit rates even than running

Figure 5. Sigcomm trace results, ONSIDE routing algorithm, ∞ network capacity

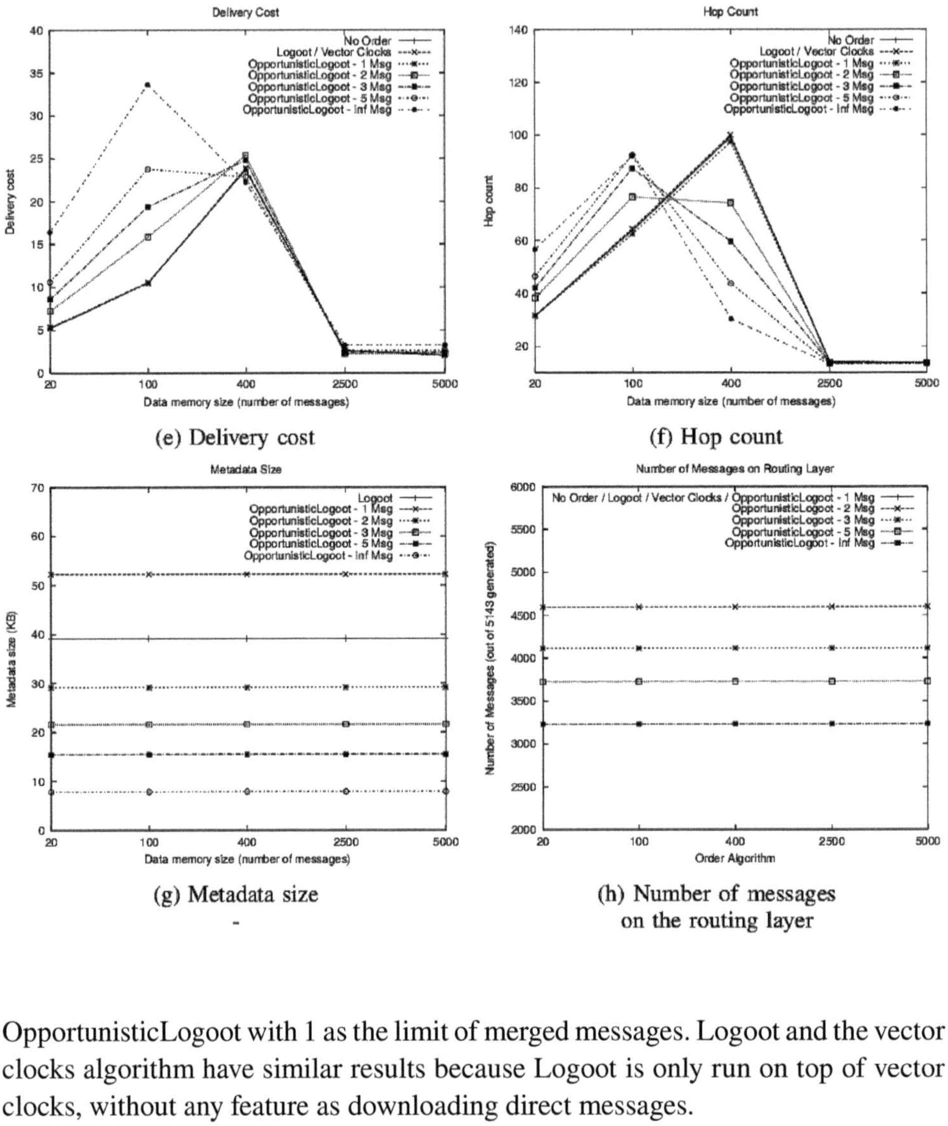

(e) Delivery cost

(f) Hop count

(g) Metadata size

(h) Number of messages on the routing layer

OpportunisticLogoot with 1 as the limit of merged messages. Logoot and the vector clocks algorithm have similar results because Logoot is only run on top of vector clocks, without any feature as downloading direct messages.

When it comes to metadata size (Figure 5g), we get almost the same results for any memory size, with variations under 0.5%, if any. Due to this reason, we will use an average size for all memory values, only differentiating them by the order algorithm, though, in case of higher variations, we shall explain each of them separately. Variations can appear because, as we change the memory size, the routing algorithm can take the decision to download new messages, which transforms them into immutable messages. Making messages immutable earlier can lead to an increase in the numbers of identifiers, which results in an increase of metadata size.

The original Logoot takes around 39 KB for all chat messages, while the total payload size (a.k.a. the total size of the chat messages) is 174 KB, resulting in 22.5% metadata size in regards to payload size. OpportunisticLogoot with 1 message as merge limit takes 52 KB, due to the extra members range left limit and range right limit, resulting in an increase of 7.5% (i.e. 30% metadata size compared to payload size). But once we have 2 messages as the merge limit, we reduce the metadata size to 29 KB, which is 16.5% compared to payload size. For 3 messages, the metadata takes 21.5 KB (i.e. 12.5%), while for 5 messages, the metadata size is 15.5 KB (i.e. 9%) As for ∞, the metadata takes around 8 KB (i.e. 4.5%). The metadata size of the vector clocks or the metadata size of the routing algorithm are not included here, as we are not interested in comparing them against algorithms that are able to ensure total order (we only consider the identifiers size). As a conclusion, in case the messages have a high payload size, it would be better to use the original Logoot with only downloading the missing messages enabled than running our algorithm with 1 message merge limit, though in all other cases, our algorithm performed better.

Figures 6 and 7 contain the results for running the order algorithms while the network capacity is limited to 1 message per second. We can observe that the results are similar, just that most metrics have a degraded performance when comparing to ∞ network capacity (the y scale is different for these experiments). The reasons for which OpportunisticLogoot performs better in most cases are the same as for ∞ network capacity. The metadata size is similar to the ∞ case due to using the same mobility trace, as it depends on the contacts between the nodes (chat messages can be merged while they are still mutable, which is until the base message is firstly downloaded), and the total number of generated chat messages, parameters that have mostly remained the same.

Figures 8, 9, 10 and 11 contain the UPB results. The total number of generated chat messages is 1786, while the number of extra messages is 0. We can generate only 1786 messages (the equivalent of 5 chat discussions) out of a maximum of 3343 messages contained by the 10 chat discussions, because UPB has only 5 topics in total. As chat discussions are assigned per channel, and a channel is represented by a topic (a.k.a. interest), we can only have 5 chats in total, without leaving any channels for extra messages. As for hit rate, delivery latency, delivery cost and hop count, the results are similar to Sigcomm. We observe that, when increasing the limit of messages that can be merged, almost all metrics improve due to downloading missing messages and reducing the total number of messages at the routing layer. When it comes to metadata size, results have a similar distribution as Sigcomm, just that the numbers are different due to fewer chat messages and different contacts. In total we have 102 KB of chat messages payload. For Logoot, we have around 20.5 KB of metadata size, which is 20.5% of metadata size compared to payload size. For OpportunisticLogoot with 1 message as merge limit, we have 28 KB of

Figure 6. Sigcomm trace results, ONSIDE routing algorithm, 1 message / second network capacity

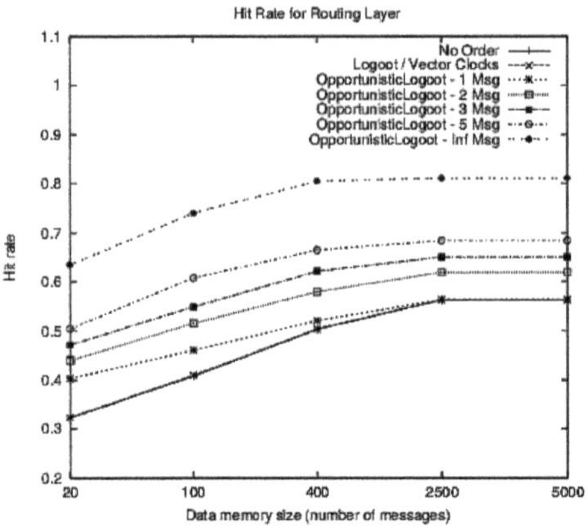

(a) Hit rate for routing layer

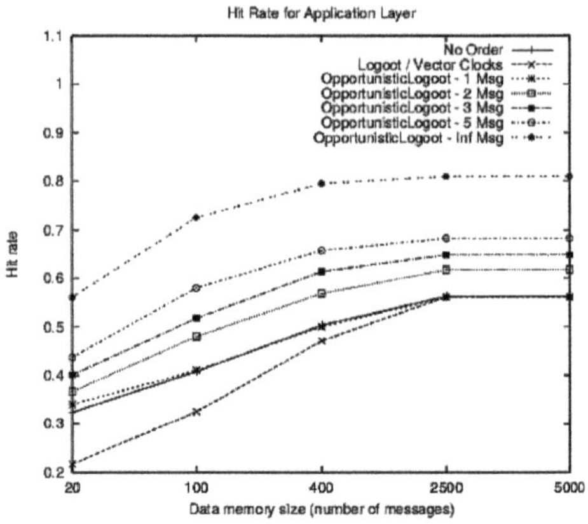

(b) Hit rate for application layer

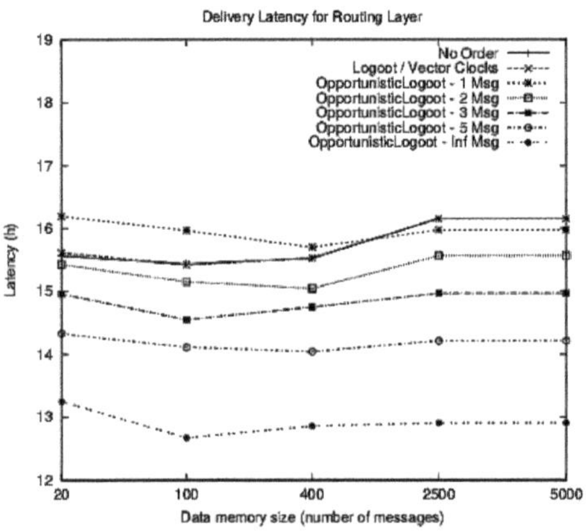

(c) Delivery latency for routing layer

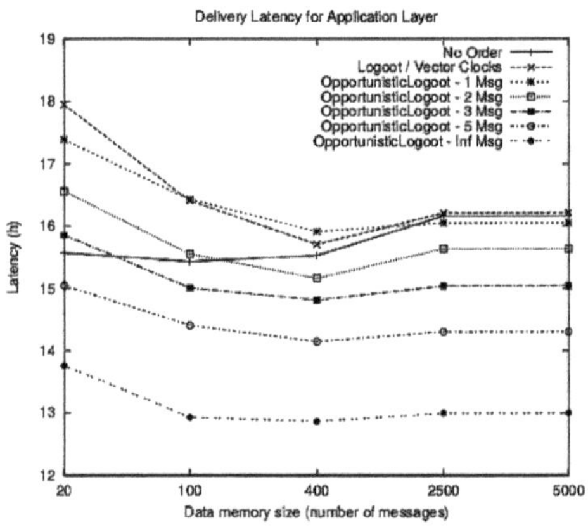

(d) Delivery latency for application layer

metadata size (i.e. 28%), while with 2 messages it takes 15.5 KB (i.e. 15.5%). With 3 messages as merge limit it takes around 12 KB (i.e. 12%), while a merge limit of 5 messages leads to 9.5 KB of metadata (i.e. 9.5%). Lastly, for ∞, the metadata takes 6.5 KB (i.e. 6.5%).

As for running the algorithms with 1 message / second network capacity, even if they have a degradation in performance for hit rate compared to ∞ network capacity (see *y* scale), the differences between Logoot, the vector clocks algorithm, and ONSIDE routing algorithm without any order, versus our proposed algorithm, tend to be the same as in the case of the ∞ network capacity, just as for Sigcomm.

Figure 7. Sigcomm trace results, ONSIDE routing algorithm, 1 message / second network capacity

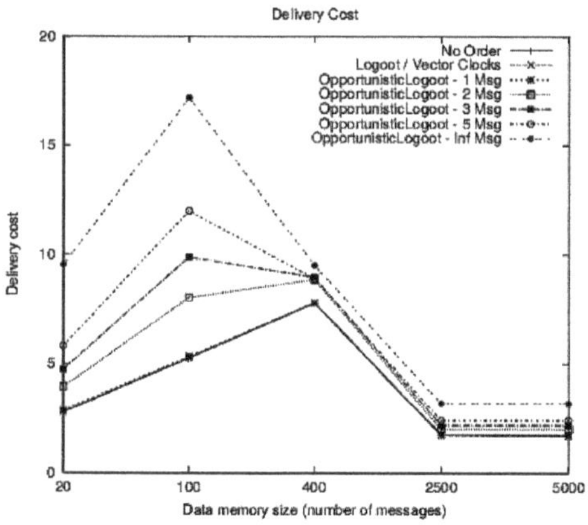

(e) Delivery cost

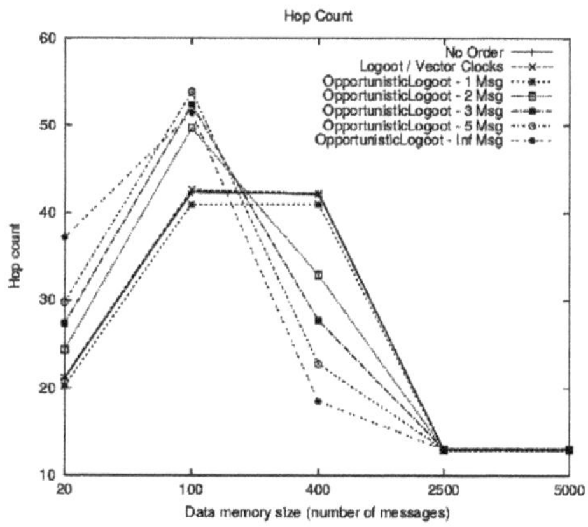

(f) Hop count

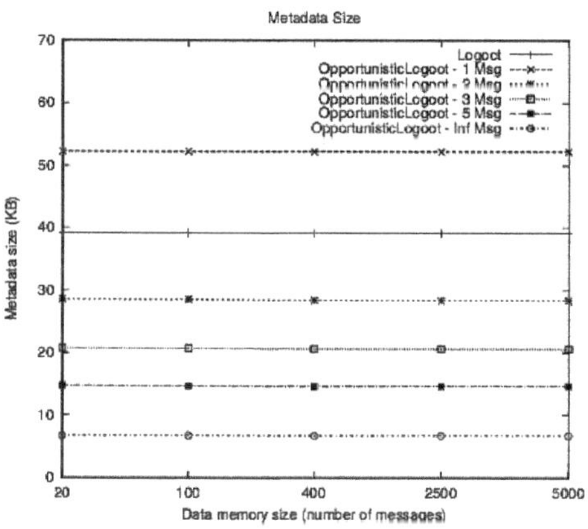

(g) Metadata size

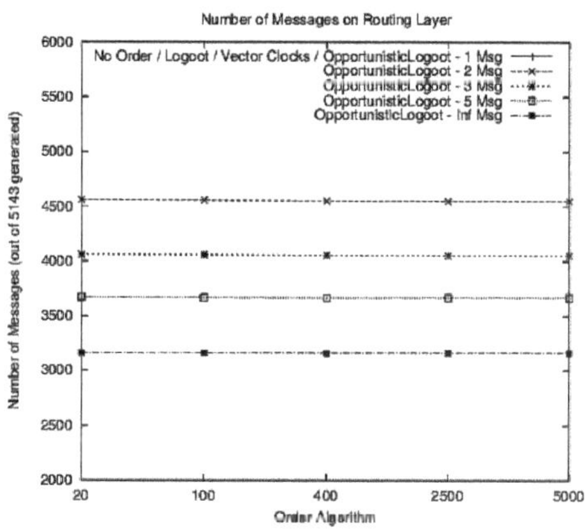

(h) Number of messages on the routing layer

Lastly, Figures 12, 13, 14 and 15 contain the results for Infocom. The total number of generated messages is 7393, out of which 3343 are chat messages, and 4050 are extra messages. Though for Sigcomm and UPB, our algorithm with two or more messages merge limit outperforms in almost all metrics even the routing algorithm without any order, for Infocom, LogootOpportunistic is not able to perform better than ONSIDE without any order in case of the hit rate for application layer when the memory is 20 or 100 messages. Though, for 1 message / second network capacity and 400 messages per memory, the results for the application layer hit rate are clearly in our favor. We still obtain better results than the vector clocks algorithm and the original Logoot in all cases. The high difference in hit rate at application

Figure 8. UPB trace results, ONSIDE routing algorithm, ∞ network capacity

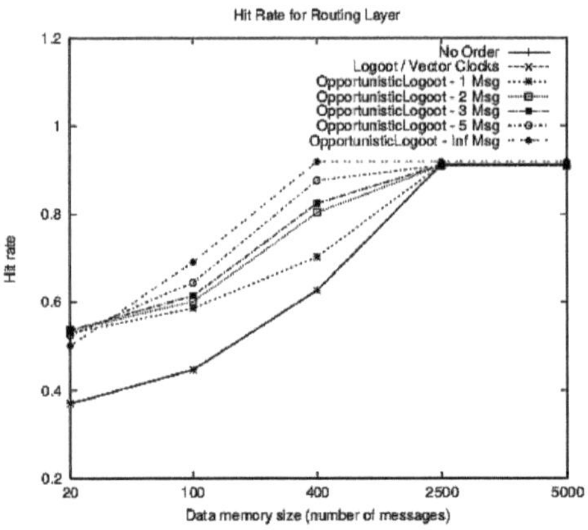

(a) Hit rate for routing layer

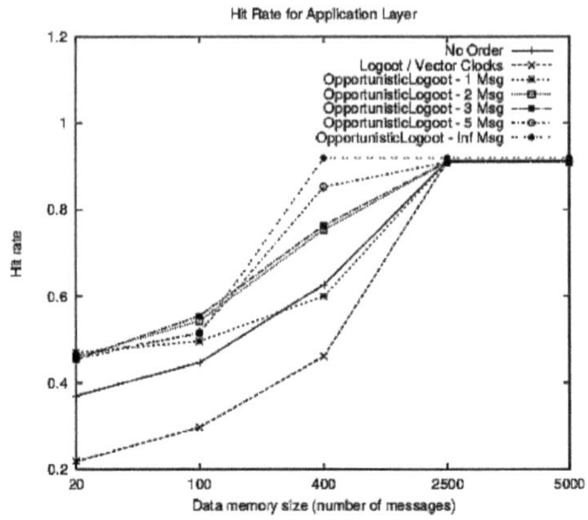

(b) Hit rate for application layer

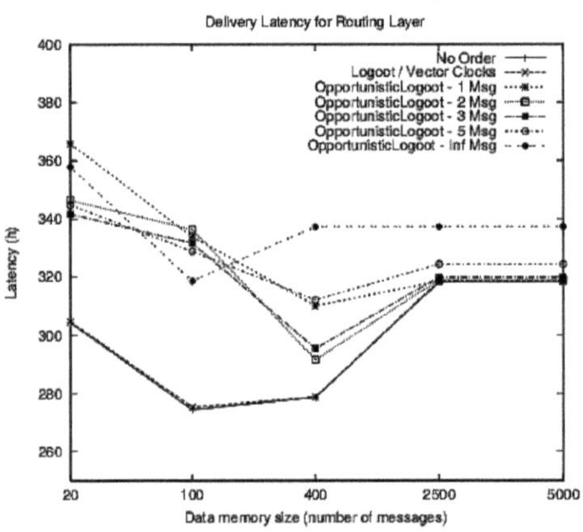

(c) Delivery latency for routing layer

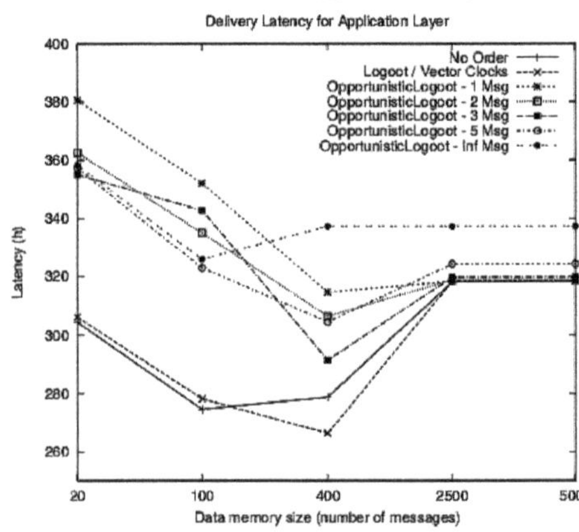

(d) Delivery latency for application layer

layer between no order and algorithms that require causal order (i.e. our proposed algorithm), plus the lower save rates of metadata size, lead to a lower performance in the case of a memory of 20 or 100 messages.

The fact that we have less savings in regard to metadata size is the result of a higher number of contacts that lead to downloading messages before they reach the merge limit, even though the total number of chat messages is the same as for Sigcomm. For the ∞ network capacity, Logoot and LogootOpportunistic with 1 message merge limit have the same metadata size as for Sigcomm, which is 39 KB for the former (i.e. 22.5% metadata size compared to payload size), and 52 KB for the

Figure 9. UPB trace results, ONSIDE routing algorithm, ∞ network capacity

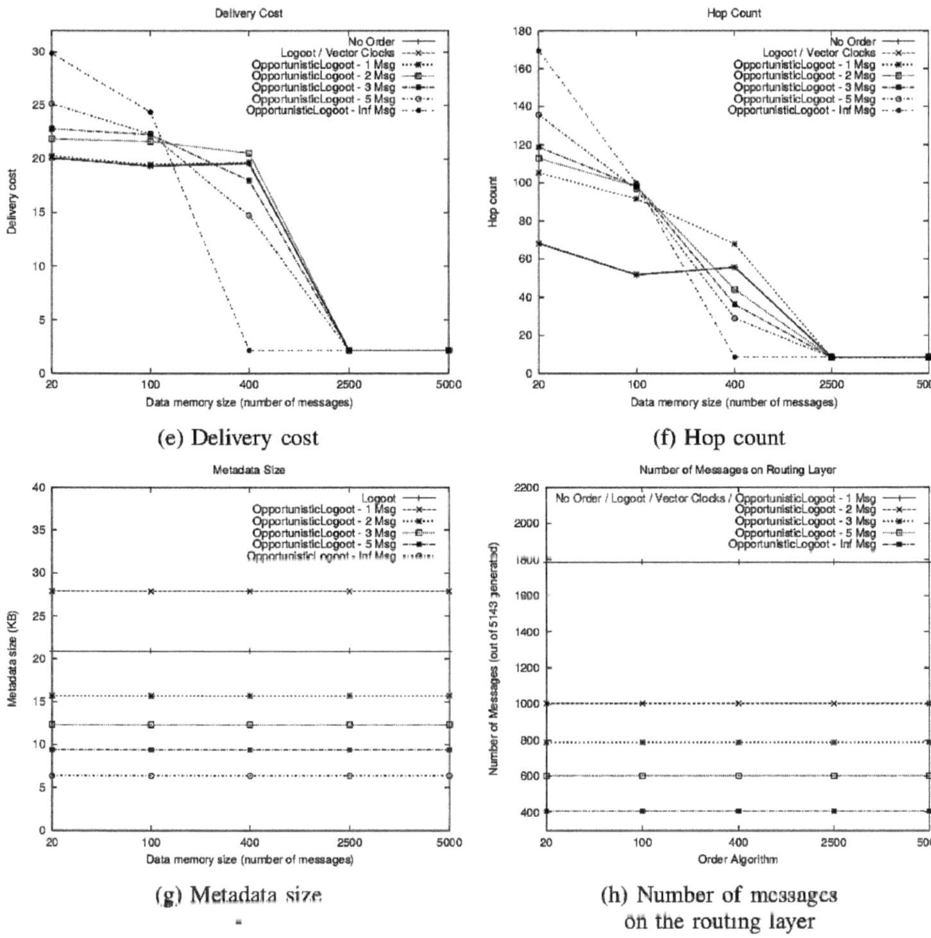

(e) Delivery cost

(f) Hop count

(g) Metadata size

(h) Number of messages on the routing layer

later (i.e. 30%). But our proposed algorithm has a metadata size of only 38.8 KB (i.e. 22.3%) when the merge limit is 2 messages, 35 KB (i.e. 20%) when the merge limit is 3 messages, 32 KB (i.e. 18.5%) when the merge limit is 5 messages, and 29 KB (i.e. 16.5%) when the merge limit is ∞. As the savings are a lot less compared to Sigcomm, we end up with a higher number of messages at the routing layer, which leads to smaller improvements in terms of performance compared to the original Logoot.

As for 1 message / second network capacity, we end up with a lower size for the metadata. These savings lead to better hit rates, due to the decrease in the

Figure 10. UPB trace results, ONSIDE routing algorithm, 1 message / second network capacity

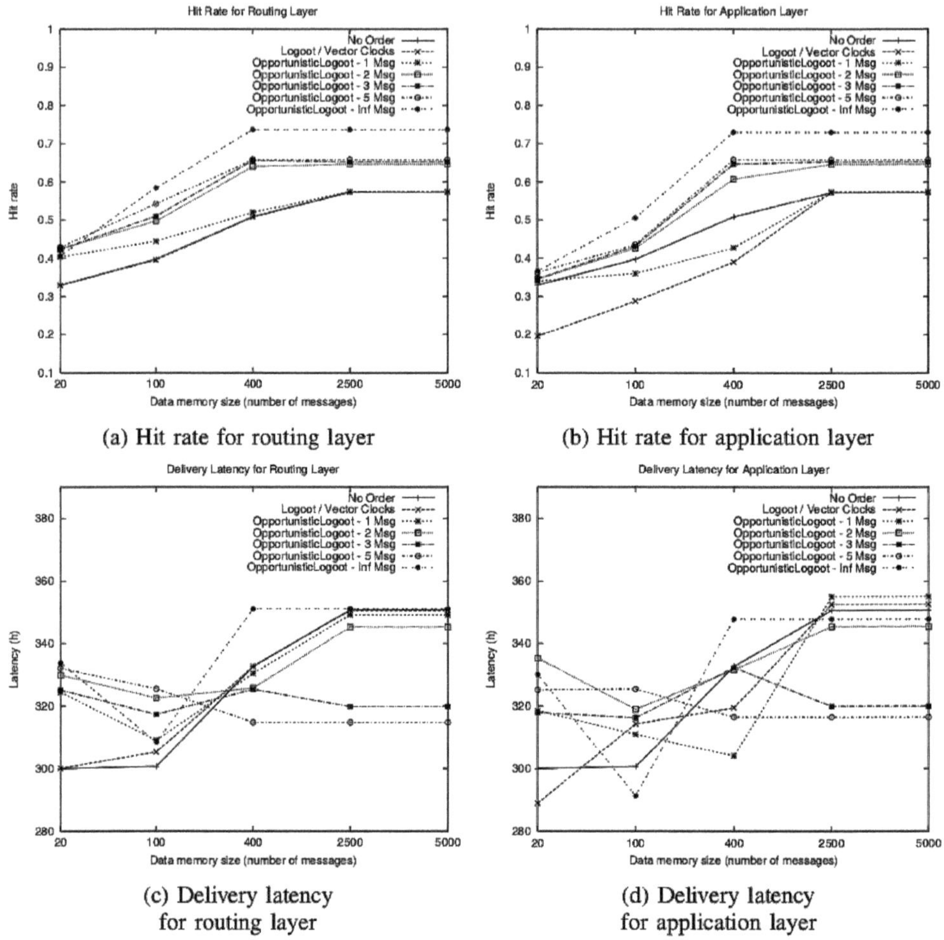

(a) Hit rate for routing layer

(b) Hit rate for application layer

(c) Delivery latency for routing layer

(d) Delivery latency for application layer

amount of messages at the routing layer. The interesting aspect is that, in the case of Infocom compared to Sigcomm and UPB, not only is the metadata size quite different between ∞ and 1 message / second for the network capacity, but we get a distinguishable difference when going from a memory of 20 messages to a memory of 5000 messages. For Logoot and OpportunisticLogoot with 1 message as merge limit, we have the same values for the metadata size (i.e. 39 KB and 52 KB), but for OpportunisticLogoot with 2 messages as merge limit, we obtain around 37.25

Figure 11. UPB trace results, ONSIDE routing algorithm, 1 message / second network capacity

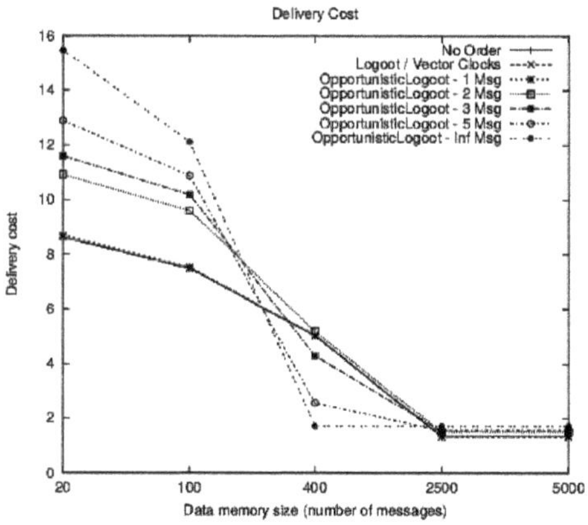

(e) Delivery cost

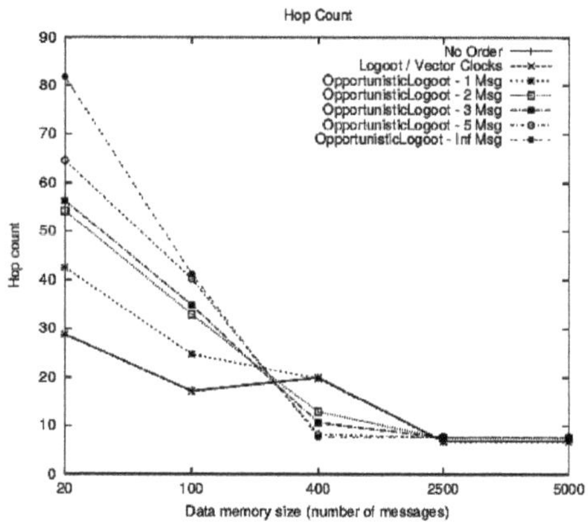

(f) Hop count

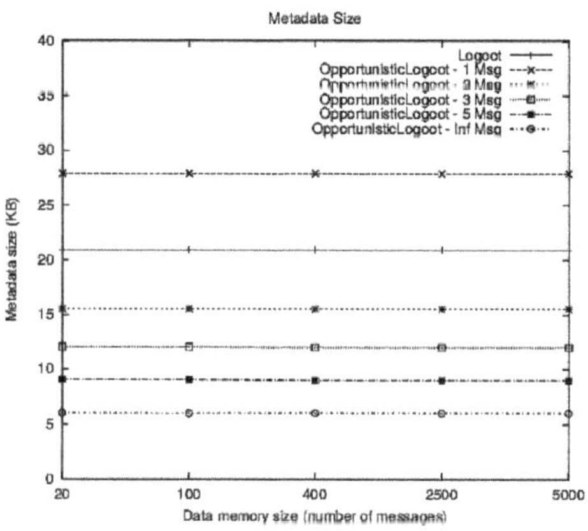

(g) Metadata size

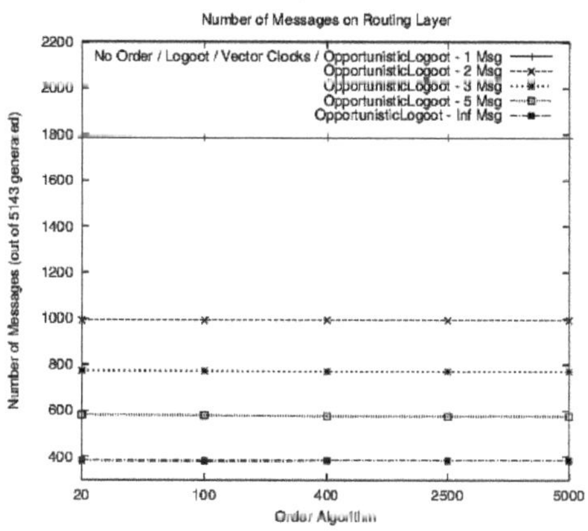

(h) Number of messages on the routing layer

KB for a memory of 20 messages, 37.15 KB for a memory of 100 messages, 36,7 KB for a memory of 400 messages, and 35.7 KB for a memory of 2500 and 5000 messages, leading to 20.5 - 21.5% metadata size compared to payload size. For a merge limit of 3 messages, we have 32.55 KB for a memory of 20 messages, 32.45 KB for a memory of 100 messages, 31.85 KB for a memory of 400 messages, and 30.8 KB for a memory of 2500 and 5000 messages (i.e. 17.7%-20.5%). As for a merge limit of 5 messages, the values are: 29.1 KB for a memory of 20 messages, 29 KB for a memory of 100 messages, 28.3 KB for a memory of 400 messages, and 27

Figure 12. Infocom trace results, ONSIDE routing algorithm, ∞ network capacity

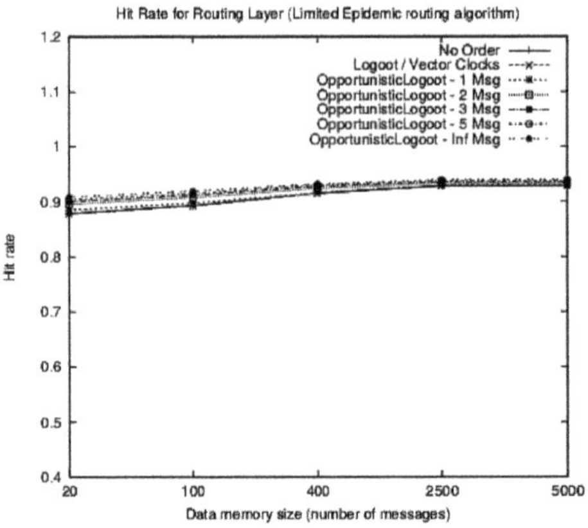

(a) Hit rate for routing layer

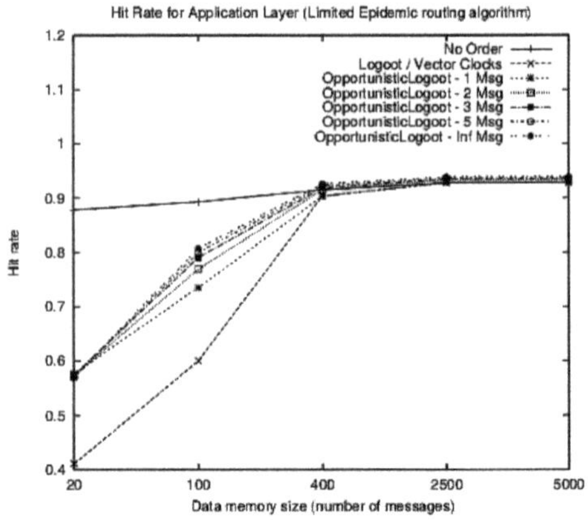

(b) Hit rate for application layer

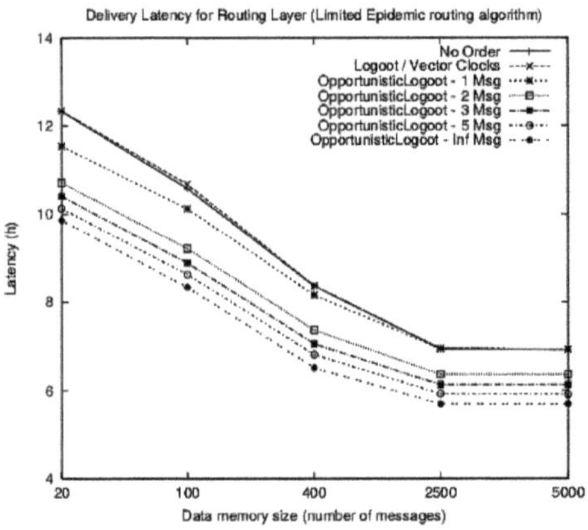

(c) Delivery latency
for routing layer

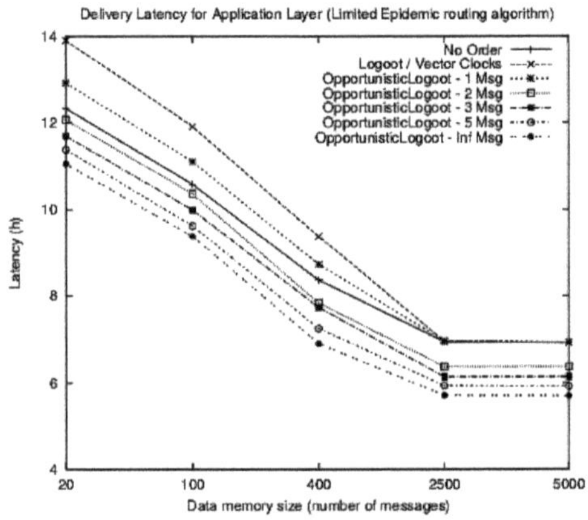

(d) Delivery latency
for application layer

KB for a memory of 2500 and 5000 messages (i.e. 15.5% - 16,7%). Lastly, for∞ as merge limit, we get 25.7 KB for 20 messages, 25.5 KB for 100 messages, 24.7 KB for 400 messages, and 23.4 KB for 2500 and 5000 messages (i.e. 13,5% - 14,75%).

The variations that appear above between memory values are due to the nature of the Infocom trace, which has a higher number of contacts, and the actual limitation of the network capacity to 1 message / second. By increasing the size of the memory, the routing algorithm gets to download more messages, either for the current node, or for other nodes. But this also means that more of the network capacity is spent on the routing algorithm and not on the direct download of missing messages,

Figure 13. Infocom trace results, ONSIDE routing algorithm, ∞ network capacity

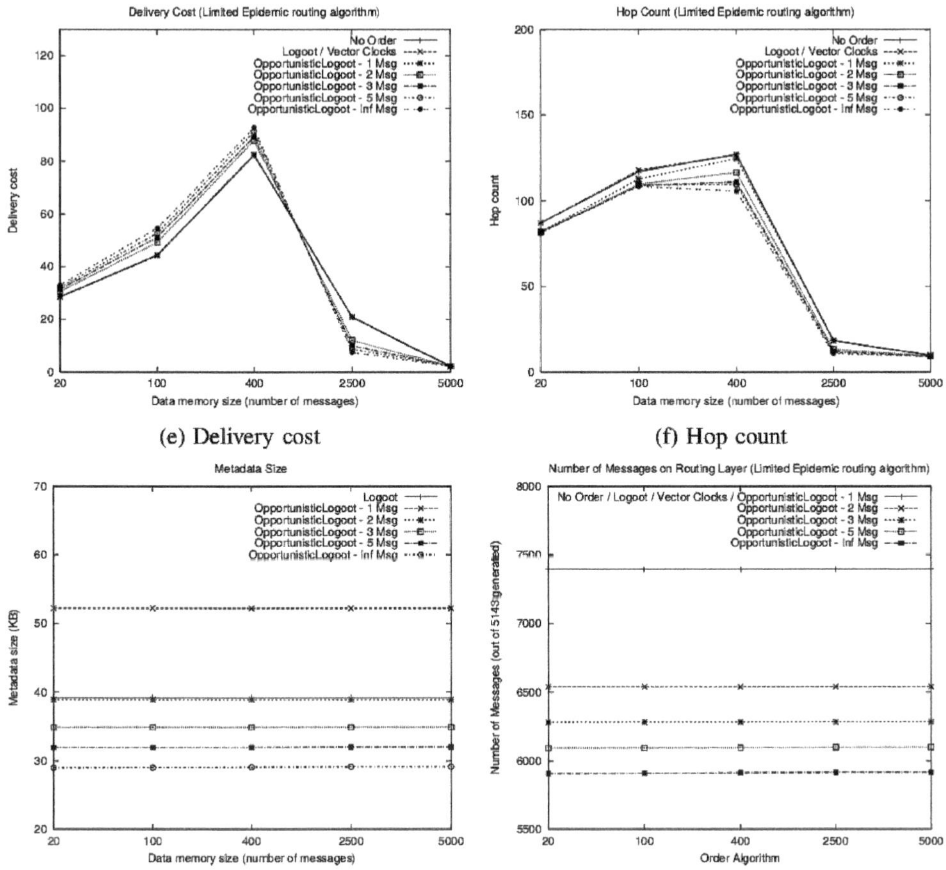

(e) Delivery cost

(f) Hop count

(g) Metadata size

(h) Number of messages on the routing layer

which is the last phase of a contact. As these former messages are intended for the contacted node (see Section IV-F), while the messages at the routing layer can have as destination other nodes, reducing the network capacity to 1 message / second from ∞ leads to fewer messages that end up as immutable, which results in a lower metadata size as the merge operation is not interrupted. Though we would expect to see a similar behavior for Sigcomm and UPB traces, those traces have a lower amount of contacts (i.e. fewer direct contacts for downloading missing messages), which reduces the impact of the network capacity.

Figure 14. Infocom trace results, ONSIDE routing algorithm, 1 message / second network capacity

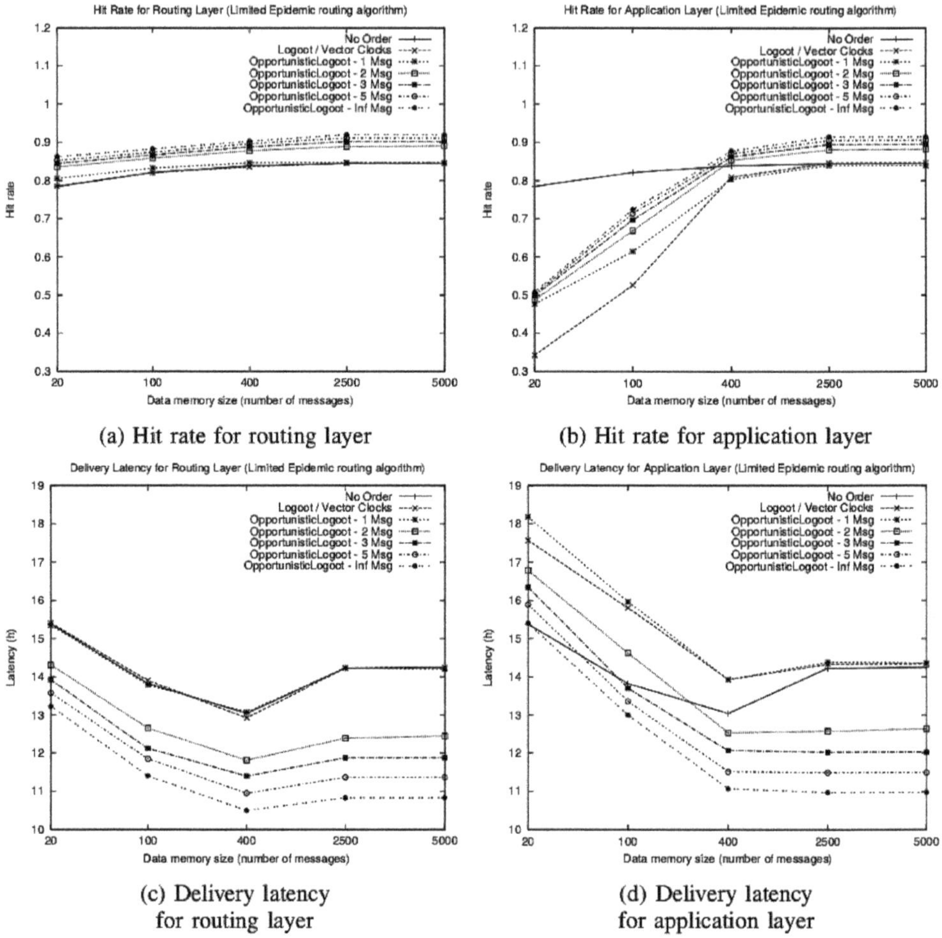

(a) Hit rate for routing layer

(b) Hit rate for application layer

(c) Delivery latency for routing layer

(d) Delivery latency for application layer

VI. CONCLUSION

In this chapter, we have proven that causal order and total ordering can be achieved in ONs. The experiments from Section III prove that the hit rate after messages respect causality is comparable to the reception hit rate in case of low memory, while it can reach the same values as the reception hit rate in case the mobile devices have enough memory. The same applies for delivery latency. The algorithm presented in Section IV ensures eventual consistency in ONs in case of direct contacts, while it reduces the metadata size of the messages as proved by the experiments from Section V.

Figure 15. Infocom trace results, ONSIDE routing algorithm, 1 message / second network capacity

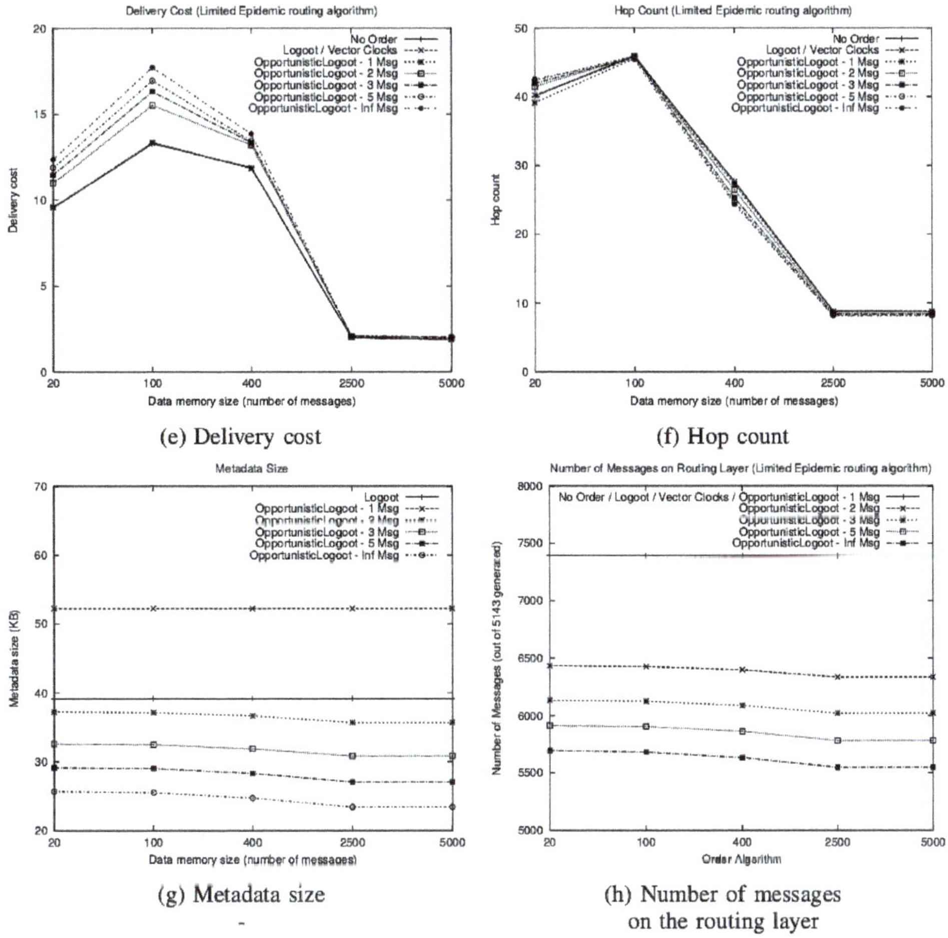

(e) Delivery cost

(f) Hop count

(g) Metadata size

(h) Number of messages on the routing layer

It is important to note that the experiments have limitations. The mobility traces we have used do not represent all the types of ONs as they have been obtained only in academic environments. Another problem is that the chat traces used in Section V were created in a server-client architecture, and modeled to support decentralized nodes as in ONs, which is not equivalent of how users will really use the chat application. Unfortunately, we are not aware of any distributed chat traces created especially for ONs where messages are not exchanged in real time.

As future work, we plan to add experiments that test both insertion and deletion for the OpportunisticLogoot algorithm we have developed. For now we have only tested the insert operation at the end of the contents by using real chat traces, but we are now working on a model to include traces of Wikipedia documents that can also test deletions. As the base element for Wikipedia documents is the character, we expect the reduction of metadata size to be higher than in the case of the chat traces. Afterwards, we plan to expand the causal order algorithm to support the direct download of missing messages on direct contact, just as the OpportunisticLogoot algorithm and implement causal barriers for a more detailed comparison. Another topic we want to tackle is security in ONs. Nodes that have not receive access to a share document, or that are not part of a chat, should not be able to see the contents of those applications messages, but only carry them to their destinations.

ACKNOWLEDGMENT

The work presented in this chapter is supported by national project MobiWay, Project PN-II-PT-PCCA-2013-4-0321 no. 16/2014.

REFERENCES

Andre, L., Martin, S., Oster, G., & Ignat, C.-L. (2013). Supporting adaptable granularity of changes for massive-scale collaborative editing. In *Collaborative Computing: Networking, Applications and Worksharing (Collaboratecom), 2013 9th International Conference Conference on*, (pp. 50–59). doi:10.4108/icst.collaboratecom.2013.254123

Birman, K., & Joseph, T. A. (1987). Reliable communication in the presence of failures. *ACM Transactions on Computer Systems*, *5*(1), 47–76. doi:10.1145/7351.7478

Birman, K., Schiper, A., & Stephenson, P. (1991). Lightweight causal and atomic group multicast. *ACM Transactions on Computer Systems*, *9*(3), 272–314. doi:10.1145/128738.128742

Boldrini, C., Conti, M., & Passarella, A. (2008). Contentplace: Social-aware data dissemination in opportunistic networks. In *Proceedings of the 11th International Symposium on Modeling, Analysis and Simulation of Wireless and Mobile Systems, MSWiM '08*. doi:10.1145/1454503.1454541

Ciobanu, R.-I., Dobre, C., & Cristea, V. (2012). Social aspects to support opportunistic networks in an academic environment. In Ad-hoc, Mobile, and Wireless Networks (LNCS), (vol. 7363, pp. 69-82). Springer Berlin Heidelberg. doi:10.1007/978-3-642-31638-8_6

Ciobanu, R.-I., Marin, R.-C., Dobre, C., Cristea, V., & Mavromoustakis, C. X. (2014). Onside: Socially-aware and interest-based dissemination in opportunistic networks. In *Network Operations and Management Symposium* (NOMS). IEEE.

Conti, M., Giordano, S., May, M., & Passarella, A. (2010). From opportunistic networks to opportunistic computing. *Comm. Mag.*, *48*(9), 126–139. doi:10.1109/MCOM.2010.5560597

Daileda, C. (2014). *How People in Hong Kong Can Communicate if Cell Networks Go Down*. Available http://mashable.com/2014/09/29/hong-kong-cell-network-chat/

Ellis, C. A., & Gibbs, S. J. (1989). Concurrency control in groupware systems. *SIGMOD Record*, *18*(2), 399–407. doi:10.1145/66926.66963

Fidge, C. J. (1988). Timestamps in message-passing systems that preserve the partial ordering. *Proceedings of the 11th Australian Computer Science Conference*, *10*(1), 56-66.

Fitzpatrick, A. (2014). *Hong Kong's Protesters Don't Need the Internet to Chat With One Another*. Available http://time.com/3449812/hong-_kong-_protesters-_firechat/

Gilbert, S., & Lynch, N. (2002). Brewer's conjecture and the feasibility of consistent, available, partition-tolerant web services. *SIGACT News*, *33*(2), 51–59. doi:10.1145/564585.564601

Google. (2015a). *Google Inc. Android one*. Available http://www.android.com/one/

Google. (2015b). *Google Inc. Google docs*. Available http://www.google.com/docs/about/

Hui, P., & Crowcroft, J. (2007). *Bubble Rap: Forwarding in small world DTNs in ever decreasing circles*. Technical Report UCAM-CL-TR-684, University of Cambridge, Computer Laboratory.

Kärkkäinen, T., & Ott, J. (2014). Shared content editing in opportunistic networks. In *Proceedings of the 9th ACM MobiCom Workshop on Challenged Networks, CHANTS '14*.

Lamport, L. (1998). The part-time parliament. *ACM Transactions on Computer Systems*, *16*(2), 133–169. doi:10.1145/279227.279229

Letia, M., Preguiça, N. M., & Shapiro, M. (2009). *Crdts: Consistency without concurrency control*. CoRR, abs/0907.0929

Li, D., & Li, R. (2008). An approach to ensuring consistency in peer-to-peer real-time group editors. *Computer Supported Cooperative Work*, *17*(5-6), 553–611. doi:10.1007/s10606-005-9009-5

Li, R., Li, D., & Sun, C. (2004). A time interval based consistency control algorithm for interactive groupware applications. In *Parallel and Distributed Systems, 2004. ICPADS 2004. Proceedings. Tenth International Conference on*, (pp. 429–436).

Marin, R.-C., Dobre, C., & Xhafa, F. (2012). Exploring predictability in mobile interaction. In EIDWT, (pp. 133–139). doi:10.1109/EIDWT.2012.29

Mattern, F. (1989). Virtual time and global states of distributed systems. In *Parallel and Distributed Algorithms* (pp. 215–226). North-Holland.

Open Garden. (2015). *Firechat*. Available http://opengarden.com/firechat

Oster, G., Urso, P., Molli, P., & Imine, A. (2006). Data consistency for p2p collaborative editing. In *Proceedings of the 2006 20th Anniversary Conference on Computer Supported Cooperative Work, CSCW '06*.

Pelusi, L., Passarella, A., & Conti, M. (2006). Opportunistic networking: Data forwarding in disconnected mobile ad hoc networks. *IEEE Communications Magazine*, *44*(11), 134–141. doi:10.1109/MCOM.2006.248176

Pietiläinen, A.-K., Oliver, E., LeBrun, J., Varghese, G., & Diot, C. (2009). Mobiclique: Middleware for mobile social networking. In *Proceedings of the 2Nd ACM Workshop on Online Social Networks, WOSN '09*.

Prakash, R., Raynal, M., & Singhal, M. (1997). An adaptive causal ordering algorithm suited to mobile computing environments. *Journal of Parallel and Distributed Computing*, *41*(2), 190–204. doi:10.1006/jpdc.1996.1300

Schiper, A., Eggli, J., & Sandoz, A. (1989). A new algorithm to implement causal ordering. In *Proceedings of the 3rd International Workshop on Distributed Algorithms*. London, UK: Springer-Verlag. doi:10.1007/3-540-51687-5_45

Shapiro, M., Preguiça, N., Baquero, C., & Zawirski, M. (2011). Conflict-free replicated data types. In *Proceedings of the 13th International Conference on Stabilization, Safety, and Security of Distributed Systems, SSS'11*. Berlin: Springer-Verlag. doi:10.1007/978-3-642-24550-3_29

Shapiro, M., & Preguiça, N. M. (2007). *Designing a commutative replicated data type*. CoRR, abs/0710.1784

Shen, H., & Sun, C. (2002). Flexible notification for collaborative systems. In *Proceedings of the 2002 ACM Conference on Computer Supported Cooperative Work, CSCW '02*. doi:10.1145/587078.587090

Skeen, D., & Stonebraker, M. (1983). A formal model of crash recovery in a distributed system. *Software Engineering. IEEE Transactions on*, *SE-9*(3), 219–228.

Socievole, A., Yoneki, E., De Rango, F., & Crowcroft, J. (2013). Opportunistic message routing using multi-layer social networks. In *Proceedings of the 2nd ACM Workshop on High Performance Mobile Opportunistic Systems, HP-MOSys '13*. doi:10.1145/2507908.2507923

Sun, C., & Ellis, C. (1998). Operational transformation in real-time group editors: Issues, algorithms, and achievements. In *Proceedings of the 1998 ACM Conference on Computer Supported Cooperative Work, CSCW '98*. doi:10.1145/289444.289469

Vahdat, A., & Becker, D. (2000). *Epidemic routing for partially-connected ad hoc networks. Technical report*. Duke University.

Vidot, N., Cart, M., Ferrié, J., & Suleiman, M. (2000). Copies convergence in a distributed real-time collaborative environment. In *Proceedings of the 2000 ACM Conference on Computer Supported Cooperative Work, CSCW '00*. doi:10.1145/358916.358988

Weiss, S., Urso, P., & Molli, P. (2008). *Logoot: a P2P collaborative editing system*. Research Report RR-6713, INRIA, France.

Yu, W. (2012). A string-wise crdt for group editing. In *Proceedings of the 17th ACM International Conference on Supporting Group Work, GROUP '12*. doi:10.1145/2389176.2389198

Compilation of References

Aggarwal, V., Feldmann, A., & Scheideler, C. (2007). Can ISPs and P2P users cooperate for improved performance? *Computer Communication Review*, *37*(3), 29–40. doi:10.1145/1273445.1273449

Agrawal, S., Roman, R., Das, M. L., Mathuria, A., & Lopez, J. (2012). A Novel Key update Protocol in Mobile Sensor Networks. In *Proceedings of International Conference on Information Systems Security* (LNCS), (vol. 7671, pp. 194-207). Berlin: Springer. doi:10.1007/978-3-642-35130-3_14

Ahmed, U., & Hussain, F. B. (2011). Energy efficient Routing Protocol for Zone Based Mobile Sensor Networks. *Wireless Communications and Mobile Computing Conference (IWCMC), 2011 7th International*. doi:10.1109/IWCMC.2011.5982691

Aitsaadi, N., Achir, N., Boussetta, K., & Pujolle, G. A. (2009). WSN Deployment Method for Monitoring Geographically Irregular Distributed Events. *Sensors (Basel, Switzerland)*, *9*(3), 1625–1643. doi:10.3390/s90301625 PMID:22573977

Akyildiz, I. F., Lee, W. Y., Vuran, M. C., & Mohanty, S. (2006). NeXt generation/dynamic spectrum access/cognitive radio wireless networks: A survey. *Computer Networks: The International Journal of Computer and Telecommunications Networking*, *50*(13), 2127–2159. doi:10.1016/j.comnet.2006.05.001

Akyildiz, I. F., Su, W., Sankarasubramaniam, Y., & Cayirci, E. (2002). Wireless Sensor Networks: A Survey. *Computer Networks*, *38*(4), 393–422. doi:10.1016/S1389-1286(01)00302-4

Ali, M. (2005). MMAC: A Mobility-Adaptive, Collision-Free MAC protocol for wireless sensor network. In *Proceedings of 24th International Conference on Performance, Computing and Communication Conference*. IEEE IPCC.

Alliance, O. (2010). *OSGi alliance*. Available at http://www.osgi.org/

Anand Paul, Y.C. (2012). Parallel Reconfigurable Computing-Based Mapping Algorithm for Motion Estimation in Advance Video Coding. *ACM Transactions on Embedded Computing Systems*, *11*(S2), 40.

Anastasi, G., Conti, M., Di Francesco, M., & Passarekka, A. (2009). Energy conservation in WSN: A Survey. *Ad Hoc Networks*, 537–568. doi:10.1016/j.adhoc.2008.06.003

Andre, L., Martin, S., Oster, G., & Ignat, C.-L. (2013). Supporting adaptable granularity of changes for massive-scale collaborative editing. In *Collaborative Computing: Networking, Applications and Worksharing (Collaboratecom), 2013 9th International Conference Conference on*, (pp. 50–59). doi:10.4108/icst.collaboratecom.2013.254123

Andreou, A., Mavromoustakis, C. X., Mastorakis, G., Bourdena, A., & Pallis, E. (2014). Adaptive Heuristic-Based P2P Network Connectivity and Configuration for Resource Availability. In Resource Management in Mobile Computing Environments (pp. 221-240). Springer International Publishing.

Aslam, M., Shah, T., Javaid, N., Rahim, A., Rahman, Z., & Khan, Z. A. (June 2012). CEEC: Centralized energy efficient clustering a new routing protocol for WSNs. *Conference on Sensor, Mesh and Ad Hoc Communications and Networks (SECON), 20129th Annual IEEE Communications Society* (pp. 103 - 105). Seoul: IEEE.

Bai, X., Santosh, K., Dong, X., Yun, Z., & Ten, H. L. (2006). Deploying Wireless Sensors to Achieve Both Coverage and Connectivity. In *Proceedings of the 7th ACM International Symposium on Mobile ad hoc networking and computing MobiHoc'06*. doi:10.1145/1132905.1132921

Bai, X., Xuan, D., Yun, Z., Lai, T. H., & Jia, W. (2008). Complete Optimal Deployment Patterns for Full-Coverage and *k*-Connectivity (*k* · 6) Wireless Sensor Networks. In *Proceedings of the 9th ACM international symposium on Mobile ad hoc networking and computing*. doi:10.1145/1374618.1374672

Bechkit, W., Challal, Y., & Bouabdallah, A. (2013). A New Class of Hash-Chain based Key Pre-distribution Schemes for WSN. *Journal of Computer Communications*.

Bennett, J., & Lanning, S. (2007, August). The netflix prize. In *Proceedings of KDD cup and workshop* (Vol. 2007, p. 35).

Bergbreiter, S., & Pister, K. S. J. (2003). CotsBots: An Off-the-Shelf Platform for Distributed Robotics. In *Proceeding of the IEEE International Conference on Intelligent Robots and Systems*. Las Vegas, NV: IEEE. doi:10.1109/IROS.2003.1248878

Biagioni, E. S., & Sasaki, G. (2003). Wireless Sensor Placement for Reliable and Efficient Data Collection. In *Proceedings of the 36th Annual Hawaii International Conference on System Sciences*. doi:10.1109/HICSS.2003.1174290

Bilstrup, K. (2007). *A survey regarding wireless communication standards intended for a high-speed vehicle environment. Technical Report IDE 0712*. Sweden: Halmstad University.

Birman, K., & Joseph, T. A. (1987). Reliable communication in the presence of failures. *ACM Transactions on Computer Systems*, *5*(1), 47–76. doi:10.1145/7351.7478

Birman, K., Schiper, A., & Stephenson, P. (1991). Lightweight causal and atomic group multicast. *ACM Transactions on Computer Systems*, *9*(3), 272–314. doi:10.1145/128738.128742

Blundo, C., De Santis, A., Herzberg, A., Kutten, S., Vaccaro, U., & Yung, M. (1993). Perfectly-secure Key Distribution for Dynamic Conferences. In *Advances in Cryptology* (LNCS), (vol. 740, pp. 471-486). Berlin: Springer.

Boldrini, C., Conti, M., & Passarella, A. (2008). Contentplace: Social-aware data dissemination in opportunistic networks. In *Proceedings of the 11th International Symposium on Modeling, Analysis and Simulation of Wireless and Mobile Systems, MSWiM '08*. doi:10.1145/1454503.1454541

Borcoci, E., Negru, D., & Timmerer, C. (2010, June). A novel architecture for multimedia distribution based on content-aware networking. In *Communication Theory, Reliability, and Quality of Service (CTRQ), 2010 Third International Conference on* (pp. 162-168). IEEE. doi:10.1109/CTRQ.2010.35

Bourdena, A., Mavromoustakis, C. X., Kormentzas, G., Pallis, E., Mastorakis, G., & Yassein, M. B. (2014). A resource intensive traffic-aware scheme using energy-aware routing in cognitive radio networks. *Future Generation Computer Systems*, *39*, 16–28. doi:10.1016/j.future.2014.02.013

Bourdena, A., Pallis, E., Kormentzas, G., & Mastorakis, G. (2013). Efficient radio resource management algorithms in opportunistic cognitive radio networks. *Transactions on Emerging Telecommunications Technologies*.

Buettner, M., Yee, G. V., Anderson, E., & Han, R. (2006). X-MAC: a short preamble MAC protocol for duty-cycled wireless sensor networks. *4th International conference on Embedded networked sensor systems (SenSys '06)*. ACM.

Burguillo-Rial, J. C., Costa-Montenegro, E., Gil-Castineira, F., & Rodriguez-Hernandez, P. (2008). Performance analysis of IEEE 802.11p in urban environments using a multi-agent model. In *Proceedings of the 19th IEEE International Symposium on Personal, Indoor and Mobile Radio Communications (PIMRC)*, (pp. 1–6).

Capkun, S., & Hubaux, J. (2005). Secure Positioning of Wireless Devices with Application to Sensor Networks. In *Proceedings of the Annual Joint Conference of the IEEE Computer and Communications Societies* (vol. 3, pp. 1917-1928). IEEE. doi:10.1109/INFCOM.2005.1498470

Cardellini, V., Colajanni, M., & Yu, P. S. (2003). Request redirection algorithms for distributed web systems. Parallel and Distributed Systems. *IEEE Transactions on*, *14*(4), 355–368.

Cayotte, E., & Buchow H. (2009). *Who dies of what in Europe before the age of 65*. Eurostat statistics.

Chakrabarti, S., & Mishra, A. (2001). QoS issues in Ad Hoc wireless networks. *IEEE Communications Magazine*, *39*(2), 142–148. doi:10.1109/35.900643

Chakrabarty, K., Iyengar, S. S., Qi, H., & Cho, E. (2002). Grid Coverage for Surveillance and Target Location in Distributed Sensor Networks. *IEEE Transcations on Computerers*, *51*(12), 1448–1453. doi:10.1109/TC.2002.1146711

Chan, H., Perrig, A., & Song, P. (2003). Random Key Pre-distribution Schemes for Sensor Networks. In *Proceedings of IEEE Symposium on Security and Privacy*, (pp. 197-213). IEEE.

Chen, H., Chen, W., Wang, Z., Wang, Z., & Li, Y. (2014). Mobile Beacon Based Wormhole Attackers Detection and Positioning in Wireless Sensor Networks.International Journal of Distributed Sensor Networks.

Chen, Y. F., Huang, Y., Jana, R., Jiang, H., Rabinovich, M., Wei, B., & Xiao, Z. (2007, June). When is P2P technology beneficial for IPTV services. In *Proceedings of the 17th International Workshop on Network and Operating System Support for Digital Audio and Video* (pp. 04-05).

Chen, B., Jamieson, K., Balakrishnan, H., & Morris, R. (2002). SPAN: An energy efficient coordination algorithm for topology maintenance in ad hoc wireless network. *ACM Wireless Network*, *8*(5), 481–494. doi:10.1023/A:1016542229220

Chen, G., Li, C., Ye, M., & Wu, J. (2007). An unequal cluster-based routing protocol in wireless sensor networks. *Wireless Networks*, *15*(2), 193–207. doi:10.1007/s11276-007-0035-8

Chen, J., Li, S., & Sun, Y. (2007). Novel Deployment Schemes for Mobile Sensor Networks. *Sensors (Basel, Switzerland)*, *7*(11), 2907–2919. doi:10.3390/S7112907

Chen, Y., Qiu, L., Chen, W., Nguyen, L., & Katz, R. H. (2003). Efficient and adaptive Web replication using content clustering. *Selected Areas in Communications. IEEE Journal on*, *21*(6), 979–994.

Cheung, Y.-, Li, M., Ng, M., & Huang, J. (2008). Agglomerative Fuzzy K-Means Clustering Algorithm with Selection of Number of Clusters. *IEEE Transactions on Knowledge and Data Engineering*, *20*(11), 1519–1534. doi:10.1109/TKDE.2008.88

Chih-Lin, I., & Pollini, G. P. (1994). The tree-search resources auction multiple access (TRAMA) protocol for wireless personal communications. *Vehicular Technology Conference, 1994 IEEE 44th*. doi:10.1109/VETEC.1994.345276

Chong, C. Y., & Kumar, S. P. (2003). Sensor Networks: Evolution, Opportunities, and Challenges. *Proceedings of the IEEE*, *91*(8), 1247–1256. doi:10.1109/JPROC.2003.814918

Ciobanu, R.-I., Dobre, C., & Cristea, V. (2012). Social aspects to support opportunistic networks in an academic environment. In Ad-hoc, Mobile, and Wireless Networks (LNCS), (vol. 7363, pp. 69-82). Springer Berlin Heidelberg. doi:10.1007/978-3-642-31638-8_6

Ciobanu, R.-I., Marin, R.-C., Dobre, C., Cristea, V., & Mavromoustakis, C. X. (2014). Onside: Socially-aware and interest-based dissemination in opportunistic networks. In *Network Operations and Management Symposium* (NOMS). IEEE.

Concolato, C., Dufourd, J. C., Le Feuvre, J., Park, K., & Song, J. (2012). Communicating and migratable interactive multimedia documents. *Multimedia Tools and Applications*, *61*(3), 599–622. doi:10.1007/s11042-011-0805-2

Conti, M., Giordano, S., May, M., & Passarella, A. (2010). From opportunistic networks to opportunistic computing. *Comm. Mag.*, *48*(9), 126–139. doi:10.1109/MCOM.2010.5560597

Coskun, V. (2008). Relocating Sensor Nodes to Maximize Cumulative Connected Coverage in Wireless Sensor Networks. *Sensors (Basel, Switzerland)*, *8*(4), 2792–2817. doi:10.3390/s8042792

Cui, T., Chen, L., Ho, T., Low, S. H., & Andrew, L. L. H. (2007). Opportunistic Source Coding for Data Gathering in Wireless Sensor Networks. In *Proceedings of the IEEE Mobile Ad-hoc and Sensor Systems (MASS)*. doi:10.1109/MOBHOC.2007.4428675

Daileda, C. (2014). *How People in Hong Kong Can Communicate if Cell Networks Go Down*. Available http://mashable.com/2014/09/29/hong-kong-cell-network-chat/

Das, S. R., Perkins, C. E., & Royer, E. M. (2000). Performance comparison of two on-demand routing protocols for ad hoc networks. *INFOCOM 2000. Nineteenth Annual Joint Conference of the IEEE Computer and Communications Societies. Proceedings*. IEEE. doi:10.1109/INFCOM.2000.832168

Dawood, M. S., Ahila, L., Sadasivam, S., & Athisha, G. (2012). Image Compression in Wireless sensor networks- A survey. *International Journal of Applied Information Systems*, *1*(9), 11–15. doi:10.5120/ijais12-450227

Dhillon, S. S., & Chakrabarty, K. (2003). Sensor Placement for Effective Coverage and Surveillance in Distributed Sensor Networks. In *Proceedings of the IEEE Wireless Communications and Networking*. doi:10.1109/WCNC.2003.1200627

Dimitriou, C. D., Mavromoustakis, C. X., Mastorakis, G., & Pallis, E. (2013, June). On the performance response of delay-bounded energy-aware bandwidth allocation scheme in wireless networks. In *2013 IEEE International Conference on Communications (ICC 2013), Workshops*, (pp. 631-636). doi:10.1109/ICCW.2013.6649310

Dowell, L. J., & Bruno, M. L. (2001). Connectivity of Random Graphs and Mobile Networks: Validation of Monte Carlo Simulation Results. In *Proceedings of the 2001 ACM symposium on Applied computing (SAC '01)*. doi:10.1145/372202.372280

Eichler, S. (2007). Performance evaluation of the IEEE 802.11p wave communication standard. In *Proceedings of the 66th IEEE Vehicular Technology Conference (VTC)*, (pp. 2199–2203).

El-Hoiydi, A., Decotignie, J. D., Enz, C., & Le Roux, E. (2003). Poster abstract: wiseMAC, an ultra-low power MAC protocol for the wiseNET wireless sensor network. *1st International conference on Embedded networked sensor systems (SenSys '03)*. ACM.

Ellis, C. A., & Gibbs, S. J. (1989). Concurrency control in groupware systems. *SIGMOD Record, 18*(2), 399–407. doi:10.1145/66926.66963

Eltoweissy, M., Moharrum, M., & Mukkamala, R. (2006). Dynamic Key Management in Sensor Networks. *IEEE Communications Magazine, 44*(4), 122–130. doi:10.1109/MCOM.2006.1632659

Eschenauer, L., & Gligor, V. (2002). A Key Management Scheme for Distributed Sensor Networks. In *Proceedings of the ACM Conference on Computer and Communication Security*, (pp. 41-47). ACM. doi:10.1145/586110.586117

ETSI. (2006). *282 001: Telecommunications and Internet converged Services and Protocols for Advanced Networking (TISPAN).* NGN Signalling Control Protocol.

Eude, N., Ducourthial, B., & Shawky, M. (2005). *Enhancing ns-2 simulator for high mobility ad hoc networks in Car-to-Car communication context.* The 7th IFIP International Conference on Mobile and Wireless Communications Networks, Morocco.

European Telecommunications Standards Institute (2011). *Final draft ETSI ES 302 663 V1.2.0, Intelligent Transport Systems (ITS); Access layer specification for Intelligent Transport Systems operating in the 5 GHz frequency band.*

Faye, S., Chaudet, C., & Demeure, I. (2012). A Distributed Algorithm for Multiple Intersections Adaptive Traffic Lights Control using a Wireless Sensor Networks. In *Proceedings of the 8th ACM International Conference on emerging Networking EXperiments and Technologies (CoNEXT)*, (pp. 13-18).

Festag, A., Fussler, H., Hartenstein, H., Sarma, A., & Schmitz, R. (2004). Fleet-Net: Bringing Car-to-Car Communication into the Real World. In *Proceedings of the 10th World Congress on ITS.*

Fidge, C. J. (1988). Timestamps in message-passing systems that preserve the partial ordering. *Proceedings of the 11th Australian Computer Science Conference, 10*(1), 56-66.

Fitzpatrick, A. (2014). *Hong Kong's Protesters Don't Need the Internet to Chat With One Another.* Available http://time.com/3449812/hong-_kong-_protesters-_firechat/

Fu, X., Yang, Wang, & Sun. (2014). Novel deployment scheme based on three dimensional voverage model for WSN, TSWJ. Hindawi.

Giannetsos, T., & Dimitriou, T. (2012). LDAC: A Localized and Decentralized Algorithm for Efficiently Countering Wormholes in Mobile Wireless Networks. *Journal of Computer and System Sciences, 80*(3), 618–643. doi:10.1016/j.jcss.2013.06.015

Gilbert, S., & Lynch, N. (2002). Brewer's conjecture and the feasibility of consistent, available, partition-tolerant web services. *SIGACT News, 33*(2), 51–59. doi:10.1145/564585.564601

Gkantsidis, C., Mihail, M., & Zegura, E. (2003). The Markov Chain Simulation Method for Generating Connected Power Law Random Graphs. In *Proceedings of the 5th SIAM Workshop on Algorithm Engineering and Experiments (ALENEX).*

Gong, S., Zuo, X., Mei, L., & Zhao, R. (2014). Enhanced Key Management Scheme Based on Random Key Pre-distribution for Wireless Sensor Networks. In Future Information Technology (LNEE), (vol. 309, pp. 343-348). Springer. doi:10.1007/978-3-642-55038-6_52

Google. (2015a). *Google Inc. Android one*. Available http://www.android.com/one/

Google. (2015b). *Google Inc. Google docs*. Available http://www.google.com/docs/about/

Grafling, S., Mahonen, P., & Riihijarvi, J. (2010). Performance evaluation of IEEE 1609 wave and IEEE 802.11p for vehicular communications. In *Proceedings of the 2nd International Conference on Ubiquitous and Future Networks (ICUFN)*, (pp. 344–348). doi:10.1109/ICUFN.2010.5547184

Guan, R., Shi, X., Marchese, M., Yang, C., & Liang, Y. (2011). Text Clustering with Seeds Affinity Propagation. *IEEE Transactions on Knowledge and Data Engineering*, *23*(04), 627–637. doi:10.1109/TKDE.2010.144

Guo, S., Leung, V., & Qian, Z. (2010). A Permutation-Based Multi-Polynomial Scheme for Pair wise Key Establishment in Sensor Networks. In *Proceedings of IEEE International Conference on Communications*, (pp. 1-5). IEEE.

Gupta, A., Liskov, B., & Rodrigues, R. (2004, March). Efficient Routing for Peer-to-Peer Overlays. In *NSDI*.

Haas, Z. J., & Tabrizi, S. (1998). On some challenges and design choices in Ad Hoc communications. In IEEE MILCOM98.

Handy, M., Haase, M., & Timmermann, D. (2002). Low energy adaptive clustering hierarchy with deterministic cluster-head selection.*4th International Workshop on Mobile and Wireless Communications Network* (pp. 368 - 372). IEEE. doi:10.1109/MWCN.2002.1045790

He, J., Chaintreau, A., & Diot, C. (2009). A performance evaluation of scalable live video streaming with nano data centers. *Computer Networks*, *53*(2), 153–167. doi:10.1016/j.comnet.2008.10.014

He, X., Niedermeier, M., & Meer, H. (2013). Dynamic Key Management in Wireless Sensor Networks: A Survey. *Journal of Network and Computer Applications*, *36*(2), 611–622. doi:10.1016/j.jnca.2012.12.010

Home Gateway Initiative. (2007). *Home Gateway Technical Requirement*. Author.

Hu & Evans. (2004). *Localization for Mobile Sensor Networks*. Tenth Annual International Conference on Mobile Computing and Networking (ACM MobiCom 2004).

Huang, Y., Fu, T. Z., Chiu, D. M., Lui, J., & Huang, C. (2008, August). Challenges, design and analysis of a large-scale p2p-vod system. *Computer Communication Review*, *38*(4), 375–388. doi:10.1145/1402946.1403001

Hui, P., & Crowcroft, J. (2007). *Bubble Rap: Forwarding in small world DTNs in ever decreasing circles*. Technical Report UCAM-CL-TR-684, University of Cambridge, Computer Laboratory.

IEA. (2005). *CO_2 Emissions from fuel combustion*. IEA.

IEEE 1609.0. (2013). *IEEE Guide for Wireless Access in Vehicular Environments (WAVE) – Architecture*. IEEE.

IEEE 802.11p. (2010). *IEEE Standard for Information Technology-Telecommunications and Information Exchange Between Systems-Local and Metropolitan Area Networks-Specific Requirements - Part II: Wireless LAN Medium Access Control (MAC) and Physical Layer (PHY) Specifications, Amendment 6: Wireless Access in Vehicular Environment*. IEEE Computer Society.

INDRA. (2008). *Annual Report 2008*. Retrieved from http://www.informeanual2008.indra.es/

Ishizuka, M., & Aida, M. (2004). Performance Study of Node Placement in Sensor Networks. In *Proceedings24th International Conference on Distributed Computing Systems Workshops* (pp. 598–603). doi:10.1109/ICDCSW.2004.1284093

Iyengar, R., Kar, K., & Banerjee, S. (2005) Low-coordination Topologies for Redundancy in Sensor Networks. In *Proceedings of the. MobiHoc'05*. doi:10.1145/1062689.1062732

Jabbar, S., Butt, A. E., Najm-us-Sehr, & Minhas, A. A. (2011). TLPER: Threshold Based Load Balancing Protocol For Energy Efficient Routing In WSN. *The 13th International Conference on Advanced Communication Technology (ICACT'11)*. Seoul, South Korea: IEEE.

Jabbar, S., Aziz, M. Z., Minhas, A. A., & Hussain, D. (2010). *PTAL: Power Tuning Anchors Localization Algorithm For Wireless Ad-Hoc Micro Sensors Network*. Bradford, UK: IEEE.

Jabbar, S., Iram, R., Minhas, A. A., Shafi, I., Khalid, S., & Ahmad, M. (2013). Intelligent Optimization Of Energy Aware Routing In Wireless Sensor Network Through Bio-Inspired Computing: Survey And Future Directions. *International Journal of Distributed Sensor Networks*, *2013*, 13. doi:10.1155/2013/421084

Jabbar, S., Minhas, A. A., Paul, A., & Rho, S. (2014). MCDA: Multilayer Cluster Designing Algorithm For Network Lifetime Improvement Of Homogenous Wireless Sensor Networks. *The Journal of Supercomputing*, 11227.

Jabbar, S., Minhas, A. A., Rahseed, T., & Rho, S. (2014, December). Heuristic Approach for Stagnation Free Energy Aware Routing in Wireless Sensor Network. *Ad Hoc Sensor Wireless Network*, *0*, 1–25.

Jang, H. C., & Feng, W. C. (2010). Network status detection-based dynamic adaptation of contention window in IEEE 802.11p. In *Proceedings of the 71st IEEE Vehicular Technology Conference (VTC)*, (pp. 1–5).

Jing, A. I., & Alhussein, A. (2006). Coverage by directional Sensors in randomly deployed wireless sensor networks. *J.Comb. Optimum*, *11*, 21–41.

Joshi, S., & Jaiswal, A. K., & Tyagi, P. K. (2013). A Novel Analysis of T Mac and S Mac Protocol for Wireless Sensor Networks Using Castalia. *International Journal of Soft Computing and Engineering, 2*(6).

Kadas, G., & Chatzimisios, P. (2011). Collaborative Efforts for Safety and Security in Vehicular Communication Networks. In *Proceedings of the 15th Panhellenic Conference on Informatics (PCI)*, (pp. 117-121). doi:10.1109/PCI.2011.75

Kadas, G., & Chatzimisios, P. (2012). The Role of Roadside Assistance in Vehicular Communication Networks: Security, Quality of Service and Routing issues. In R. Daher & A. Vinel (Eds.), *Roadside Networks for Vehicular Communications: Architectures, Applications and Test Fields* (pp. 1–37). IGI Global.

Kaddachi, M. L., Makkaoui, L., Soudani, A., Lecuire, V., & Moureaux, J. M. (2011, December). *FPGA-based image compression for low-power Wireless camera Sensor Networks.* Paper Presented at Next Generation Networks and Services (NGNS), 2011 3rd International Conference on. doi:10.1109/NGNS.2011.6142537

Kangasharju, J., Ross, K. W., & Turner, D. A. (2007, May). Optimizing file availability in peer-to-peer content distribution. In *INFOCOM 2007.26th IEEE International Conference on Computer Communications*. IEEE. doi:10.1109/INFCOM.2007.229

Kapthikeyan, A., Shankar, T., Srividhya, V., Sarkar, S., & Gupte, A. (2013). Energy Efficient Distributed Image Compression Using JPEG2000 in Wireless Sensor Networks (WSNS*). Journal of Theoretical and Applied Information Technology*, *47*(2).

Karkazis, P., Leligou, H. C., & Zahariadis, T. (2012). Geographical Routing in Wireless Sensor Networks. In *Proceedings of the 5th International Conference on Telecommunications and Multimedia (TEMU)*, (pp. 19-24). doi:10.1109/TEMU.2012.6294717

Kärkkäinen, T., & Ott, J. (2014). Shared content editing in opportunistic networks. In *Proceedings of the 9th ACM MobiCom Workshop on Challenged Networks, CHANTS '14*.

Karn, P. (1990, September). *MACA-a new channel access method for packet radio*. Paper presented at ARRL/CRRL Amateur Radio 9th Computer Networking Conference.

Khairnar, V., & Kotecha, K. (2013). Performance of Vehicle-to-Vehicle Communication using IEEE 802.11p in Vehicular Ad-hoc Network Environment. *International Journal of Network Security & Its Applications*, *5*(2), 143–170. doi:10.5121/ijnsa.2013.5212

Khan, M., Gansterer, W., & Haring, G. (2007). Congestion Avoidance and Energy Efficient Routing Protocol for Wireless Sensor Networks with a Mobile Sink. *Journal of Networks*, *2*(6), 42–49. doi:10.4304/jnw.2.6.42-49

Khan, S. U. (2007) Approximate Optimal Sensor Placements in Grid Sensor Fields. In *Proceedings of the IEEE 65th Vehicular Technology Conference, VTC2007*. doi:10.1109/VETECS.2007.63

Kilkki, K. (2008). Quality of Experience in Communications Ecosystem. *J. UCS*, *14*(5), 615–624.

Koo, S. G., Lee, C. G., & Kannan, K. (2004, October). A genetic-algorithm-based neighbor-selection strategy for hybrid peer-to-peer networks. In *Computer Communications and Networks, 2004. ICCCN 2004. Proceedings. 13th International Conference on* (pp. 469-474). IEEE. doi:10.1109/ICCCN.2004.1401710

Kumar, S., Singhal, D., & Garimella, R. M. (2012). *Cognitive Wireless Sensor Networks in Intelligent Sensor Networks: The Integration of Sensor Networks, Signal Processing and Machine Learning*. Retrieved from https://books.google.co.in/books?id=ZmLNBQAAQBAJ&printsec=frontcover

Kumar, S. (2004). Medium Access Control protocols for Ad Hoc wireless networks. *Survey (London, England)*, 326–329.

Lamport, L. (1998). The part-time parliament. *ACM Transactions on Computer Systems*, *16*(2), 133–169. doi:10.1145/279227.279229

Laoutaris, N., Rodriguez, P., & Massoulie, L. (2008). ECHOS: Edge capacity hosting overlays of nano data centers. *Computer Communication Review*, *38*(1), 51–54. doi:10.1145/1341431.1341442

Lazos, L., & Poovendran, R. (2004). SeRLoc: Secure Range-Independent Localization for Wireless Sensor Networks. In *Proceedings of the Third ACM workshop on Wireless security*, (pp. 21-30). ACM.

Lazos, L., & Poovendran, R. (2006). HiRLoc: High-resolution Robust Localization for Wireless Sensor Networks. IEEE Journal on Selected Areas in Communications, 24(2), 233-246.

Lazos, L., Poovendran, R., & Capkun, S. (2005). Rope: Robust Position Estimation in Wireless Sensor Networks. In *Proceedings of the Fourth International Conference on Information Processing in Sensor Networks*, (pp. 324-331). Academic Press.

Lecuire, V., Duran-Faundez, C., & Krommenacker, N. (2008). Energy-efficient image transmission in sensor networks. *International Journal of Sensor Networks*, *4*(1-2), 37–47. doi:10.1504/IJSNET.2008.019250

Lee, J.-S., & Cheng, W.-L. (2012). Fuzzy-Logic-Based Clustering Approach for Wireless Sensor Networks Using Energy Predication. *IEEE Sensors Journal*, *12*(9), 2891–2897. doi:10.1109/JSEN.2012.2204737

Lee, K., & Lee, H. (2012). A Self-Organized and Smart-Adaptive Clustering and Routing Approach for Wireless Sensor Networks. *International Journal of Distributed Sensor Networks*, 13.

Leligou, H. C., Chatzimisios, P., Sarakis, L., Orphanoudakis, T., Karkazis, P., & Zahariadis, T. (2014). An IEEE 802.11p compliant system prototype supporting road safety and traffic management applications. *International Journal of Wireless Networks and Broadband Technologies*, *3*(1), 1–17. doi:10.4018/ijwnbt.2014010101

Leligou, H. C., Karkazis, P., Hatziefremidis, A., Orphanoudakis, T., & Zahariadis, T. (2013). Implementation of road safety and traffic management applications over mobile opportunistic systems. In *Proceedings of the 17th WSEAS International Conference on Communications (CSCC)*, (pp. 163-168).

Leontiadis, I., Marfiazy, G., Mack, D., Pauyx, G., Mascolo, C., & Gerla, M. (2011). On the Effectiveness of an Opportunistic Traffic Management System for Vehicular Networks. *IEEE Transactions on Intelligent Transportation Systems*, *12*(4), 1537–1548. doi:10.1109/TITS.2011.2161469

Letia, M., Preguiça, N. M., & Shapiro, M. (2009). *Crdts: Consistency without concurrency control*. CoRR, abs/0907.0929

Li, R., Li, D., & Sun, C. (2004). A time interval based consistency control algorithm for interactive groupware applications. In *Parallel and Distributed Systems, 2004. ICPADS 2004. Proceedings. Tenth International Conference on*, (pp. 429–436).

Liben-Nowell, D., Balakrishnan, H., & Karger, D. (2002, July). Analysis of the evolution of peer-to-peer systems. In *Proceedings of the twenty-first annual symposium on Principles of distributed computing* (pp. 233-242). ACM. doi:10.1145/571825.571863

Li, C., Ye, M., Chen, G., & Wu, J. (2005). An Energy-Efficient Unequal Clustering Mechanism for Wireless Sensor Networks.*IEEE International Conference on Mobile Adhoc and Sensor Systems Conference*. Washington, DC: IEEE.

Li, D., & Li, R. (2008). An approach to ensuring consistency in peer-to-peer real-time group editors. *Computer Supported Cooperative Work*, *17*(5-6), 553–611. doi:10.1007/s10606-005-9009-5

Lim, J. S., & Oppenheim, A. V. (1987). Advanced Topics in Signal Processing. Englewood Cliffs, NJ: Prentice Hall.

Lin, F. Y. S., & Chiu, P. L. (2005). A Near-Optimal Sensor Placement Algorithm to Achieve Complete Coverage/ Discrimination in Sensor Networks. *IEEE Communications Letters*, *9*, 43–45.

Liu, D., & Ning, P. (2003). Establishing Pair wise Keys in Distributed Sensor Networks. In *Proceedings of the ACM Conference on Computer and Communications Security*, (pp. 52-61). ACM.

Liu, Y., & Liu, Y. (2008). A robust routing algorithm with fair congestion control in WSN. *Computer Communication and Networks, 2008, ICCCN '08'.Proceedings of 17th International Conference*. doi:10.1109/ICCCN.2008.ECP.167

Liu, D., Ning, P., Liu, A., Wang, C., & Du, W. (2008). Attack Resistant Location Estimation in Wireless Sensor Networks. *ACM Transactions on Information and System Security*, *11*(4), 1–39. doi:10.1145/1380564.1380570

Magoni, D., & Hoerdt, M. (2005). Internet core topology mapping and analysis. *Computer Communications*, *28*(5), 494–506. doi:10.1016/j.comcom.2004.09.002

Maher, R., Berre, Hnaien, Snoussi, & Khoukhi. (2014). A Branch and Bound Algorithm for critical Grid coverage problem WSN. *IJDSN*, 1-9.

Mahmud, M., Rahman, M., & Akhtar, M. (2012). Improvement of K-means clustering algorithm with better initial centroids based on weighted average.*7th International Conference on Electrical & Computer Engineering* (pp. 647 - 650). Dhaka: IEEE. doi:10.1109/ICECE.2012.6471633

Manhas, E. B., Jr., Brante, G., Souza, R. D., & Pellenz, M. E. (2012, April). *Energy-Efficient Cooperative Image Transmission over Wireless Sensor Networks.* Presented at of Wireless Communication and Networking Conference (WCNC), 2012 IEEE, Shanghai, China. doi:10.1109/WCNC.2012.6214121

Marin, R.-C., Dobre, C., & Xhafa, F. (2012). Exploring predictability in mobile interaction. In EIDWT, (pp. 133–139). doi:10.1109/EIDWT.2012.29

Markakis, E. K., Pallis, E., Skianis, H., & Zaxaropoulos, V. (2012). Dynamic TCP Window Size Adaptation for Improving Network Performance of P2P-Aware Interactive Broadcasting Environments. *Journal of Communication and Computer*, *9*, 1395–1403.

Mastorakis, G., Bourdena, A., Mavromoustakis, C. X., Pallis, E., & Kormentzas, G. (2013, July). An energy-efficient routing protocol for ad-hoc cognitive radio networks. In Future Network and Mobile Summit (IEEE FutureNetworkSummit), 2013 (pp. 1-10).

Mastorakis, G., Bourdena, A., Mavromoustakis, C. X., Pallis, E., & Kormentzas, G. (2013, July). An energy-efficient routing protocol for ad-hoc cognitive radio networks. In Future Network and Mobile Summit (IEEE FutureNetworkSummit), 2013 (pp. 1-10). IEEE.

Mastorakis, G., Mavromoustakis, C. X., Bourdena, A., Kormentzas, G., & Pallis, E. (2013, September). Maximizing energy conservation in a centralized cognitive radio network architecture. In *2013 IEEE 18th International Workshop on Computer Aided Modeling and Design of Communication Links and Networks (IEEE CAMAD 2013),* (pp. 175-179). doi:10.1109/CAMAD.2013.6708112

Mattern, F. (1989). Virtual time and global states of distributed systems. In *Parallel and Distributed Algorithms* (pp. 215–226). North-Holland.

Mavromoustakis, C. X., Mousicou, P., Papanikolaou, K., Mastorakis, G., Bourdena, A., & Pallis, E. (2015). Dynamic Cloud Resource Migration for Efficient 3D Video Processing in Mobile Computing Environments. In Novel 3D Media Technologies (pp. 119-134). Springer New York.

Mavromoustakis, C., Andreou, A., Mastorakis, G., Papadakis, S., Bourdena, A., & Stratakis, D. (2014, August). Energy Consumption Optimization through Pre-scheduled Opportunistic Offloading in Wireless Devices. In *EMERGING 2014,The Sixth International Conference on Emerging Network Intelligence* (pp. 22-28).

Mavromoustakis, C., Dimitriou, C. D., & Mastorakis, G. (2013). On the Real-Time Evaluation of Two-Level BTD Scheme for Energy Conservation in the Presence of Delay Sensitive Transmissions and Intermittent Connectivity in Wireless Devices. *International Journal on Advances in Networks and Services, 6*(3-4), 148-162.

Mavromoustakis, C. X., Bourdena, A., Mastorakis, G., Pallis, E., & Kormentzas, G. (2014). An energy-aware scheme for efficient spectrum utilization in a 5G mobile cognitive radio network architecture. *Telecommunication Systems*, 1–13.

Mavromoustakis, C. X., Dimitriou, C., Mastorakis, G., & Pallis, E. (2013). Real-Time Performance Evaluation of F-BTD scheme for optimized QoS Energy Conservation in Wireless Devices. *Proceedings of IEEE GlobeCom, 2013*, 1156–1161.

Mavromoustakis, C. X., & Karatza, H. D. (2004). *Adaptive Energy Conservation Model using Dynamic Caching for Wireless Devices*. IEEE. doi:10.1109/SIMSYM.2004.1299490

Mavromoustakis, C. X., Mastorakis, G., Bourdena, A., & Pallis, E. (2014). *Energy efficient resource sharing using a traffic-oriented routing scheme for cognitive radio networks*. IET Networks.

Mavromoustakis, C. X., Pallis, E., & Mastorakis, G. (2014). *Resource Management in Mobile Computing Environments. Modeling and Optimization in Science and Technologies*. Springer International Publishing.

MICA2 Mote Datasheet. (n.d.). Retrieved from http://www.xbow.com/Products/Product_pdf_files/Wireless_pdf/6020-0042-01_A_MICA2.pdf

Miller, L. E. (2001). Distribution of Link Distances in a Wireless Network. *Journal of Research of the National Institute of Standards and Technology, 106*(2), 401–412. doi:10.6028/jres.106.017

Miorandi, D., Sicari, S., Pellegrini, F. D., & Chlamtac, I. (2012). Internet of things: Vision, applications and research challenges. *Ad Hoc Networks, 10*(7), 1497–1516. doi:10.1016/j.adhoc.2012.02.016

Mi, Q., Stankovic, J., & Stoleru, R. (2012). Practical and Secure Localization and Key Distribution for Wireless Sensor Networks. *Ad Hoc Networks, 10*(6), 946–961. doi:10.1016/j.adhoc.2011.12.008

Misic, J., Badawy, G., Rashwand, S., & Misic, V. B. (2010). Tradeoff issues for cch/sch duty cycle for IEEE 802.11p single channel devices. In *Proceedings of IEEE Global Telecommunications Conference (GLOBECOM)*, (pp. 1-6).

Mousicou, P., Mavromoustakis, C. X., Bourdena, A., Mastorakis, G., & Pallis, E. (2013, November). Performance evaluation of dynamic cloud resource migration based on temporal and capacity-aware policy for efficient resource sharing. In *Proceedings of the 2nd ACM workshop on High performance mobile opportunistic systems* (pp. 59-66). doi:10.1145/2507908.2507917

Murray, C., & Lopez, A. (1996). *The global burden of disease* (Vol. 1). Cambridge, MA: Harvard University Press.

Muruganathan, S. d., Ma, D. C., Bhasin, R., & Fapojuwo, A. (2005). A Centralized Energy-Efficient Routing Protocol for Wireless Sensor Networks. *IEEE Radio Communications*.

Naeimi, S., Ghafghazi, H., Chow, C.-O., & Ishii, H. (2012). A Survey on the Taxonomy of Cluster-Based Routing Protocols for Homogeneous Wireless Sensor Networks. *Sensors (Basel, Switzerland)*, 12. PMID:22969350

Nantulya, V. M., & Reich, M. R. (2002). The neglected epidemic: Road traffic injuries in developing countries. *British Medical Journal*, *324*(7346), 1139–1141. doi:10.1136/bmj.324.7346.1139 PMID:12003888

Nasri, M., Helali, A., Sghaier, H., & Maared, H. (2010, October). *Energy-Efficient Wavelet Image Compression in Wireless Sensor Network.* Paper Presented In Communication in wireless environments and ubiquitous systems: New challenges (ICWUS), 2010 international conference on, Sousse, Tunisia.

Nasri, M., Helali, A., Sghaier, H., & Maaref, H. (2011, December). *Energy conservation for image transmission over wireless sensor networks.* Presented at Microelectronics (ICM), 2011 International Conference on. doi:10.1109/ICM.2011.6177376

NEC Laboratories. (2013). *Car2X Communication SDK 2 Userguide v2.3.3.*

Nitnaware, D., & Verma, A. (2010). Energy Constraint Node Cache Based Routing Protocol for AdHoc Network. *International Journal of Wireless & Mobile Networks*, *1*(1), 77–86.

Niu, Y., Gao, D., & Gao, S. (2012). A Robust Localization in Wireless Sensor Networks against Wormhole Attack. Networks, 7(1), 187-194. doi:10.4304/jnw.7.1.187-194

Nygren, E., Sitaraman, R. K., & Sun, J. (2010). The akamai network: A platform for high-performance internet applications. *Operating Systems Review*, *44*(3), 2–19. doi:10.1145/1842733.1842736

O'Rourke, J. (1987). *Art Gallery Theorems and Algorithms.* New York, NY: Oxford University Press.

Oh, K., & Woo, S. (2007). Improved energy efficiency of DMAC with periodic full sleep cycle for WSN with heavy traffics. *Intelligent Sensors, Sensors Networks and Information, 2007, ISSNIP 2007,3rd International Conference.* .4496824 doi:10.1109/ISSNIP.2007

Open Garden. (2015). *Firechat.* Available http://opengarden.com/firechat

Oster, G., Urso, P., Molli, P., & Imine, A. (2006). Data consistency for p2p collaborative editing. In *Proceedings of the 2006 20th Anniversary Conference on Computer Supported Cooperative Work, CSCW '06.*

Panorama of Transport. (2009). *Eurostat EC.*

Pathan, A. M. K., & Buyya, R. (2007). *A taxonomy and survey of content delivery networks*. Grid Computing and Distributed Systems Laboratory, University of Melbourne, Technical Report.

Pathan, M., Buyya, R., & Vakali, A. (2008). Content delivery networks: State of the art, insights, and imperatives. In *Content Delivery Networks* (pp. 3–32). Springer Berlin Heidelberg. doi:10.1007/978-3-540-77887-5_1

Paul, A. (2013). Dynamic Power Management for Ubiquitous Network Devices. *Advance Science Letters*, *19*(7), 2046–2049. doi:10.1166/asl.2013.4634

Pelusi, L., Passarella, A., & Conti, M. (2006). Opportunistic networking: Data forwarding in disconnected mobile ad hoc networks. *IEEE Communications Magazine*, *44*(11), 134–141. doi:10.1109/MCOM.2006.248176

Pietiläinen, A.-K., Oliver, E., LeBrun, J., Varghese, G., & Diot, C. (2009). Mobiclique: Middleware for mobile social networking. In *Proceedings of the 2Nd ACM Workshop on Online Social Networks, WOSN '09*.

Pincus, M. (2011). *Intelligent Transportation Systems Benefits, Costs, Deployment, and Lessons Learned Desk Reference: 2011 Update*. US Department of Transportation, Research and Innovative Technology Administration.

Polastre, J., Hill, J., & Culler, D. (2004). Versatile low power media access for wireless sensor networks. *2nd international conference on Embedded networked sensor systems (SenSys '04)*. ACM.

Pradhan, S. S., & Ramchandran, K. (2003). Distributed source coding using syndromes (DISCUS): Design and construction. *IEEE Transactions on Information Theory*, *49*(3), 626–643. doi:10.1109/TIT.2002.808103

Prakash, R., Raynal, M., & Singhal, M. (1997). An adaptive causal ordering algorithm suited to mobile computing environments. *Journal of Parallel and Distributed Computing*, *41*(2), 190–204. doi:10.1006/jpdc.1996.1300

Qiu, L., Padmanabhan, V. N., & Voelker, G. M. (2001). On the placement of web server replicas. In INFOCOM 2001. Twentieth Annual Joint Conference of the IEEE Computer and Communications Societies. *Proceedings of the IEEE*, *3*, 1587–1596.

Raid, A. M., Khedr, W. M., El-dosuky, M. A., & Ahmed, W. (2014). *Jpeg Image Compression Using Discrete Cosine Transform- A Survey*. arXiv preprint arXiv:1405.6147

RamMurthy, G., Naveen Reddy, G., & Ravi Shankar Varma, A.RamMurthy. (2011). Modified Distributed Laxity Based Priority Scheduling Scheme. *Wireless Personal Communications*, *58*(3), 627–636. doi:10.1007/s11277-010-0126-1

Rasmussen, K., Capkun, K., & Cagalj, M. (2007). SecNav: Secure Broadcast Localization and Time Synchronization in Wireless Networks. In *Proceedings of the ACM International Conference on Mobile Computing and Networking*, (pp. 310-313). ACM.

Royer, E. M., & Toh, C. K. (1999). A review of current routing protocols for Ad Hoc mobile wireless networks. *IEEE Personal Commun.*, *6*(2), 46–55. doi:10.1109/98.760423

Samet, H. (2008). K-Nearest Neighbor Finding Using Max Nearest Dist. *IEEE Transactions on Pattern Analysis and Machine Intelligence*, *30*(02), 243–252. doi:10.1109/TPAMI.2007.1182 PMID:18084056

Saucez, D., Donnet, B., & Bonaventure, O. (2008). *Idips: Isp-driven informed path selection*. Academic Press.

Savarese, C., Rabaey, J., & Beutel, J. (2001). Locationing in Distributed Ad-hoc Wireless Sensor Networks. In *Proceedings of IEEE International Conference on Acoustics, Speech, and Signal Processing*, (pp. 2037-2040). IEEE.

Schiper, A., Eggli, J., & Sandoz, A. (1989). A new algorithm to implement causal ordering. In *Proceedings of the 3rd International Workshop on Distributed Algorithms*. London, UK: Springer-Verlag. doi:10.1007/3-540-51687-5_45

Shapiro, M., & Preguiça, N. M. (2007). *Designing a commutative replicated data type*. CoRR, abs/0710.1784

Shapiro, M., Preguiça, N., Baquero, C., & Zawirski, M. (2011). Conflict-free replicated data types. In *Proceedings of the 13th International Conference on Stabilization, Safety, and Security of Distributed Systems, SSS'11*. Berlin: Springer-Verlag. doi:10.1007/978-3-642-24550-3_29

Shen, H., & Sun, C. (2002). Flexible notification for collaborative systems. In *Proceedings of the 2002 ACM Conference on Computer Supported Cooperative Work, CSCW '02*. doi:10.1145/587078.587090

Singh, S., & Raghavendra, C. S. (1998). PAMAS-power aware multi-access protocol with signalling for Ad Hoc Networks. *SIGCOMM Comput. Commun. Rev.*, *28*(3), 5–26. doi:10.1145/293927.293928

Skeen, D., & Stonebraker, M. (1983). A formal model of crash recovery in a distributed system. *Software Engineering. IEEE Transactions on*, *SE-9*(3), 219–228.

Socievole, A., Yoneki, E., De Rango, F., & Crowcroft, J. (2013). Opportunistic message routing using multi-layer social networks. In *Proceedings of the 2nd ACM Workshop on High Performance Mobile Opportunistic Systems, HP-MOSys '13*. doi:10.1145/2507908.2507923

Sodagar, I. (2011). The MPEG-DASH Standard for Multimedia Streaming Over the Internet. *IEEE MultiMedia*, *18*(4), 62–67. doi:10.1109/MMUL.2011.71

Sohraby, K., Minoli, D., & Znati, T. (2007). *Wireless Sensor Networks: Technology, Protocols and Applications*. Hoboken, NJ: John Wiley & Sons. doi:10.1002/047011276X

Sokullu, & Donertas. (2008). *Combined effects of mobility, congestion and contention on network performance for IEEE 802.15.4 Based Networks*. Computer and Information Science, 2008, ISCIS '08', 23rd International Symposium on Digital Object Identifier.

Srinivasan, A., & Wu, J. (2007). A Survey on Secure Localization in Wireless Sensor Networks. In The Encyclopedia of Wireless and Mobile Communications. CRC Press.

Sriporamanont, T. (2006). *Wireless Sensor Network Simulator.* (Master's Thesis). Halmstad University, Sweden.

Sun, C., & Ellis, C. (1998). Operational transformation in real-time group editors: Issues, algorithms, and achievements. In *Proceedings of the 1998 ACM Conference on Computer Supported Cooperative Work, CSCW '98.* doi:10.1145/289444.289469

Takaishi, D., Nishiyama, H., Kato, N., & Miura, R. (2014, September). Towards Energy Efficient Big Data Gathering in Densely Distributed Sensor Networks. *IEEE Transactions on Emerging Topics in Computing*, *2*(3), 388–397. doi:10.1109/TETC.2014.2318177

Tanenbaum, A. S. (2003). *Computer Networks* (4th ed.). Pearson Education Inc.

Tang, X., & Xu, J. (2004, March). On replica placement for QoS-aware content distribution. In *INFOCOM 2004. Twenty-third AnnualJoint Conference of the IEEE Computer and Communications Societies* (Vol. 2, pp. 806-815). IEEE.

Tang, C., & Raghavendra, C. S. (2004). *Compression Techniques for Wireless Sensor Networks.* Book Wireless Sensor Networks Kluwer Academic.

Tao, L., & Yu, F. (2009). ECODA- Enhanced congestion detection and avoidance for multiple class of traffic in sensor network. *Communication, 2009, APCC 2009,15th Asia-Pacific Conference on.* doi:10.1109/APCC.2009.5375498

The Network Simulator. (n.d.). *NS-2.* Retrieved from http://www.isi.edu/nsnam/ns

Toh, C.-K. (2002). *Ad Hoc Mobile Wireless Networks: Protocols and Systems.* Prentice-Hall PTR.

Tran, M., & Tavanapong, W. (2005, November). Peers-assisted dynamic content distribution networks. In *Local Computer Networks, 2005. 30th Anniversary. The IEEE Conference on* (pp. 123-131). IEEE. doi:10.1109/LCN.2005.102

Tsai, A. C., Paul, A., Wang, J. C., & Wang, J. F. (2007). Efficient Intra Prediction in H.264 based on Intensity Gradient approach. *IEEE International Symposium on Circuits and Systems* (ISCAS 2007), (pp. 3952-3955). doi:10.1109/ISCAS.2007.378665

Tsai, H. C., Chen, B. W., Wang, J. F., & Paul, A. (2013). Enhanced long-rang Personal Identification based Multimodal Information of Human Features. *Multimedia Tools and Applications.*

Tubaishat, M., & Madria, S. (2003). Sensor Networks: An Overview. *IEEE Potential Magazine*, *22*(2), 20–23. doi:10.1109/MP.2003.1197877

Vahdat, A., & Becker, D. (2000). *Epidemic routing for partially-connected ad hoc networks. Technical report.* Duke University.

Valancius, V., Laoutaris, N., Massoulié, L., Diot, C., & Rodriguez, P. (2009, December). Greening the internet with nano data centers. In *Proceedings of the 5th international conference on Emerging networking experiments and technologies* (pp. 37-48). ACM. doi:10.1145/1658939.1658944

Vandenberghe, W., Moerman, I., & Demeester, P. (2011). Approximation of the IEEE 802.11p Standard Using Commercial Off-The-Shelf IEEE 802.11a Hardware. In *Proceedings of IEEE 11th International Conference on ITS Telecommunications (ITST)*, (pp. 21-26).

Vidot, N., Cart, M., Ferrié, J., & Suleiman, M. (2000). Copies convergence in a distributed real-time collaborative environment. In *Proceedings of the 2000 ACM Conference on Computer Supported Cooperative Work, CSCW '00*. doi:10.1145/358916.358988

Wang, S. Y., Chou, C. L., Liu, K. C., Ho, T. W., Hung, W. J., & Huang, C. F., … Lin, C. C. (2009). Improving the channel utilization of IEEE 802.11p/1609 networks. In *Proceedings of Wireless Communications and Networking Conference (WCNC)*, (pp. 1–6).

Wang, Y., Ahmed, A., Krishnamachari, B., & Psounis, K. (2008). IEEE 802.11p performance evaluation and protocol enhancement. In *Proceedings of IEEE International Conference on Vehicular Electronics and Safety (ICVES)*, (pp. 317–322).

Wang, S. S., & Chen, Z. P. (2013). LCM: A Link-Aware Clustering Mechanism for Energy-Efficient Routing in Wireless Sensor Networks. *IEEE Sensors Journal*, *13*(2), 728–736. doi:10.1109/JSEN.2012.2225423

Wang, Y. C., Hu, C. C., & Tseng, Y. C. (2005). Efficient Deployment Algorithms for Ensuring Coverage and Connectivity of Wireless Sensor Networks. In *Proceedings of the First International Conference on Wireless Internet*. doi:10.1109/WICON.2005.13

Warberg, A., Larsen, J., & Jorgensen, R. M. (2008). *Green wave traffic optimization – a survey. Technical Report*. Technical University Denmark.

Webopedia. (n.d.). *The 7 Layers of the OSI Model*. Retrieved September 24, 2010, from http://www.webopedia.com/quick_ref/OSI_Layers.asp

Weiss, S., Urso, P., & Molli, P. (2008). *Logoot: a P2P collaborative editing system*. Research Report RR-6713, INRIA, France.

Wikipedia. (n.d.). *Network Layer*. Retrieved February 18, 2014, from http://en.wikipedia.org/wiki/Network_layer

Wu, M., & Chen, C. W. (2006). Multiple Bit Stream Image Transmission over Wireless Sensor Networks. IEEE & Wiley Interscience. doi:10.1002/9780471784173.ch13

Xian, C. (2012). Fault-tolerant monitor placement for out of band WSN monitoring. *Ad Hoc Networks*, *10*, 62–74.

Xiao, Y., Rayi, V., Sun, B., Du, X., Hu, F., & Galloway, M. (2007). A Survey of Key Management Schemes in Wireless Sensor Networks. Computer Communications, 30, 2314-2341. doi:10.1016/j.comcom.2007.04.009

XiaoxingGuo., M. R., & Frater, M. J. R. (2006). A Propagation-delay-tolerant Collision Avoidance Protocol for Underwater Acoustic Sensor Networks. In *Proc. MTS/IEEE OCEANS*.

Xie, H., Yang, Y. R., Krishnamurthy, A., Liu, Y. G., & Silberschatz, A. (2008, August). P4P: Provider portal for applications. *Computer Communication Review*, *38*(4), 351–362. doi:10.1145/1402946.1402999

Xing, G., Wang, X., Zhang, Y., Lu, C., Pless, R., & Gill, C. (2002). Integrated coverage and connectivity configuration of energy conservation in sensor network. *ACM Transactions on Sensor Networks*, *1*(1), 36–72. doi:10.1145/1077391.1077394

Xiong, Z., Liveris, A. D., & Cheng, S. (2004). Distributed source coding for sensor networks. *IEEE Signal Processing Magazine*, *21*(5), 80–94. doi:10.1109/MSP.2004.1328091

Xiuchao, W. (2004). *Simulate 802.11b Channel within NS2*. Retrieved from http://cir.nus.edu.sg/reactivetcp/report/80211ChannelinNS2_new.pdf

Xu, Y., Hiedemann, J., & Estrin, D. (2001). Geography informed Energy Conservation for Ad Hoc Routing. In *Proceeding of the Seventh annual ACM lst IEEE International Conference on Mobile computing and Networking* (MobiCom). doi:10.1145/381677.381685

Xu, D., Kulkarni, S. S., Rosenberg, C., & Chai, H. K. (2006). Analysis of a CDN–P2P hybrid architecture for cost-effective streaming media distribution. *Multimedia Systems*, *11*(4), 383–399. doi:10.1007/s00530-006-0015-3

Yadav, R., Varma S., & Malaviya, N. (2009). A survey of MAC protocols for Wireless Sensor Networks. *UbiCC Journal, 4*(3).

Yang, M., & Fei, Z. (2003, May). A model for replica placement in content distribution networks for multimedia applications. In *Communications, 2003. ICC'03. IEEE International Conference on* (Vol. 1, pp. 557-561). IEEE.

Yang, P.-T., & Lee, S. (2012). A Distributed Reclustering Hierarchy Routing Protocol Using SocialWelfare in Wireless Sensor Networks. *International Journal of Distributed Sensor Networks*, *2012*, 14. doi:10.1155/2012/681026

Yang, S., Tseng, H., Wu, E. H.-K., & Chen, G. (2005). Utilization based duty cycle tuning MAC protocol for wireless sensor networks. *Global Telecommunications Conference*, (vol. 6, pp. 32-62).

Yang, X., & Vaidya, N. H. (2002). Priority Scheduling in Wireless Ad Hoc Networks. *Proceedings of the 3rd ACM international symposium on Mobile ad hoc networking & computing (MobiHoc'02)*. doi:10.1145/513800.513809

Ye, W. (2002). *SCADDS: Scalable Coordination Architectures for Deeply Distributed Systems.* Retrieved July 1, 2014, from http://www.isi.edu/scadds/projects/smac/

Yin, H., Liu, X., Zhan, T., Sekar, V., Qiu, F., Lin, C., & Li, B. et al. (2010). Livesky: Enhancing cdn with p2p. *ACM Transactions on Multimedia Computing, Communications, and Applications*, *6*(3), 16. doi:10.1145/1823746.1823750

Younis, O., & Fahmy, S. (2004). HEED: A hybrid energy-efficient, distributed clustering approach for ad hoc wireless networks. *IEEE Trans. On Mobile Computing*, *3*(4), 660–669. doi:10.1109/TMC.2004.41

Yousef, K. M., Al-Karaki, J. N., & Shatnawi, A. M. (2010). Intelligent traffic light flow control system using wireless sensors networks. *Journal of Information Science and Engineering*, *26*(3), 753–768.

Yu, J., Qi, Y., Wang, G., Guo, Q., & Gu, X. (2011). An Energy-Aware Distributed Unequal Clustering Protocol for Wireless Sensor Networks. *International Journal of Distributed Sensor Networks*, 2011.

Yu, W. (2012). A string-wise crdt for group editing. In *Proceedings of the 17th ACM International Conference on Supporting Group Work, GROUP '12*. doi:10.1145/2389176.2389198

Yu, Z., Liu, Y., & Cai, Y. (2008). Design of an Energy-Efficient Distributed Multi-level Clustering Algorithm for Wireless Sensor Networks. *4th International Conference on Wireless Communications, Networking and Mobile Computing* (pp. 1 - 4). Dalian: IEEE. doi:10.1109/WiCom.2008.839

Zahariadis, T., Trakadas, P., Leligou, H. C., Maniatis, S., & Karkazis, P. (2013). A novel trust-aware geographical routing scheme for wireless sensor networks. *Wireless Personal Communications*, *69*(2), 805–826. doi:10.1007/s11277-012-0613-7

Zaidi, Z. R., & Mark, B. L. (2004). Mobility Estimation for Wireless Networks Based on an Autoregressive Model. In *Proc. IEEE Globecom 2004*. doi:10.1109/GLOCOM.2004.1378999

Zaidi, Z. R., Mark, B. L., & Roshan, K. (2004). A Two-tier Representation of Node Mobility in Ad Hoc Networks. In *Proc. IEEE Sensor and Ad Hoc Communications and Networks (SECON)*. Santa Clara, CA: IEEE. doi:10.1109/SAHCN.2004.1381913

Zeinalipour-Yazti, D., Papadakis, H., Georgiou, C., & Dikaiakos, M. D. (2008). *Mobile Sensor Network Data Management. Parallel Processing Letters Journal.*

Zhan, C., Shen, L., Hadi, M. A., & Gan, A. (2008). *Understanding the Characteristics of Secondary Crashes on Freeways.* Paper presented at the 87th Annual Meeting of the Transportation Research Board, Washington DC.

Zhang, T., He, J., & Yu, H. (2012). Secure Localization in Wireless Sensor Networks with Mobile Beacons.International Journal of Distributed Sensor Networks.

Zhang, H., & Hou, J. C. (2005). Maintaining Sensing Coverage and Connectivity in Large Sensor Networks. *Ad Hoc Sensor Wireless Networks*, *1*, 89–124.

Zhang, X. Y., Zhang, Q., Zhang, Z., Song, G., & Zhu, W. (2004). A construction of locality-aware overlay network: mOverlay and its performance.Selected Areas in Communications. *IEEE Journal on*, *22*(1), 18–28. doi:10.1109/JQE.2003.820842

Zhu, S., Setia, S., & Jajodia, S. (2003). LEAP: Efficient Security Mechanisms for Large-scale Distributed Sensor Networks. In *Proceedings of the ACM Conference on Computer and Communications Security*, (pp. 62-72). ACM.

Zorbas, D., Razafindralambo, T., & Lilli, I. (2012). *Wireless Sensor Network Redeployment under Target Coverage Constraint*. New Technology, Mobility and Security (NTMS), 2012 5th International Conference on.

Zorbas, D., & Douligeris, C. (2011). Connected coverage in wsns based on critical targets. *Computer Networks*, *55*(6), 1412–1425. doi:10.1016/j.comnet.2010.12.023

Zou, L., Lu, M., & Xiong, Z. (2005). A Distributed Algorithm for the Dead End Problem of Location Based Routing in Sensor Networks. *Vehicular Technology, IEEE Transactions on, 54*(4), 1509 – 1522.

About the Contributors

Naveen Chilamkurti is currently acting head of the Department of Computer Science and Information Technology, La Trobe University, Melbourne, Victoria, Australia. He is also the inaugural Editor-in-Chief of the *International Journal of Wireless Networks and Broadband Technologies* launched in July 2011. He is currently serving as a Technical Editor for *IEEE Wireless Communications,* Associate Technical Editor for *IEEE Communications Magazine and in editorial board for IEEE Transactions on Vehicular Technology.* He has published about 180 journal and conference papers since 2001. His current research areas include intelligent transport systems, Cyber Security, wireless multimedia, Vehicular cloud, Mobile security, wireless sensor networks, and so on. He is also an Associate Editor and Guest Editor for Wiley, Elsevier, Springer and Hindawi Journals. He was also Keynote speaker for number of conferences, including QSHINE 2016. He also won best paper award as a co-author in IEEE 2015 Best Land Transportation Paper Award from the IEEE Vehicular Technology Society. He is also recipient of Australian Fellowship for research visit to India in 2012. He also edited book published by Springer, Elsevier and IGI Global publications.

* * *

Athina Bourdena received her B.Sc. from the Department of Applied Informatics and Multimedia of Technological Educational Institute of Crete, Greece in 2007. She obtained her M.Sc. in Digital Communications and Networks from the Department of Digital Systems of the University of Piraeus, Greece in 2009 and her Ph.D. diploma in the field of cognitive radio in November 2013 from the Department of Information & Communication Systems Engineering of University of the Aegean in Greece. From 2009 until now she works as a teaching assistance of the Technological Educational Institute of Crete and from 2014 until now as a post-doctoral fellow at the Department of Computer Science of the University of Nicosia in Cyprus. Her research interests are in the fields of cognitive radio networks, mobile networks and digital interactive television systems. She has published more than 35 publication articles in refereed scientific journals and international conferences.

Joachim Bruneau-Queyreix received his MSc. Degree in Telecommunications in 2014 from the Institut Polytechnique de Bordeaux, France. He has also spent 6 months studying research in networking and security at the University of Santa Barbara, California, USA. He is actually pursuing a PhD degree at the University of Bordeaux in the field of multiple sources and multiple paths for multimedia streaming over overlay and cloud networks. He is the main author of paper entitled "Home-Box based collaborative caching strategy: An asset for Content Delivery Networks" published in the IEEE TEMU 2014 conference, for which he received the best paper award. He is a member of IEEE.

Radu-Ioan Ciobanu is PhD student within the Computer Science Department, in University Politehnica of Bucharest, Romania. The PhD topic is on Routing and Dissemination in Mobile Opportunistic Networks. His general interests regard distributed and collaborative systems, artificial intelligence and algorithms. His current research is focused on models and techniques for representing and reasoning on mobility support for opportunistic networking and applications.

Mihail Costea finished his Bachelor studies within the University POLITEHNICA of Bucharest. His main research areas are collaborative systems, distributed systems and algorithms. He is currently enrolled for Master studies at University POLITEHNICA of Bucharest.

Manik Lal Das received his Ph.D. degree from Indian Institute of Technology, Bombay. He is an Associate Professor in Dhirubhai Ambani Institute of Information and Communication Technology, Gandhinagar, India. He is a member of IEEE and Cryptology Research Society of India. His research interests include Cryptography and Information Security.

Ciprian Dobre has scientific and scholarly contributions in the field of large scale distributed systems concerning monitoring (MonALISA), data services (PRO, DataCloud@Work), high-speed networking (VINCI, FDT), large scale application development (EGEE III, SEE-GRID-SCI), evaluation using modeling and simulation (MONARC 2, VNSim). Dr. Ciprian Dobre was awarded a PhD scholarship from California Institute of Technology (Caltech, USA), and another one from Oracle. His results received two CENIC Awards, and three Best Paper Awards, and were published in 6 books, 10 articles in major international peer-reviewed journal, and over 60 articles in well-established international conferences and workshops (these articles received more than 150 citations). He is local project coordinator for national projects 'CAPIM - Context-Aware Platform using Integrated Mobile Services', and 'TRANSYS – Models and Techniques for Traffic Optimizing in Urban Environments'.

Tata Jagannadha Swamy is Research Scholar in Department of Electronics and Communication Engineering, JNT University, Hyderabad, India. He obtained B.Tech Degree from Acharya Nagarjuna University in 2001, Master of Engineering in Systems and Signal Processing from Osmania University in 2006. His research interests are in the domain of Cognitive Radio Networks, Wireless Sensor Networks, Wireless Communication and Digital Signal Processing and Probablistic nature of signals. He published and presented several research papers in International Journals and conferences.

Radu-Corneliu Marin is PhD student within the Computer Science Department, in University Politehnica of Bucharest, Romania. The PhD topic is on Mobile Opportunistic Computing. His general interests regard distributed and collaborative systems, artificial intelligence and algorithms. His current research is focused on models and techniques for representing and reasoning on mobility for code execution in smartphone communities.

Evangelos Markakis holds a PhD from the department of Information & Communication Systems Engineering of the University of the Aegean. Currently he acts as the Senior Network Infrastructure engineer of the Largest Hospital in Crete maintaining a network of 1.2K+ PCs, 3k+ Users and as a Research Associate in a Post-Doctoral position at PASIPHAE laboratory for the project FP7/ICT-619520 "T-NOVA". His expertise are in the fields of distributed systems and P2P applications, design of large scale Heterogeneous Networks, Next Generation Networks, as well as in network management and virtualisation techniques, including SDN/NFV concepts. He has actively participated in more than 10 EU FP6/FP7 funded projects (IST/ICT/Health/Security) and 20 national research programs from 2003. He has more than 30 publications in international journals, conference and book chapters, in the above areas. He is a Member of IEEE.

George Mastorakis graduated from the Department of Electrical & Electronic Engineering of University of Manchester Institute of Science and Technology (UMIST), UK in July 2000. He obtained his M.Sc. in Telecommunications from the Department of Electrical & Electronic Engineering of University College London (UCL), UK in November 2001 and his Ph.D. diploma in the field of interactive digital television in September 2008 from the Department of Information & Communication Systems Engineering, Faculty of Sciences of University of the Aegean in Greece.

He is serving as an Assistant Professor and a research associate at Technological Educational Institute of Crete, teaching theoretical and laboratory discipline courses/ modules in the field of data networks, wireless networks, principles of digital television networks and interactive digital terrestrial television. He is a researcher of "Research and Development of Telecommunication Systems - PASIPHAE" Laboratory in the Department of Applied Informatics and Multimedia at Technological Educational Institute of Crete and at the same time, works as a visiting research fellow in the Centre for Technological Research of Crete, the National Centre for Scientific Research (NCSR) Demokritos and University of the Aegean in Greece. He collaborates with academic staff members of Electronics Business Intelligence Laboratory (e-Business Intelligence Lab) at Technological Educational Institute of Crete and Distributed Information Systems Laboratory in the Department of Computer Science at the University of Nicosia in Cyprus. His research interests include mobile networks, interactive marketing, new technologies and marketing, multimedia applications and services, cognitive radio networks, radio resource management, optimization algorithms, interactive digital television (interactive broadcasting), next generation networks, network management, Quality of Service (QoS) through hybrid systems and energy-efficient networks. He has participated in a large number of European research projects of the 6th and 7th European Framework (IST/ICT) Programme and national research projects of the 3rd and 4th Greek Community Support Framework Programme. Indicative research projects include "ATHENA" (FP6-IST-507312), "IMOSAN" (FP6-IST-027457), "PASIPHAE" (GSRT-KR_22) and recent ones of 7th Framework like "VITAL + +" (FP7-ICT-224287), "HURRICANE" (FP7-ICT-1-1.1-216006), "COGEU" (FP7-ICT-248560) and "ALICANTE" (FP7-ICT- 248,652). He has also served as a Technical Manager of several research projects funded by the General Secretariat for Research and Technology (GSRT) of Greek Ministry of Development. Dr. George Mastorakis has served as a peer reviewer of scientific papers in international conferences and journals, as a technical programme committee member of international conferences and workshops and as a member of scientific journals editorial boards. Additionally, he has served as an organizing and scientific committee member of the international conference series, "IEEE International Conference on Telecommunications and Multimedia" (TEMU), which is organized every two years at Technological Educational Institute of Crete. He is editor of Journal of Networks and author of more than 100 research articles in refereed journals, international conferences and edited volumes (book chapters) of scientific books, published in IEEE, Elsevier, Springer and Wiley. He is author/ editor of two Books. He is a member of Technical Chamber of Greece, IEEE and the National Accreditation Centre of Vocational Training Structures and Accompanying Support Services. Finally, he serves as evaluator of research projects proposals of the European Research Programme "Eurostars", the "Innovative Entrepreneurship" Programme and the "Digital Convergence" Operational Programme of Greece.

Constandinos X. Mavromoustakis is currently an Associate Professor at the Department of Computer Science at the University of Nicosia, Cyprus. He received a five-year dipl.Eng (BSc/BEng/MEng, KISATS approved) in Electronic and Computer Engineering from Technical University of Crete, Greece, MSc in Telecommunications from University College of London, UK, and his PhD from the department of Informatics at Aristotle University of Thessaloniki, Greece. Dr. Mavromoustakis is leading the Mobile Systems Lab. (MOSys Lab., http://www.mosys.unic.ac.cy/) at the Department of Computer Science at the University of Nicosia. Dr. Mavromoustakis has extensive experience in simulation of wireless systems, MP2P and Opportunistic Systems, Mobile Computing and Web technologies. He serves as an Associate editor International Journal of Communication Systems (IJCS), Wiley, and editorial board member of the Simulation Practice and Theory (SIMPAT) Journal, Elsevier and as member of Technical Program Committees for various International Conferences.

Rachit Mittal is a Product Architect at Sprinklr Solutions Private Limited, Gurgaon, Haryana, India. He received his B.Tech from DA-IICT, Gandhinagar, Gujarat.

Daniel Negru received his MSc. Degree in Computer Science from the University of Pierre and Marie Curie, Paris 6, in 2002 and his PhD Degree in 2006 in the field of Broadcast and Internet convergence solutions at the network and service levels. In 2002, he worked in the centre of research of Motorola, Paris, specifying and developing a complete IPv6 stack from scratch, including Mobile IPv6 and multimedia components. Between 2003 and 2006, he worked at the CNRS laboratory, Paris area – French computer science research centre that does research in the underlying technologies for the next generation global information infrastructure – and participated in several national and European projects, such as ANR RIAM NMS, IST FP6 ATHENA, IST FP6 ENTHRONE, IST FP6 ENTHRONE2, IST FP6 IMOSAN. In 2007, he became Associate Professor at ENSEIRB School of Engineers / University of Bordeaux, specializing in multimedia and networking and his main fields of research reside in the domains of mobility and multimedia over heterogeneous networks, IMS convergence, service accessibility and adaptation. He is currently working in the field of media-aware networks, including content-, context- and service-awareness at the network level and network awareness at the service level. He has coordinated the ICT FP7 ALICANTE project that tackles these fields.

Evangelos Pallis received his B.Sc. in Electronic Engineering from Technological Educational Institute of Crete in 1994, his M.Sc. in Telecommunications from University of East London in 1997 and his Ph.D. in Telecommunications from University of East London in 2002. He currently serves as an Associate Professor at Technological Educational Institute of Crete in the Department of Informatics Engineering and acts as the Director of Research and Development of Tele communication Systems Laboratory. His research interests are in the fields of wireless broadband and mobile networks and network management. He has more than 100 publications in international scientific journals, conference and workshop proceedings.

Garimella Ramamurthy obtained his B.Tech degree in Electronics and Communication Engineering from Sri Venkateswara University, Tirupati, India in 1984, M.S in Electrical Engg. from the Louisiana State University, Baton Rouge U.S.A in 1986 and Ph.D degree in Computer engineering from Purdue University, West Lafayette, U.S.A in 1989. His research interests include Wireless Sensor Networks, Adhoc Wireless Networks, Computer Architecture, Multi Dimensional Neural Networks, Perormace Evaluation, Cybermetics and Optimal Filtering. He is currently an associate professor at the International Institute of Information Technology (IIIT), Hyderabad, India. He is a member of Eta Kappa Nu, Phi Kappa Phi. IEEE and ACM. He published several papers in various Journals including IEEE, ACM transactions and presented papers in various International/national conferences. He is currently the editor of International Journal of Systemics, Cybernetics and Informatics and Research Journal of Telecommunication and Information Technology (INSINET publications), International Journal Algorithms, Computing Mathematics.

Odysseas Shiakallis is currently a Research Associate at the Department of Computer Science of the University of Nicosia. He received his B.Sc. in Computer Science at the Department of Computer Science of the University of Nicosia. He is currently performing research in the field of sensor networks (WSNs) at the Mobile Systems Lab (MoSys Lab), a laboratory under the Computer Science department at the University of Nicosia, engaged in the design, development, and evaluation of mobile networks, including sensor networks (WSNs), Mobile and Wearable Computing and related technologies, Modeling and Simulation of Mobile hybrid systems (including metaheuristic scheduling techniques and spatiotemporal scheduling techniques) and performance evaluation of high performance Mobile computing systems. He has extensive knowledge and experience in wireless and mobile systems, in complex wireless systems simulation. Odysseas Shiakallis joined the Mobile Systems Lab (MoSys Lab) in January 2013.

Index

A

B

C

D

E

G

H

I

K

L

M

N

O

P

R

S

T

U

V

W

IRMA INTERNATIONAL

Information Resources Management Association

Become an IRMA Member

Members of the **Information Resources Management Association (IRMA)** understand the importance of community within their field of study. The Information Resources Management Association is an ideal venue through which professionals, students, and academicians can convene and share the latest industry innovations and scholarly research that is changing the field of information science and technology. Become a member today and enjoy the benefits of membership as well as the opportunity to collaborate and network with fellow experts in the field.

IRMA Membership Benefits:

- **One FREE Journal Subscription**
- **30% Off Additional Journal Subscriptions**
- **20% Off Book Purchases**
- Updates on the latest events and research on Information Resources Management through the IRMA-L listserv.
- Updates on new open access and downloadable content added to Research IRM.
- A copy of the Information Technology Management Newsletter twice a year.
- A certificate of membership.

IRMA Membership $195

Scan code to visit irma-international.org and begin by selecting your free journal subscription.

Membership is good for one full year.

www.irma-international.org

www.ingramcontent.com/pod-product-compliance
Ingram Content Group UK Ltd.
Pitfield, Milton Keynes, MK11 3LW, UK
UKHW050946290926
15470UKWH00005B/1675

9 781466 699410